Insect–Fungal Associations

Insect–Fungal Associations
Ecology and Evolution

Edited by

Fernando E. Vega

Meredith Blackwell

*For Sarah Emche
Best of luck for
a long productive
career.*

Meredith Blackwell.

*For Sarah,
One of the best scientists
I have ever collaborated
with.*

FERNANDO

OXFORD
UNIVERSITY PRESS

2005

OXFORD

UNIVERSITY PRESS

Oxford   New York
Auckland   Bangkok   Buenos Aires   Cape Town   Chennai
Dar es Salaam   Delhi   Hong Kong   Istanbul   Karachi   Kolkata
Kuala Lumpur   Madrid   Melbourne   Mexico City   Mumbai   Nairobi
São Paulo   Shanghai   Taipei   Tokyo   Toronto

Copyright © 2005 by Oxford University Press, Inc.

Published by Oxford University Press, Inc.
198 Madison Avenue, New York, New York 10016

www.oup.com

Oxford is a registered trademark of Oxford University Press

Library of Congress Cataloging-in-Publication Data
Insect-fungal associations : ecology and evolution /
edited by Fernando E. Vega, Meredith Blackwell.
p. cm.
Includes bibliographical references.
ISBN 0-19-516652-3
1. Fungi as biological pest control agents.   2. Insect
pests—Biological control.   3. Fungi in agriculture.   4. Insect-fungus
relationships.   I. Vega, Fernando E.   II. Blackwell, Meredith.
SB976.F85 I57 2004

9 8 7 6 5 4 3 2 1

Printed in the United States of America
on acid-free paper

It is not only poets who are interested in puzzles. All of us live in basic mystery. Science and religion jostle one another in the shadows, throwing in each other's eyes the dust of beauty, possibilities, distant myths, and approximate truth.

Pablo Neruda
Passions and Impressions

Acknowledgments

F.E.V. thanks Pedro Barbosa, Pat Dowd, and Eric Rosenquist. Also, thanks are due to Ann Sidor, Francisco Posada, and Monica Pava-Ripoll for their superb work in the Beltsville laboratory. Special thanks to Wendy and Ian, for being so wonderful and for their love. M.B. thanks Sung-Oui Suh and undergraduate students including Christine Ackerman, Katie Brillhart, Cennet Erbil, Nhu Nguyen, and Amy Whittington, who keep her lab productive. Thanks also to James Kimbrough and Quentin Wheeler, who first showed her the insect fungi, and David Malloch, with whom she pursued the origin of euascomycetes without mycelium. Renée, Elise, David, and Esme provide support. We both thank Kirk Jensen, our initial editor at Oxford University Press, for taking on this project, and Anne Rockwood for helping us to complete it. We also thank Lisa Stallings and Rosanne Hallowell at Oxford University Press for overseeing the copy-editing and production, and Lisa Hamilton for the cover design. Thanks also to Sung-Oui Suh, who helped with proofreading.

Contents

Contributors

Duur K. Aanen, Department of Population Ecology, Zoological Institute, University of Copenhaguen, Universitetsparken 15, 2100 Copenhagen, Denmark

A. Elizabeth Arnold, Department of Biology, Duke University, Durham, North Carolina 27708, USA

Michael J. Bidochka, Department of Biological Sciences, Brock University, St. Catharines, Ontario L2S 3A1, Canada

Meredith Blackwell, Department of Biological Sciences, Louisiana State University, Baton Rouge, Louisiana 70803, USA

Jacobus J. Boomsma, Department of Population Ecology, Zoological Institute, University of Copenhaguen, Universitetsparken 15, 2100 Copenhagen, Denmark

Cameron R. Currie, Department of Ecology and Evolutionary Biology, Haworth Hall, University of Kansas, Lawrence, Kansas 66045, USA

Patrick F. Dowd, Crop Bioprotection Research Unit, U.S. Department of Agriculture, Agricultural Research Service, Peoria, Illinois 61642, USA

Naomi M. Fast, Department of Botany, University of British Columbia, 3529-6270 University Boulevard, Vancouver, British Columbia V6T 1Z4, Canada

Michael J. Furlong, Department of Zoology and Entomology, School of Life Sciences, University of Queensland, Brisbane 4072, Queensland, Australia

Thomas C. Harrington, Department of Plant Pathology 221 Bessey Hall, Iowa State University, Ames, Iowa 50011, USA

Patrick J. Keeling, Department of Botany, University of British Columbia 3529-6270 University Boulevard, Vancouver, British Columbia V6T 1Z4, Canada

Leslie C. Lewis, Corn Insects and Crop Genetics Research Unit, U.S. Department
of Agriculture, Agricultural Research Service, Ames, Iowa 50011, USA

Ulrich G. Mueller, School of Biological Sciences, University of Texas, Austin,
Texas 78712, USA

Judith K. Pell, Division of Plant and Invertebrate Ecology, Rothamsted
Research, Harpenden, Hertfordshire AL5 2JQ, England

Stephen A. Rehner, Insect Biocontrol Laboratory, U.S. Department of Agriculture,
Agricultural Research Service, Bldg. 011A, Beltsville, Maryland 20705, USA

Ted R. Schultz, Department of Systematic Biology, MRC 188, National Museum
of Natural History, Smithsonian Institution, Washington, DC 20560, USA

Cherrie L. Small, Department of Biological Sciences, Brock University,
St. Catharines, Ontario L2S 3A1, Canada

Sung-Oui Suh, Department of Biological Sciences, Louisiana State University,
Baton Rouge, Louisiana 70803, USA

Fernando E. Vega, Insect Biocontrol Laboratory, U.S. Department of Agriculture,
Agricultural Research Service, Bldg. 011A, Beltsville, Maryland 20705, USA

Alex Weir, Environmental and Forest Biology, SUNY College of Environmental
Science and Forestry, 350 Illick Hall, 1 Forestry Drive, Syracuse, New York
13210, USA

Introduction: Seven Wonders of the Insect–Fungus World

Meredith Blackwell
Fernando E. Vega

This book first arose out of a passage in Borges, out of the laughter that shattered, as I read the passage, all the familiar landmarks of my thought—*our* thought, the thought that bears the stamp of our age and our geography—breaking up all the ordered surfaces and all the planes with which we are accustomed to tame the wild profusion of existing things, and continuing long afterwards to disturb and threaten with collapse our age-old distinction between the Same and the Other. This passage quotes a "certain Chinese encyclopaedia" in which it is written that "animals are divided into: (a) belonging to the Emperor, (b) embalmed, (c) tame, (d) sucking pigs, (e) sirens, (f) fabulous, (g) stray dogs, (h) included in the present classification, (i) frenzied, (j) innumerable, (k) drawn with a very fine camelhair brush, (l) etcetera, (m) having just broken the water pitcher, (n) that from a long way off look like flies." In the wonderment of this taxonomy, the thing we apprehend in one great leap, the thing that, by means of the fable is demonstrated as the exotic charm of another system of thought, is the limitation of our own, the stark impossibility of thinking *that.*

Michel Foucault
The Order of Things

Insects and fungi share a long history of association in the common habitats where they endure similar environmental conditions. Only relatively recently did mycologists stop killing the fungus-destroying insects they collected with their specimens and entomologists realize the fungal influence on insects. Prompted by an interest in understanding the associations between the two groups of unrelated organisms, an opportunity for people with interests in the interactions between the insects and fungi arose. Quentin Wheeler, a proponent of cladistic analysis, hoped to find new disciples for Hennig among mycologists and hoped to encourage entomologists and mycologists to work together to understand the interactions among the speciose groups of organisms. A symposium organized by Wheeler and one of us (M.B.) was held at the Entomological Society of America, Eastern Branch, meeting at Syracuse and then at Ithaca, New York, in 1981. An edited book was an outcome of the gathering (Wheeler and Blackwell 1984). Some people only noticed the book when it hit the remainder tables, and oddly enough a single—although five star—Amazon.com review, from "a reader from Ithaca, NY USA" suggests, "This is an excellent book about the relationships between fungus and insects. It shows how cladistics is the ultimate method to codify biological information into a framework common to all Biology." It is encouraging that some of the chapters have been cited in this volume. In fact, a cluster of edited volumes on a similar topic published over a 10-year period more than 20 years ago indicates that there was intense interest in the interactions of these organisms (Batra 1979; Pirozynski and Hawksworth 1988; Schwemmler and Gassner 1989; Wilding et al. 1989). This is, by complete coincidence, the 20th anniversary of the Wheeler and Blackwell volume that was slightly before its time, and the many advances in the study of the insect—fungal associations call for another look at these associations.

Foucault Made a Distinction between "the Same and the Other.". . . Twenty Years Later, It Is Not the "Same."

Over the intervening 20 years cladistics has become a way of life to systematists and other biologists. Entomologists and, especially, mycologists, who previously had had few characters to analyze, are now collecting and analyzing molecular data. Over the last 20 years the encounters between entomologists and mycologists have increased, and to some extent teams of these scientists work together. Several new scientists have been trained as symbiologists.

In late 2002, one of us (F.E.V.) suggested that the time had come for a new book on insect–fungus associations. The present volume is different in content from the range of topics covered two decades ago, but that is not the major difference. The revolution has come with our ability to collect genetically based characters, DNA sequences, and other molecular markers, for use in phylogenetic analysis, for identification, and for population markers. No longer do we need classifications as that of Borges' "encyclopaedia" with animals divided into artificial categories such as belonging to the Emperor, embalmed, or that from a long way off look like flies. We can distinguish a fly from a wasp or other fly-mimic with a convergent morphology. Other difficult tasks are possible, such as distinguishing among species of *Ceratocystis* and those of *Ophiostoma*, both of which have convergent perithecial morphologies selected for success in insect-associated spore dispersal. Not only can these fungi be distinguished, but a measure of their vast genetic difference that reflects their present taxonomy is also possible. As you read many of the chapters in this volume, consider what it was like not so long ago when the organisms could not be identified or their taxonomic position established. Insect–fungus research of only a few years ago often was done in ignorance of the identity of the fungus, let alone with population-level information. Even estimated dates of associations between insects and fungi and rates of evolution during establishment of symbiotic associations can be attempted, albeit sometimes with controversy, using molecular information and grounded by the fossil record (Berbee and Taylor 1993, 2001; Lutzoni and Pagel 1997; Blackwell 2000; Heckman et al. 2001). Another recently published book (Bourtzis and Miller 2003) emphasizes insect symbioses, including some involving fungi, and perhaps we are at the front end of another cycle of volumes on insect–microbial interactions.

In this volume, the types of associations among insects and fungi are divided into two groups: interactions in which fungi act against insects and those in which fungi form mutualistic associations with insects. The division is artificial and emphasizes the fact that we know more about the effect of the fungus on the insect, but with the molecular markers at hand, we are beginning to make progress toward understanding the direct benefits to the fungus. We like to think of the interactions discussed here as the wonders of the insect–fungal association world, contrived to number seven, as described below.

First we present a section on fungi acting against insects, including discussions of parasitism. Two of the most important necrotrophic parasites (1) are important for their insect control potential in a world that becomes increasingly polluted, especially in agricultural systems. Stephen A. Rehner, Michael J. Bidochka, and Cherrie L. Small discuss *Beauveria* and *Metarhizium*, asexual ascomycetes that kill a variety of insect hosts. The use of molecular markers has advanced the field, so a better understanding of host–parasite interactions is possible. Several necrotrophic interactions that are exciting topics for research include the complexity of multipartite interactions. Michael J. Furlong and Judith K. Pell report on interactions among three groups of organisms, insect parasitoids, predators, and entomopathogenic fungi, and the insects they attack. Other complex interactions involve fungi that are hidden away within the leaves of plants. A possible beneficial role for endophytes (2) in the broad-leafed plants they inhabit without symptoms has been suggested for many years. Elizabeth Arnold and Leslie C. Lewis outline the distribution of such fungi within plants and propose a role for at least one of them, *Beauveria bassiana*.

Analysis of molecular characters helps us discover what a fungus is. Slime molds and water molds are excluded from the kingdom Fungi, but several groups previously considered as protists (e.g., *Pneumocystis* and members of the "DRIPS" phylum, including protistan fish parasites, *Dermocystidium* and *Ichthyophonus*) are now considered to be fungi. Several chapters discussed in this volume help to place parasites with reduced morphology (3), for which morphological characters did not allow them to be recognized for what they are. Naomi Fast and Patrick J. Keeling discuss the evidence that leans toward recognition of fungal roots of Microsporidia, a group of parasites of a variety of organisms including arthropods. Some evidence suggests a relationship not just to fungi, but possibly to a group of necrotrophic parasites of insects. Another group that has benefited greatly from analysis of DNA sequence characters has been the Laboulbeniales. Fungal biotrophic parasites of insects are rare, except the very successful exception of the Laboulbeniomycetes. This group was once suggested as a connection between floridean red algae and ascomycetes, and also was considered as a member of three different fungal phyla. Alex Weir and Meredith Blackwell discuss the associations of these ascomycetes with certain arthropods and assess what phylogenetic analysis tells us about these uncultivable organisms. Certain Laboulbeniomycetes provide examples of complex methods of spore dispersal by arthropods (4).

An area of insect–fungus interactions that has prospered from the use of molecular techniques is that of farming mutualisms (5). These highly developed associations occur among different groups of insects, including one in the New World and another in the Old World. Ted R. Schultz, Ulrich G. Mueller, Cameron R. Currie, and Stephen A. Rehner make use of molecular techniques to develop a new body of symbiotic theory relating to attine ants, the two unrelated basidiomycetes that they cultivate, and a parasitic fungus controlled by the antibiotics of an associated bacterium. In this volume, they compare agriculture in humans and ants and arrive at new ideas concerning the benefit accrued to the organisms involved in the interaction. Phylogenetic analyses have enlightened us about the associations between the Old World fungus-growing termites and their fungal crop. Duur K. Aanen and Jacobus J. Boomsma provide evidence that the termite–basidiomycete association, unlike that of the ants and fungi, arose once. Their work traces the migrations that the termites and fungi made together.

Certain bark and ambrosia beetles also rely on fungi as a sole source of food; although they disperse fungi, they do not actively farm them in the manner of attine ants and termites. Mycophagy (6) occurs among some bark beetles and both ascomycetes and a few basidiomycetes, and Thomas C. Harrington provides insight into the ecology and evolution of fungus-feeding bark beetles and their fungal partners. His emphasis is on closely related fungal species pairs, detected by molecular markers. The fungi have different degrees of specialization with the beetles, and the pathway of evolution of mycangial associations is clarified.

Some fungi also interact with insects by providing nutritional supplements (7). Fernando E. Vega and Patrick F. Dowd emphasize the role of yeast–insect endosymbionts in aiding the digestion and detoxification of plant material ingested by insects and provide information on the basis of such interactions. Biodiversity studies have led to the discovery of an enormous variety of true yeasts (Saccharomycetes). Sung-Oui Suh and Meredith Blackwell describe the gut of beetles as a habitat for new yeasts that will increase the known number from about 800 species to almost 1000. It is suspected that these yeasts also might provide nutritional supplements for insects.

Insects involved in fungal associations include members of the Coleoptera, Diptera, Homoptera, Hymenoptera, and Isoptera, among others. The fungi involved in interactions with insects may be clustered taxonomically, as is the case for Ascomycetes in the Hypocreales (e.g., *Beauveria*, *Metarhizium*, *Fusarium*), ambrosia fungi in the genera *Ophiostoma* and *Ceratocystis* and their asexual relatives, Laboulbeniomycetes, Saccharomycetes, and the more basal Microsporidia. Other groups, however, have only occasional members (e.g., mushrooms cultivated by attine ants and termites) in such associations. The 11 chapters included in this volume, however, are only a beginning, and there are certain to be many more than seven wonders to observe in the study of insect—fungus associations.

Literature Cited

Batra, L. R. 1979. *Insect-fungus symbiosis: Nutrition, mutualism, and commensalism.* Monclair, NJ: Allanheld, Osmun & Co.

Berbee, M. L., and J. W. Taylor. 1993. Dating the evolutionary radiations of the true fungi. *Canadian Journal of Botany* 71:1114–1127.

Berbee, M. L., and J. W. Taylor. 2001. Fungal molecular evolution: Gene trees and geologic time. In *The Mycota: A comprehensive treatise on fungi as experimental systems for basic and applied research.* Vol. 7. *Systematics and evolution*, part B, ed. D. J. McLaughlin, E. G. McLaughlin, and P. A. Lemke, pp. 229–245. Berlin: Springer Verlag.

Blackwell, M. 2000. Perspective: evolution: terrestrial life—fungal from the start? *Science* 289:1884–1885.

Bourtzis K., and T. Miller, eds. 2003. *Insect symbiosis.* Boca Raton, FL: CRC Press.

Heckman, D. S., D. M. Geiser, B. R. Eidell, R. L. Stauffer, N. L. Kardos, and S. B. Hedges. 2001. Molecular evidence for the early colonization of land by fungi and plants. *Science* 293:1129–1133.

Lutzoni, F., and M. Pagel. 1997. Accelerated evolution as a consequence of transition to mutualism. *Proceedings of the National Academy of Science of the USA* 94:11422–11427.

Martin, M. M. 1987. *Invertebrate-microbial interactions.* Ithaca, NY: Cornell University Press.

Pirozynski, K. A., and D. L. Hawksworth, eds. 1988. *Coevolution of fungi with plants and animals.* London: Academic Press.

Schwemmler, W., and G. Gassner, eds. 1989. *Insect endocytobiosis: Morphology, physiology, genetics, evolution.* Boca Raton, FL: CRC Press.

Wheeler, Q., and M. Blackwell, eds. 1984. *Fungus-insect relationships: Perspectives in ecology and evolution.* New York: Columbia University Press.

Wilding, N., N. M. Collins, P. M. Hammond, and J. F. Webber, eds. 1989. *Insect-fungus interactions.* New York: Academic Press.

Part I

Fungi Acting against Insects

1

Phylogenetics of the Insect Pathogenic Genus *Beauveria*

Stephen A. Rehner

B *eauveria* (Balsamo) Vuillemin (Ascomycota: Hypocreales) is a cosmopolitan genus of soilborne entomogenous molds. As one of the first entomopathogenic fungi to be discovered, elucidation of the role of *Beauveria* as the cause of white muscardine, a devastating disease afflicting the European silkworm industry in the 18th and 19th centuries, initiated the study of fungal insect pathology. Additionally, scientific studies by Agostino Bassi demonstrating that *Beauveria* was the infectious disease agent that caused the white muscardine disease of silkworm were important antecedents to the germ theory of disease, arguably one of the most significant theories in the history of science. Although investigation of *Beauveria* was instigated by the need to protect a beneficial domesticated insect, *Beauveria* is also an important natural pathogen of insects, its hosts including many economically important insect pests. As a result, a great deal of research on *Beauveria* has been motivated by and has focused on the applied use of *Beauveria* in insect biological control.

Since its discovery nearly two centuries ago, *Beauveria* has been found to possess several advantageous characteristics that have positioned it as one of the principal organisms used in research on fungal insect pathology (Steinhaus 1963) and in insect biological control (Dunn and Mechalas 1963; Ferron 1978; Gillespie and Moorehouse 1989; Ferron et al. 1991). First, due to its cosmopolitan distribution, easy recognition, and frequent appearance in nature, *Beauveria* is one of the most widely recognized and encountered of all entomopathogenic fungi. Second, the extremely broad host range of *Beauveria*, which is known to infect more than 700 species of insects (Goettel et al. 1990), and its wide variation in virulence toward different insect hosts, make it one of the more versatile candidate fungal entomopathogens for biological control of insects. Third, *Beauveria* is an extremely tractable organism. It is easily isolated from insect cadavers or from soil by using simple media, antibiotics, and selective agents (Beilharz et al. 1982; Chase et al. 1986), and by baiting soil with insects (Zimmerman 1986). It flourishes in the laboratory on simple media (Goettel and Inglis 1997) and can be conserved by storage in glycerol solutions at ultra-low temperatures or by freeze-drying (Humber 2001). Consequently, *Beauveria* is the best represented entomopathogenic fungus in world culture collections. Furthermore, *Beauveria* is amenable to industrial-scale production and to formulation as a mycoinsecticide (Bartlett and Jaronski 1988; Feng et al. 1994). One species, *B. bassiana*, has the ability to exist as an endophyte in corn (Bing and Lewis 1992; Wagner and Lewis 2000; chapter 4), which may lead to unique methods for crop protection.

Despite nearly two centuries of research, progress toward elucidating the evolution and genetics of *Beauveria* has just begun due to the recent development of adequate tools and genetic methods for asexual organisms. The purpose of this chapter is to review the history of systematic and genetic research on *Beauveria* and to provide a summary of recent progress in understanding its phylogeny, population genetics, and genetics.

Discovery and Characterization of *Beauveria*

Beauveria was discovered at a time when the existence of fungi and other microbes and their biological roles, particularly as agents of plant and animal diseases, were first being discovered, communicated, and debated (Ainsworth 1976). The Italian lawyer and scientist Agostino Bassi (1773–1856) first brought *Beauveria* to the attention of western science. The following account of Bassi and his work with *Beauveria* is reconstructed from Major (1944), Steinhaus (1949), Ainsworth (1956), and Porter (1973).

Trained in law and granted a doctoral degree by the University of Pavia in 1798, Bassi also pursued a comprehensive curriculum in the natural sciences, his lifelong interest. Bassi's most significant and enduring contributions to biology were his studies on the muscardine disease of silkworm larvae, which chronically afflicted the silk industries of France and Italy in the 18th and 19th centuries. The muscardine disease was characterized by the rapid death and mummification of stricken larvae, whose exteriors, under conditions of high humidity, typically became covered with a white, powdery layer. Muscardine, the English term for the disease of silkworm larvae and many other fungal diseases of insects, is probably derived from the Italian *moscardino*, which apparently derives from a perceived similarity to fruit confections, but, the etymology has additional interpretations (see Steinhaus 1949). In Italy, muscardine disease was known as *mal del segno* ("the mark disease") and also was referred to variously as *calcinaccio, calcinetto*, or *calcino* in reference to the chalky texture of the insect cadavers, or *cannellino*, a little bean.

Bassi began his studies on silkworm culture and the muscardine disease in 1807, and he continued this work over the next 30 years. The first decade of Bassi's research, by his own admission, yielded only frustration as he attempted unsuccessfully to verify prevailing beliefs that the disease developed spontaneously as a result of environmental conditions (e.g., atmosphere, diet, breeding method) and that deliquescence of the animal's tissues in the early stages of infection was associated with acidification. Bassi was particularly creative in devising ways to kill and mummify silkworm larvae that mimicked the muscardine disease, except for one essential characteristic: the silkworm larvae stricken in his experiments were unable to transmit the disease to healthy insects. By 1816, discouraged by his lack of success with this line of inquiry, Bassi speculated that the insects acquired the disease from a foreign agent.

Focusing on this idea, Bassi determined that the sole infectious agent of white muscardine was the white powdery substance frequently produced on the exterior of animals killed by the disease. He observed that the powder, when implanted into healthy silk moths on the point of a needle, reproduced the symptoms, lethal outcome, and postmortem effects of muscardine disease. Bassi demonstrated that the disease developed only in larvae exposed directly to diseased insects or in areas that previously housed infected insects. Furthermore, Bassi demonstrated that muscardine infections could be completely eliminated or reduced from moth-rearing areas by isolating healthy larvae from these environments. He also demonstrated that chemical and heat treatments of the rearing environment eliminated the infectious properties of the muscardine powder. By 1836 Bassi had concluded the muscardine agent was a living germ and, because of the minute size of its "seed," determined that the agent was a fungus, which he referred to as *Botrytis paradoxa* (Yarrow 1958).

Despite detailed analysis of the etiology of the muscardine disease, skeptics disputed Bassi's claims, holding to the notion that the muscardine disease was due to spontaneous generation. Bassi sought independent validation of his claims by repeating his experiments and submitted these results to a panel at the University of Pavia, which, upon reviewing the evidence, endorsed Bassi's methods and conclusions. Bassi summarized his observations and conclusions in two monographs, for which there is an English translation (Yarrow 1958).

Bassi's research and analysis of the muscardine disease yielded three significant results. First, the discovery of the etiology of the muscardine disease, its infectious nature, and the development of control strategies benefited the European silkworm industry. Second, Bassi's work illustrated for the first time that fungi can cause disease in insects and opened a new area of investigation in basic and applied mycology. Third, Bassi's description of muscardine disease was the first enunciation of the germ theory of disease, one of the most important discoveries in the history of science.

In spite of his detailed observations and insights into the underlying cause and true nature of the muscardine disease, Bassi is rarely credited with the discovery of the germ theory of disease because he did not provide the essential physical proofs for its basis. Even though he had access to a microscope, he did not provide a detailed description of the fungus, perhaps due to his poor eyesight. Bassi also did not isolate the fungus into pure culture for the simple reason that methods for axenic culture of microorganisms had not yet been developed. Had he been able to examine and describe the fungus in isolation from other microorganisms and demonstrate that pure cultures were infectious and capable of producing the disease, Bassi might have received credit for the germ theory instead of Pasteur, Koch, Schoenlein, and others who provided these critical details for several important human, animal, and plant diseases in the decades after Bassi's work. These scientists had access to improved microscopes for describing the minute structure of microbes and were able to grow pathogenic microorganisms in pure culture, with which they demonstrated the ability of these microbes to infect and cause disease in their hosts. Bassi's publications were cited by both Schoenlein and Pasteur (Arcieri 1956), and it is plausible that his work influenced the thinking of these later scientists.

Bassi's publications on the silkworm muscardine disease were soon followed by characterization of white muscardine disease and its causal organism. Balsamo-Crivelli (1835) formally named the muscardine fungus *Botrytis bassiana*, in honor of Bassi, facilitating scientific and technical communications about the organism and muscardine disease. Audoin (1837a,b) repeated Bassi's inoculation studies confirming the etiology of the disease; he also reported that the disease was not restricted to silkworm but also caused disease in other insect species. Johanys (1839) first grew *B. bassiana* on organic substrates. Vittadini (1852) grew *B. bassiana* in pure culture and apparently was the first to use a solid culture medium, gelatin. Furthermore, Vittadini (1853) provided additional validation that *B. bassiana* was the cause of muscardine of silkworm. Thus, within a short span of time, understanding muscardine disease of silkworms was transformed from superstitious lore to a reasonably well-characterized biological phenomenon known to be caused by a specific organism.

Taxonomy of *Beauveria*

Vuillemin (1912) formally described the genus *Beauveria*, assigning *Botrytis bassiana* as the type species, *Beauveria bassiana* in recognition of J. Beauverie (1914) and his studies on *Beauveria* and muscardine disease. Because species of *Beauveria* reproduce by production of conidia (mitospores), they have been presumed to be asexual and were traditionally classified as hyphomycetous asexual fungi (Deuteromycetes or Fungi Imperfecti), a practice no longer followed because molecular methods make it possible to place such fungi among their sexual relatives. Despite several comprehensive taxonomic analyses of *Beauveria* during the last century, there remain significant problems in the identification, taxonomy, and nomenclature of species in this genus.

Beauveria is easy to distinguish morphologically. Its most distinctive characteristics are sympodial to whorled clusters of short-globose to flask-shaped conidiogenous cells that produce a succession of one-celled, sessile, hyaline, holoblastic conidia on a progressively elongating sympodial (zig-zag) rachis (Kirk et al. 2001). *Beauveria* has been compared to other genera, including *Acrodontium*, *Isaria*, *Tritirachium* (de Hoog 1972), and *Tolypocladium* (Gams 1971) due to similarities in their conidiogenous cells (Kirk et al. 2001). von Arx (1986) synonymized *Tolypocladium* with *Beauveria*, but this proposal has generally been rejected. A comparison of available internal transcribed spacer (ITS) sequences among species in these three genera indicated that they are distinct from each other and from *Beauveria* (Rehner and Buckley unpublished ms.).

In culture, *Beauveria* species typically produce concolorous white mycelium and conidia, although some isolates may produce a yellow pigment in the older, central parts of the colony, as in *B. amorpha* and *B. velata*. Colony growth tends to be rapid, and the texture of the mycelium is typically lanate to woolly, and synnemalike projections are occasionally observed (Kirk et al. 2001). Conidia in culture can often be copious, frequently creating a chalky, mealy, or powdery appearance at the colony surface. At low magnification the surfaces of conidium-producing colonies usually are covered by myriad "spore balls," compact globose heads of conidiogenous cells and accumulated conidia, giving the mature colony surface a minute granular appearance. Many isolates excrete a red pigment into the culture medium, although not under all culture conditions (de Hoog 1972; Eyal et al. 1994).

In direct contrast to the straightforward generic delimitation of *Beauveria*, species recognition and identification are problematic because of a lack of informative morphological variation. The conidial form is the only morphological feature of *Beauveria* that has proven somewhat useful for species delimitation, and conidia vary from globose, ellipsoid, cylindrical, reniform, to vermiform and range in size from 1.8 to 6.0 μm in their greatest dimensions (fig. 1.1). The conidia of all described species of *Beauveria* are thin-walled and smooth, except for the South American species *B. velata*, which has verrucose conidia (Samson and Evans 1982).

New species of *Beauveria* were described steadily throughout the 19th and 20th centuries, complicating the nomenclature and identification of these species. In total, 49 species have been classified in *Beauveria*, approximately 22 of which are currently considered valid (Kirk 2003). Unfortunately, overlapping variation in the size and shape of conidia among many of these species has raised questions about their validity and has complicated routine species identifications.

Monographic studies of *Beauveria* have been conducted by Petch (1926), MacLeod (1954), and de Hoog (1972). Petch (1926) detailed the morphology and cultural characteristics of isolates representing eight putative species but could confidently discern only two distinctive entities based on conidial form; he referred to these as *B. bassiana* and *B. densa*. Petch (1926) further concluded that cultural features of *Beauveria* are influenced by culture conditions and thus are unreliable as taxonomic characters. In his monograph, MacLeod (1954) examined 16 specific entities. However, like Petch, he was able to differentiate just two species, *B. bassiana* and *B. tenella*, which correspond to Petch's *B. bassiana* and *B. densa*, respectively. de Hoog (1972) also came to similar conclusions as Petch and MacLeod but used the name *B. brongniartii* in place of *B. densa* (*sensu* Petch) and *B. tenella* (*sensu* MacLeod) and recognized an additional species, *B. alba*. He later transferred *B. alba* to *Engyodontium* (de Hoog 1978).

Since de Hoog's (1972) analysis, several distinctive new species from South America have been described. These include *B. verminconia* with small sickle-shaped conidia (de Hoog and Rao 1975), *B. velata* with verrucose conidia (Samson and Evans 1982), and *B. amorpha*, characterized by large, cylindric conidia (Samson and Evans 1982). The name *B. amorpha* is also used for Asian and North American isolates that produce cylindric conidia (Humber 1997). The most recently

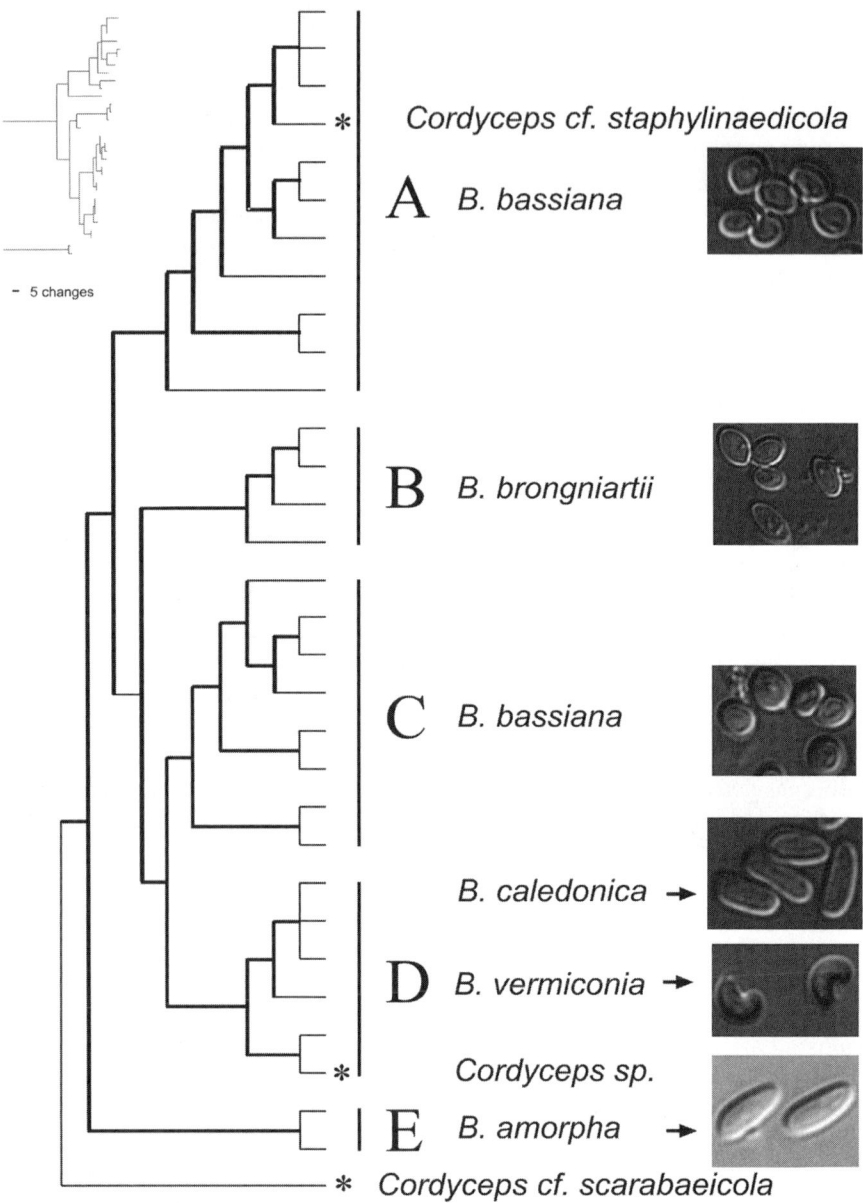

Figure 1.1. Phylogenetic relationships of *Beauveria*. Consensus cladogram summarizing several independent Bayesian and parsimony analyses. Branches with significant nodal support are indicated in bold. Clades A–E are labeled according to their morphological species identification and include micrographs of representative conidia. Inset tree in upper left is a phylogram indicating relative branch lengths.

described *Beauveria* species, *B. caledonica*, from Scotland, also produces cylindrical conidia similar to those of *B. amorpha* (Bissett and Widden 1986). Six morphological species can be discerned in *Beauveria*: *B. amorpha*, *B. bassiana*, *B. brongniartii*, *B. caledonica*, *B. velata*, and *B. vermiconia* (fig. 1.1). However, in North America and Europe, the majority of environmental isolates of *Beauveria* fall within the morphological circumscription of either *B. bassiana* or *B. brongniartii*, and this is reflected in taxonomic keys and descriptions for species identification used in these regions (Tanada and Kaya 1993; Humber 1997). Unquestionably, the greatest taxonomic confusion surrounds *B. bassiana*, for which many morphologically similar species have been described. Thus far, seven species have been placed into synonymy with *B. bassiana*, including *B. densa*, *B. doryphorae*, *B. effusa*, *B. globulifera*, *B. shiotae*, *B. stephanoderis*, and *B. sulfurescens* (Kirk 2003).

The most significant recent development in the systematics of *Beauveria* has been evidence of direct links to the teleomorph genus *Cordyceps* (Ascomycota: Hypocreales: Clavicipitaceae), a connection previously suspected but not confirmed. Shimazu et al. (1988) and Li et al. (2001) each described species of *Cordyceps* (e.g., *C. brongniartii* and *C. bassiana*) that produced *Beauveria* anamorphs (asexual states) from cultures isolated from perithecial stromata. As their specific epithets imply, the anamorphs of each species correspond morphologically to *B. brongniartii* and *B. bassiana*, respectively. Huang et al. (2002) provided additional confirmation of the *C. bassiana*–*B. bassiana* relationship by showing that *C. bassiana* was nested within a clade of *B. bassiana* isolates in phylogeny based on ITS sequences. Similarly, a phylogeny of nuclear, small subunit ribosomal RNA (SSU rRNA) sequences (Sung et al. 2001) yielded the inference that *C. scarabaeicola* arose from within *Beauveria* and is the sister to *B. caledonica*. Both anamorph–teleomorph connections have been independently corroborated in subsequent molecular phylogenetic analyses, which also revealed a third possible teleomorph connection, that of a *C. cf. scarabaeicola* specimen that was placed as the most basal branch of the *Beauveria* lineage (Rehner and Buckley in press). All three *Beauveria* teleomorphs produce morphologically similar yellow perithecial stromata. However, comparative morphological studies of the teleomorphs have not yet been made, and the extent to which characters of the teleomorphs will contribute to a better understanding of the taxonomy and systematics of this group cannot be predicted. In any case, future systematic studies in *Beauveria* must be expanded to include characters of the teleomorphs as these become known.

All *Beauveria* teleomorphs described thus far are from Asia. It is not known whether this regional concentration of teleomorphs is coincidental (i.e., that populations of the different species groups in Asia just happen to be sexual) or whether, because of the traditional interests in the medicinal properties of *Cordyceps*, Asian mycologists simply are more attuned and diligent in their search for *Cordyceps* teleomorphs, including those of *Beauveria*. Another possibility has been suggested by Bidochka and Small (chapter 2, this volume), who discuss *Cordyceps* teleomorphs of *Metarhizium* known only from Asia. This finding, coupled with high taxonomic diversity, was used to suggest an Asian center of origin for *Metarhizium*. Not enough information is available yet for *Beauveria* to draw such conclusions. It is clear, however, that consideration of cultural and molecular phylogenetic criteria is likely to play a pivotal role in the identification and verification of anamorph–teleomorph links. In view of the recent discovery of *Cordyceps* teleomorphs of *Beauveria* in Asia, the search for additional anamorph–teleomorph connections in *Beauveria* should be expanded to include other regions of the world.

Assessments of Genetic Diversity in *Beauveria*

Since the 1970s, the advent of numerous molecular techniques and associated analytical methods provided new opportunities for advancing taxonomic, phylogenetic, and population genetics of *Beauveria*. During the 1980s and 1990s, numerous studies were published in which a wide array of chemotaxonomic, biochemical, and DNA-based techniques were applied to characterize underlying patterns of genetic variation and relatedness within *Beauveria* (table 1.1). Each of the different types of molecular markers described in these studies was effective in detecting interstrain variation within *Beauveria*, usually providing a finer level of resolution than previously achieved using morphology alone. However, few of these techniques yielded insights or inferences that effectively resolved specific taxonomic or genetic relationships in *Beauveria*. In many cases use of a technique was restricted to a single

Table 1.1. Techniques used to characterize patterns of genetic variation and relatedness in *Beauveria*.

Technique	References
Isozymes	Riba et al. (1986), Poprawski et al. (1988), St. Leger et al. (1992)
Serology	Shimizu and Aizawa (1988)
Morphology/Biochemistry API zymograms, acetylesterase, casein hydrolysis	Mugnai et al. (1989)
Mitochondrial DNA restriction fragment length polymorphisms (RFLPs)	Hegedus et al. (1993b)
Large subunit ribosomal DNA (rDNA) sequencing	Rakotonirainy et al. (1991)
RFLPs	Kosir et al. (1991), Maurer et al. (1997)
Electrophoretic karyotyping	Pfeifer and Khachatourians (1993)
Large subunit rDNA intron sequences	Neuvéglise and Brygoo (1994), Neuvéglise et al. (1996)
RFLP and nucleotide sequences of internal transcribed spacer (ITS) regions	Neuvéglise et al. (1994)
PCR–single-strand conformation polymorphism of taxon-specific markers	Hegedus and Khachatourians (1993a, 1996)
Random amplified polymorphic DNA (RAPD) markers	Bidochka et al. (1994), Cravanzola et al. (1997), Maurer et al. (1997)
Morphology and RAPD markers	Glare and Inwood (1998)
ITS RFLP	Coates et al. (2002)

exploratory study (e.g., serology, single-strand conformation polymorphisms, isozymes, karyotyping, chemotaxonomy, ribosomal DNA introns, mitochondrial restriction fragment length polymorphisms [RFLP]); thus the usefulness of these methods cannot be considered to have been thoroughly evaluated. Overall, it is difficult to synthesize or generalize from the results of these studies due to the disparate types of characters and isolates sampled. In any case, most of these early methods have been superseded by the use of nucleotide sequence data, microsatellite markers, amplified fragment length polymorphism (AFLP), and single nucleotide polymorphism (SNP) markers for analyzing genetic relationships in fungi. Nevertheless, these studies define the transition from a reliance on morphology to acceptance of genetic approaches as a basis for determining genetic and taxonomic relationships in *Beauveria*.

Phylogenetics of *Beauveria*

Past attempts to order relationships in *Beauveria* have rarely been explicitly phylogenetic in focus, but have instead emphasized taxonomic identification and, particularly, the application of molecular-based markers for discrimination of individual isolates. As a result, few of the central questions about the phylogeny of *Beauveria* have been specifically investigated. These questions include: (1) What is the evolutionary origin of *Beauveria*? (2) Is *Beauveria* monophyletic? (3) What is the sister group to *Beauveria*? (4) Are the morphological species in *Beauveria* monophyletic, and how are they related to one another? (5) Are geographically widespread species composed of multiple cryptic species? and (6) What *Cordyceps* teleomorphs are most closely related to *Beauveria*, and are any derived from within *Beauveria*? Only recently have these questions begun to be addressed; however, significant progress toward answers has already been achieved, suggesting that a comprehensive phylogeny for *Beauveria* will soon be available.

The first explicit molecular phylogenetic hypothesis for the evolutionary origin of *Beauveria* was the SSU rRNA phylogeny of clavicipitaceous fungi and other pyrenomycetes by Sung et al. (2001). In this phylogeny, *Beauveria* was shown to form a monophyletic group derived within the Sordariomycetes (Hypocreales: Clavicipitaceae) (Eriksson et al. 2003). The origin of *Beauveria* appears to have been recent, based on low SSU rRNA sequence divergence between *B. bassiana* and *B. caledonica* (<1%), which, at this time, is the deepest divergence known within the genus. Additional support for a recent origin of *Beauveria* is a lack of evidence for saturation at synonymous sites of protein-coding genes, introns, and noncoding sequences between these two species (Rehner unpublished data). Molecular dating of fungal divergences is complicated by a lack of fossils and the genes used (Berbee and Taylor 1995, 2001; Heckman et al. 2001; Kasuga et al. 2002). Thus, *Beauveria* is likely to have originated during the last 75–100 million years (my), with a minimum estimate of 25 my for the origin of *B.* cf. *bassiana* based on an amber-preserved fossil (Poinar and Thomas 1984). Future progress toward dating the origin and major divergences in *Beauveria* from molecular data will most likely rely on analyses of synonymous positions of protein-coding genes, introns, and noncoding sequences and, it is hoped, the discovery of additional fossil material. As was mentioned above, an important inference obtained from the SSU rRNA analysis is the demonstration of close relationships between *Beauveria* and the sexual genus *Cordyceps*, such as the relatively close placement of *Cordyceps militaris* to the *Beauveria* clade and the placement of *C. scarabaeicola* within the *Beauveria* clade as sister to *B. caledonica* (Sung et al. 2001). These findings are consistent with hypotheses that at least some lineages within *Beauveria* remain sexual.

Phylogenetic relationships within *Beauveria* have been inferred from analyses of two nuclear encoded loci, the ribosomal ITS1-5.8S-ITS2 and elongation factor 1-alpha (EF1-α; Rehner and Buckley in press). Taxon sampling in this study included five morphological species of *Beauveria*, including *B. amorpha*, *B. bassiana*, *B. brongniartii*, *B. caledonica*, and *B. vermiconia*. Three *Cordyceps* species, *C. staphylinaedicola*, *C. bassiana*, and *C. scarabaeicola*, each of which has been reported to produce a *Beauveria* anamorph, were also included. In addition, the analysis included a geographic survey of *B. bassiana* isolates to assess phylogenetic structure within this globally distributed species. In the analysis, five principal clades (A–E) were resolved that correspond closely, with one exception, to the five morphological species discussed above (fig. 1.1).

A surprising and significant finding from the molecular phylogenetic analyses of EF1-α and ITS was that *B. bassiana* isolates were partitioned into two unrelated clades, A and C (fig. 1.1); thus, *B. bassiana* is polyphyletic. Additional corroboration for the polyphyly of *B. bassiana* has been obtained from phylogenetic reconstructions with six other nuclear genes (Rehner unpublished data). Past inability to differentiate groups A and C is understandable; isolates from both clades produce indistinguishable small, globose to subglobose conidia (clade A conidia 2.3–3.2 μm; clade C conidia, 2.1–2.9 μm), the only reliable morphological criterion by which *B. bassiana* is defined and identified. Pattern of host affiliation is likewise uninformative for distinguishing clades A and C; both groups are apparently generalist entomopathogens and have wide, overlapping host ranges (Rehner and Buckley in press).

The discovery that *B. bassiana* is polyphyletic further complicates taxonomic and biological understanding of this important species. No type specimen of *B. bassiana* is known to exist; thus a neotype, either from clade A or clade C, needs to be selected. Although original collection location and host source are sometimes used as a basis for selecting neotypes, these criteria are of limited value in the case of *B. bassiana* because clades A and C are sympatric in Europe where *B. bassiana* was originally discovered and because *B. bassiana* was originally described from domestic silkworms. However, clade A has been demonstrated to have a global distribution (Rehner and Buckley in press) and thus is likely to be familiar to the greatest number of people working with *B. bassiana*. Also, the commercial biocontrol formulations such as Mycotrol and Botanigard (Emerald Biotech, Butte, Montana) and Boverin (from former USSR), are placed in clade A (Rehner unpublished data). Together, these facts suggest that selection of clade A would probably be the most appropriate source for designating a neotype for *B. bassiana*. Molecular diagnosis currently is the only method that reliably differentiates clades A and C (Rehner unpublished data). DNA sequence differences are used routinely to distinguish certain fungal taxa, although the formal delimitation of taxa on the basis of nucleotide data is not yet a common practice in fungal systematics. Formal description of new taxa that incorporates nucleotide information is not, however, without precedent. Recently, molecular characters were used in the circumscription of the cryptic species *Cryptococcus posadasii* (Fisher et al. 2002). This may also be necessary in the case of *B. bassiana*.

The discovery that *B. bassiana* is polyphyletic should raise concern that important differences may exist in the ecology, virulence, toxicology, or other important biological characteristics of clades A and C. Detailed studies are therefore needed to determine whether any significant biological differences exist between these two clades, particularly with regard to their ecology and role as entomopathogens.

Clade B, the sister to clade A (fig. 1.1) contains isolates accessioned in the Agricultural Research Service Collection of Entomopathogenic Fungal Cultures (ARSEF) primarily under the name of *B. brongniartii*, a species characterized by ellipsoid to subcylindrical conidia, 3.3–4.8 × 2.1–2.5 μm in size. Originally described from Europe by Petch (1926), *B. brongniartii* has a Eurasian distribution (Rehner and Buckley in press). Lineage diversity detected with EF1-α suggests that *B. brongniartii* is a complex of regionally endemic cryptic species (Rehner unpublished data). *Beauveria brongniartii* is commonly affiliated with Coleoptera and is used as a biocontrol agent against the European cockchafer, *Melolontha melolontha* (Keller et al. 1989) in Europe and against cerambicid beetles in Japan (Biolisa, Nitto Denko Corporation, Japan). A teleomorph, *Cordyceps brongniartii*, was described from Japan (Shimazu et al. 1988), but this anamorph–teleomorph link has not yet been corroborated with either molecular or cultural data.

Clade D (fig. 1.1) includes isolates that, with one exception, produce cylindrical and often slightly curved conidia 3.5–5.2 × 1.5–3.1 μm in size. European isolates with these conidia characteristics are placed in *B. caledonica* (Bissett and Widden 1988). Asian and South American isolates with cylindrical conidia accessioned in ARSEF are most frequently identified as *B. amorpha*. However, authentic *B. amorpha* (ARSEF 2251) from South America is placed apart in clade E. The distribution of cylindrical conidia in both clades D and E may indicate at least two origins for cylindrical-shaped conidia in *Beauveria*. The one isolate in clade D that does not produce cylindrical conidia is the South American species, *B. vermiconia*, which has tightly curled, sickle-shaped conidia (de Hoog and Rao 1972); however, the single available culture, ARSEF 2922, has apparently lost the ability to form conidia. Isolates in clade D have been cultured from Coleoptera and from soil (Bissett and Widden 1988; de Hoog and Rao 1972), but too few isolates in this clade have been examined to draw conclusions about their pattern of host affiliation. An Asian teleomorph identified as *Cordyceps* cf. *scarabaeicola* was placed in clade D, documenting that at least some species in this clade are sexual (Rehner and Buckley in press).

Clade E (fig. 1.1) contains what is likely to be authentic *B. amorpha*, a South American species that occurs on Coleoptera and Hymenoptera. Morphologically, *B. amorpha* is characterized by nearly cylindrical conidia 3.8–5.2 × 1.7–2.5 μm and a tendency for yellow pigment to accumulate in older parts of cultures. No teleomorphs have yet been placed within this lineage.

Additional lineage diversity was consistently detected within each of the five major clades resolved in the *Beauveria* molecular phylogeny; thus, traditional morphological species concepts have conservatively underestimated species diversity in this genus. Although molecular phylogenetic sampling is preliminary, data obtained thus far suggest that the number of independent terminal groups, if recognized as phylogenetic species, would likely significantly increase species number estimates in *Beauveria*. Furthermore, preliminary data from highly polymorphic microsatellite markers indicate that there may be additional hierarchical structure below that resolved by nuclear gene phylogenies (Rehner unpublished data).

Cryptic diversity has been analyzed in greatest detail in the *B. bassiana* clade A (fig. 1.2). The extensive cladistic structure, the low sequence divergence (0–1.5%), and the disjunct distributions of closely related lineages in clade A are consistent with a recent radiation that coincided with frequent intercontinental dispersal. In several cases, terminal clades corresponded to geographically discrete groups, suggesting that allopatric divergence is an important mode of speciation in this group. This finding appears to be true for the other species groups, but sampling in these lineages is more limited. Although EF1-α provides exceptional insight into the phylogenetic history of clade A and other species complexes in *Beauveria*, more detailed taxon sampling, phylogenetic analyses of additional genes in combination with population genetic markers will be needed to diagnose and resolve species relationships *Beauveria*.

Host Specificity

Perhaps one of the most persistent questions regarding the ecology of *Beauveria* is whether species of *Beauveria* are host specific. Conclusions from previous analyses are ambiguous. Several studies report a correlation between genetic marker data (isozyme, random amplified polymorphic DNA [RAPD], ITS, RFLP) and host affiliation (Poprawski et al. 1988; Neuvéglise et al. 1994; Viaud et al. 1996; Cravanzola et al. 1997; Berretta et al. 1998; Piatti et al. 1998). In contrast, other studies using identical techniques but different sets of isolates found no statistically significant correlation between genotype and host affiliation (St. Leger et al. 1992; Maurer et al. 1997; Urtz and Rice 1997; Coates et al. 2002). In the molecular phylogeny of Rehner and Buckley (in press) (figs. 1.1 and 1.2), only clade B (*B. brongniartii*) and clade D (*B. caledonica, B. vermiconia*, and several unidentified isolates) showed specificity at the level of insect order (e.g., Coleoptera). In contrast, host affiliations are extremely diverse in *B. bassiana* clades A and C, which attack seven and six insect orders, respectively. Permutation tests showed that the actual pattern of host affiliation was indistinguishable from a random distribution (Rehner and Buckley in press). At a minimum, the data suggest that prolonged, intimate coevolution does not occur between *B. bassiana* clade A and particular insect host groups. However, in previous studies where a correlation between *Beauveria* genotypes and host affiliation have been reported, sampling was often conducted at local geographic scales (Berretta et al. 1998; Castrillo and Brooks 1998)

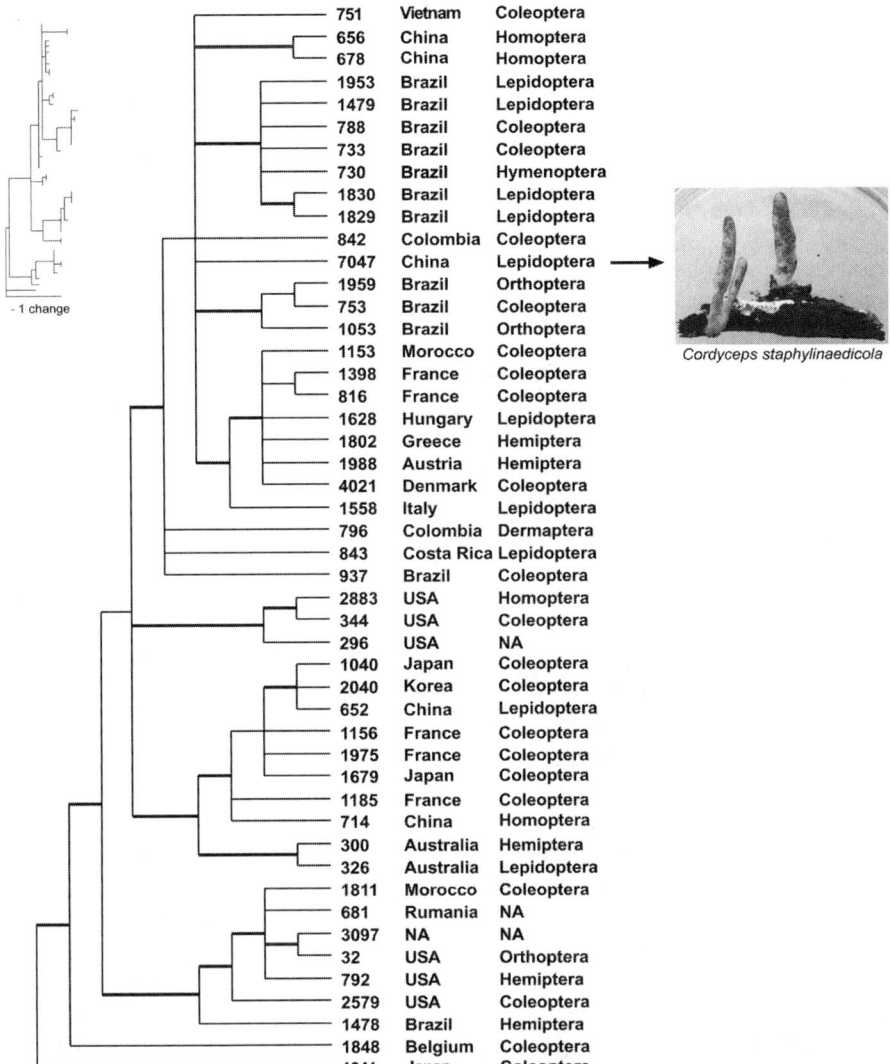

751	Vietnam	Coleoptera
656	China	Homoptera
678	China	Homoptera
1953	Brazil	Lepidoptera
1479	Brazil	Lepidoptera
788	Brazil	Coleoptera
733	Brazil	Coleoptera
730	Brazil	Hymenoptera
1830	Brazil	Lepidoptera
1829	Brazil	Lepidoptera
842	Colombia	Coleoptera
7047	China	Lepidoptera
1959	Brazil	Orthoptera
753	Brazil	Coleoptera
1053	Brazil	Orthoptera
1153	Morocco	Coleoptera
1398	France	Coleoptera
816	France	Coleoptera
1628	Hungary	Lepidoptera
1802	Greece	Hemiptera
1988	Austria	Hemiptera
4021	Denmark	Coleoptera
1558	Italy	Lepidoptera
796	Colombia	Dermaptera
843	Costa Rica	Lepidoptera
937	Brazil	Coleoptera
2883	USA	Homoptera
344	USA	Coleoptera
296	USA	NA
1040	Japan	Coleoptera
2040	Korea	Coleoptera
652	China	Lepidoptera
1156	France	Coleoptera
1975	France	Coleoptera
1679	Japan	Coleoptera
1185	France	Coleoptera
714	China	Homoptera
300	Australia	Hemiptera
326	Australia	Lepidoptera
1811	Morocco	Coleoptera
681	Rumania	NA
3097	NA	NA
32	USA	Orthoptera
792	USA	Hemiptera
2579	USA	Coleoptera
1478	Brazil	Hemiptera
1848	Belgium	Coleoptera
1041	Japan	Coleoptera

- 1 change

Cordyceps staphylinaedicola

Figure 1.2. Phylogenetic relationships of *Beauveria bassiana* clade A. Consensus cladogram summarizing several independent Bayesian and parsimony analyses. Branches with significant nodal support are indicated in bold. Branch tips are labeled according to Agricultural Research Service Collection of Entomopathogenic Fungal Cultures accession number, geographic origin, and insect host. Inset tree in upper left is a phylogram of a single most parsimonious tree indicating relative branch lengths; inset photograph is *Cordyceps staphylinaedicola* (courtesy Bo-Huang, Anhui, China), a teleomorph with a *B. bassiana*-type anamorph.

15

or isolates were sampled from a single, highly abundant insect host with a wide geographic distribution (Neuvéglise et al. 1994).

Population Genetics of *Beauveria*

Little information is currently available about allelic variation, allelic associations, modes of reproduction, population differentiation, or gene flow in *Beauveria*. Population genetics investigations in *Beauveria* are complicated by the absence of criteria for accurate diagnosis of species and populations and by a lack of well-characterized, simple, independent, and robust genetic markers that both facilitate species diagnosis and resolve intraspecific relationships. Previous population-level investigations in *B. bassiana* and *B. brongniartii* have been conducted with dominant genetic markers such as RFLP (Neuvéglise et al. 1994; Viaud et al. 1996) or RAPDs (Cravanzola et al. 1997; Berretta et al. 1998; Piatti et al. 1998). Although these studies revealed extensive underlying genetic variation, the types of markers and methods of data analysis were not informative to the determination of population structure.

To avoid confounding population and phylogenetic comparisons, it is essential to correctly assign isolates to populations. In view of past difficulties in defining biologically meaningful species concepts in morphologically depauperate organisms such as *Beauveria*, it is no surprise that the recognition of populations has also proven elusive. The use of multiple-gene genealogies has been advocated for the diagnosis of cryptic species (Avise and Ball 1990; Taylor et al. 2000), an approach that has proven extremely effective in the study of cryptic diversification patterns in fungi (Koufopanou et al. 1997, 2001; James et al. 2001; Kroken and Taylor 2001; O'Donnell et al. 2002), including *Beauveria* (Rehner and Buckley in press). However, this approach requires intensive effort and ultimately may not provide the resolution needed to diagnose the transition between codiverging phylogenetic structure of past speciation events to the genetic patterns generated by evolutionary processes within populations (Taylor et al. 2000). Nonetheless, multigene phylogenies are extremely valuable to population genetics analyses inasmuch as they help focus attention on the least inclusive clades that are likely to encompass population entities. In addition to multilocus phylogenetic diagnosis, independent approaches for diagnosing phylogenetic species and documenting population genetic parameters are likely to be necessary.

Robust population inference requires an abundance of discrete, highly polymorphic, potentially independently segregating markers capable of detecting significant patterns of population genetic structure, illuminating mode(s) of genetic transmission, and resolving genetic relationships among individuals within and between populations (McEwen et al. 2000). Markers currently used that best satisfy these criteria are microsatellites (Weber 1990; Field and Wills 1996) and SNPs (Brookes 1999; Kuhner et al. 2000; Sunnucks 2000). Mutation rates estimated for microsatellites and SNPs differ by several orders of magnitude (e.g., $\approx 10^{-4}$ versus 10^{-8}–10^{-9}, respectively; Brumfield et al. 2003). Theoretical and empirical data suggest that SNPs may be preferable to microsatellites for population inference (Brumfield et al. 2003). However, because of the earlier discovery of microsatellite markers, the development of methods for direct isolation of microsatellite markers that do not require prior sequence information, and the characteristically high level of polymorphisms encountered at these loci, microsatellites are currently the most widely used population genetic markers.

Microsatellite markers, amenable to analysis under standard population genetic criteria, have only recently been developed for *Beauveria*. Enkerli et al. (2001) described 10 microsatellite markers for *B. brongniartii*. Distance analysis of the microsatellite allelic data showed that European *B. brongniartii* isolates from *M. melolontha* larvae formed a cluster distinct from non–European *B. brongniartii* isolates and other species of *Beauveria* (Enkerli et al. 2001). Explicit population genetic analyses with these data were not conducted, so it was not determined whether the population was recombining. However, the high diversity of multilocus haplotypes and corresponding low frequency of repeated genotypes in the population sample, the absence of a correlation between genetic relatedness and geography, and the relatively long terminal branches characterizing isolates within the *B. brongniartii* ingroup suggest that the *B. brongniartii* population may be undergoing recombination.

A single microsatellite marker that can be amplified by polymerase chain reaction (PCR) from *B. bassiana, B. brongniartii, B. amorpha,* and *B. caledonica* was described by Coates et al. (2002). No significant association between microsatellite genotype and either the host or geographic origin was detected in a global survey of *B. bassiana* (Coates et al. 2002). However, this analysis was conducted without knowledge of phylogenetic relationships of the isolates surveyed. Thus, it is possible that more than one cryptic species was sampled. Additionally, because only a single locus was analyzed, the data are uninformative about recombination processes.

Microsatellite markers recently have been developed for the globally distributed *B. bassiana* clade A characterized by Rehner and Buckley (2003), although detailed population genetic analyses have not yet been conducted. With the development of microsatellite markers for *B. bassiana* and *B. brongniartii*, a common set of tools and standardized approaches are now available for investigating species histories, species boundaries, and population genetics in these cryptic species complexes. Additionally, the markers will also have practical uses for tracking the fate of strains released for biological control of insects, and at the same time they will enable monitoring of the impact of environmental releases to indigenous communities of *Beauveria*.

Genetics of *Beauveria*

The genetics of *Beauveria* is not well characterized. *Beauveria* has long been assumed to be asexual, reproducing exclusively by conidia. Consequently, classical methods of Mendelian genetic analysis that depend on a sexual cycle for genetic exchange have not been developed for any species of *Beauveria*. Attempts at genetic analyses in *Beauveria* have relied on methods for promoting recombination by the parasexual cycle, a mechanism of mitotic recombination (Pontecorvo et al. 1953; Pontecorvo 1956). The parasexual cycle occurs in vegetative heterokaryons by the following sequence of events: (1) anastomosis of genetically different mycelia; (2) formation of transient diploid nuclei by fusion of different types of nuclei, followed by mitotic proliferation; (3) formation of mitotic chiasmata between homologous chromosomes and chromosomal exchange; (4) restoration of haploidy by reductive aneuploidization; and (5) partitioning of recombinant haploid daughter nuclei by conidiogenesis. Although it is difficult to monitor the process in nature, parasexual recombination can be manipulated experimentally.

Parasexual recombination was first demonstrated in *Beauveria* by Paccola-Meirelles and Azevedo (1991), who recovered prototrophic recombinants from crosses between auxotrophic parental strains. Using a similar approach, Bello and Paccola-Meirelles (1998) used segregant analysis of progeny produced in parasexual crosses to assign several auxotrophic markers and a mutation-conferring resistance to benomyl to four linkage groups. Parasexuality also has been demonstrated to occur in many other species of filamentous ascomycetes, including related entomopathogenic species of *Metarhizium* (Tinline and Noviello 1971; Al-Aidroos 1980; Messias and Azevedo 1980; Bagagli et al. 1991; Leal-Bertioli et al. 2000) and *Paecilomyces* (Cantone and Vandenberg 1998).

Protoplast fusions between auxotrophic strains also have been used to achieve parasexual recombination in *Beauveria* (Paccola-Meirelles and Azevedo 1994), a strategy that potentially increases the rate of recovery of parasexual recombinants. Couteaudier et al. (1996) described interspecific prototrophic recombinants from protoplast fusions between *B. bassiana* and *B. sulfurescens*. RFLP markers from both parents were observed in some progeny, indicating that these progeny were either recombinants or partial diploids. Additionally, one progeny showed enhanced lethality to *Ostrinia nubilalis*, although its RFLP genotype matched that of only one parent.

Thus far the use of parasexual genetics in *Beauveria* has not progressed significantly beyond demonstrating proof of the concept, and consequently it cannot be predicted if, when, or to what extent parasexual genetics will contribute to an understanding of *Beauveria* genetics. Similarly, predictions that parasexual genetics can be used to create novel strains with enhanced characteristics for use in biological control, although demonstrated in principle (Couteaudier et al. 1996), have not been fully realized. Future investigation of parasexuality in *Beauveria* will benefit from the development of molecular markers that can establish genetic linkage relationships and the distribution and frequency of crossover points during mitotic recombination. In the foregoing studies of parasexual recombination in *Beauveria*, the parental strains did not readily form heterokaryons, and so it was necessary first to create complementing auxotrophic markers to force heterokaryon formation. To date the causal basis for the barriers to hyphal fusion and heterokaryon formation has not been investigated in *Beauveria*. However, intraspecific barriers to heterokaryosis are known to be widespread in filamentous fungi, a phenomenon known as vegetative incompatibility (also known as heterokaryon or somatic incompatibility; Glass and Kuldau 1992; Leslie 1993, 1996; Glass and Saupe 2002) and appear to be an evolutionarily conserved system of non–self-recognition in sexual fungi (Wu et al. 1998).

Vegetative incompatibility has been described for *B. bassiana* (see also chapter 2). Use of complementary nitrate nonutilizing mutants delineated 14 vegetative compatibility groups (VCGs) among 25 strains of *B. bassiana* from diverse geographic origins (Couteaudier and Viaud 1997). VCG groupings correlated with geographic origin, although one group, VCG-1, occurred at several distant sites in Europe, indicating it is a widely distributed genotype. VCG groups were isolated from a single host insect species, indicating host specificity. Individual insect species, however, could be infected by more than one VCG group (e.g., *Sitona discoideus* [Coleoptera: Curculionidae] was host to VCGs 5, 6, 7, and 8 in France and Morocco). Members of the same VCG were more closely related to each other than to those of other VCGs; however, no data were presented on genetic relationships among VCGs. The proximate geographic, host, and genetic similarities were attributed to shared VCG membership, but no explanation was presented for the underlying cause of the observed VCG diversity. In light of recent evidence for sexuality in *B. bassiana* (Li et al. 2001), it is possible that the VCG diversity, as described by Couteaudier and Viaud (1997), is the signature of past sexual recombination. Apart from direct observation of *Beauveria* teleomorphs, detailed knowledge of population genetic structure within and between VCGs of *Beauveria* will be essential to determine the relationship between vegetative incompatibility and recombination. Additionally, a better understanding of mitotic recombination is necessary to determine whether it can be distinguished from sexual recombination and whether it indeed occurs in nature. Vegetative incompatibility also has been reported in the related entomopathogenic hyphomycetes *Metarhizium anisopliae* (Bidochka et al. 2000; chapter 2, this volume), *Verticillium lecanii* (Korolev and Gindin 1999), and *Paecilomyces fumosoroseus* (Riba and Ravelojoana 1984; Cantone and Vandenberg 1998), indicating that non–self-recognition systems are conserved in these putatively mitotic fungi.

The accumulating evidence that at least some *Beauveria* species are sexual may lead to development of methods for Mendelian genetic analysis in these fungi. Indeed, progress toward *in vitro* fruiting *Cordyceps staphylinidicola*, a teleomorph of *B.* cf. *bassiana*, has already been achieved in two instances (M. Shimazu, pers. comm.; J.-M. Sung, pers. comm.). However, each of these fruitings originated from cultures isolated from perithecial stromata; thus it is possibile that these isolates are homothallic. Considerable work remains before the reproductive biology of *Beauveria* is fully characterized and tractable genetic systems are developed. An important step toward achieving these goals is the isolation of ascospore progeny from teleomorph collections, as these will likely provide the most direct route to the development of methods for *in vitro* mating and sexual morphogenesis in *Beauveria*.

Chromosome number and nuclear genome size have been determined for several different isolates of *B. bassiana*. Pfeifer and Khachatourians (1993) determined the electrophoretic karyotype of a single *B. bassiana* strain and observed eight chromosomes and estimated a nuclear genome size of 40.6 Mb. This is in agreement with the findings of Viaud et al. (1996), who used both telomere probes and electrophoretic karyotypes and estimated that there are seven or eight chromosomes per haploid genome, depending on the strain. Individual chromosome sizes ranged between 1 and 7.7 Mb in size, with complete genome size estimates varying between 34.1 and 44.1 Mb among the seven strains examined. These genome size estimates are similar to those reported for other pyrenomycetes, including the closely related hypocrealean entomogenous fungi *Paecilomyces fumosoroseus* (Shimizu et al. 1993) and

Metarhizium anisopliae (Shimizu et al. 1992), and the endophyte *Epichloe typhina* (Kuldau et al. 1999), all of which have haploid chromosome counts of six to eight chromosomes and nuclear genome sizes ranging between 30 and 40 Mb.

Conclusion

Despite past and current limitations to genetic analysis in *Beauveria*, the outlook for developing a comprehensive understanding of the genetics of *Beauveria* in the near future is extremely promising. First, increasing knowledge of *Beauveria* species histories through molecular phylogenetic analysis is likely to improve criteria for species recognition and identification dramatically, which will lead to a more detailed understanding of the biodiversity and geographic distributions of this ecologically ubiquitous group of entomopathogenic fungi. Also, the development of robust phylogenies will provide new interpretative tools that can aid in understanding the distribution and evolution of important biological characteristics of *Beauveria* such as host range, pathogenicity and virulence, and production of toxins and other metabolites, to name only a few. Second, with the development of population genetic markers, the determination of whether individual species are clonal or recombining will define the methods of genetic analysis that can be developed for these organisms. Third, comprehensive genomic sequencing efforts can quickly inform knowledge of gene diversity in *Beauveria* by comparison to available genomic sequences from other insect and plant pathogenic ascomycetes, thus facilitating the identification of common sets of functional and regulatory genes involved in pathogenesis. Additionally, analysis of gene expression through microarray technologies will expedite the identification of genes whose expression is specifically associated with entomopathogenesis, and this is probably the most direct approach to understanding the molecular basis for variation in pathogenicity and virulence commonly observed among isolates of *Beauveria*. Finally, methods for functional analysis of genes, such as transformation-mediated gene disruption for *Beauveria*, also are needed. Fortunately, the haploid genome and conidia greatly simplify these objectives. Ultimately, an improved understanding of the mechanisms and genetics of traits integral to entomopathogenesis are most likely to inform the future use of *Beauveria* as a biocontrol agent of pest insects.

Literature Cited

Ainsworth, G. C. 1956. Agostino Bassi 1773–1856. *Nature* 177:255–257.
Ainsworth, G. C. 1976. *Introduction to the history of mycology*. Cambridge: Cambridge University Press.
Al-Aidroos, K. 1980. Demonstration of a parasexual cycle in the entomopathogenic fungus *Metarhizium anisopliae*. *Canadian Journal of Genetics and Cytology* 22:309–314.
Arcieri, G. P. 1956. Agostino Bassi in the history of medical thought. Florence: Olschki.
Audoin, V. 1837a. Recherches anatomiques et physiologiques sur la maladie contagieuse qui attaque les vers à soie, et qu'on désigne sous le nom de Muscardine. *Annales des Sciences Naturelles* 8:229–245.

Audoin, V. 1837b. Nouvelles expériences sur la nature de la maladie contagieuse qui attaque les vers à soie et qu'on désigne sous le nom de Muscardine. *Annales des Sciences Naturelles* 8:257–270.

Avise, J. C., and R. M. Ball. 1990. Principles of genealogical concordance in species concepts and biological taxonomy. *Oxford Surveys in Evolutionary Biology* 7:45–67.

Bagagli, E., M. C. Valadares, and J. L. Azevedo. 1991. Parameiosis in the entomopathogenic fungus *Metarhizium anisopliae* (Metsch.) Sorokin. *Revista Brasileira de Genética* 14:261–271.

Balsamo-Crivelli, G. 1835. Ossevazione sopra una nuova specie di Mucedinea del genere *Botrytis*. *Biblioteca Italiana Ossia Giornale di Letteratura Scienze ed Arti* 39: 125.

Bartlett, M. C., and S. T. Jaronski. 1988. Mass Production of entomogenous fungi for biological control of insects. In *Fungi in biological control systems*, ed. M. N. Burge, pp. 61–85. Manchester: Manchester University Press.

Beauverie J. 1914. Les Muscardines. Le genre *Beauveria*. *Revue Génerale de Botanique* 26:81–157.

Beilharz, V. C., D. G. Parbery, and H. J. Swart. 1982. Dodine: A selective agent for certain soil fungi. *Transactions of the British Mycological Society* 79:507–511.

Bello, V. A., and L. D. Paccola-Meirelles. 1998. Localization of auxotrophic and benomyl resistance markers through the parasexual cycle in the *Beauveria bassiana* (Bals.) Vuill. entomopathogen. *Journal of Invertebrate Pathology*. 72:119–125.

Berbee, M. L., and J. W. Taylor. 1995. From 18S ribosomal sequence data to evolution of morphology among the fungi. *Canadian Journal of Botany* 73 (suppl. 1):S677–S683.

Berbee, M. L., and J. W. Taylor. 2001. Fungal molecular evolution: gene trees and geologic time. In *The Mycota*, ed. D. J. McLaughlin, E. McLaughlin, and P. A. Lemke, pp. 229–246. Berlin: Springer.

Berretta, M. F., R. E. Lecuona, R. O. Zandomeni, and O. Grau. 1998. Genotyping isolates of the entomopathogenic fungus *Beauveria bassiana* by RAPD with fluorescent labels. *Journal of Invertebrate Pathology* 71:145–150.

Bidochka M. J., M. A. McDonald, R. J. St. Leger, and D. W. Roberts. 1994. Differentiation of species and strains of entomopathogenic fungi by random amplification of polymorphic DNA (RAPD). *Current Genetics* 25:107–113.

Bidochka M. J., M. J. Melzer, T. M. Lavender, and A. M. Kamp. 2000. Genetically related isolates of the entomopathogenic fungus *Metarhizium anisopliae* harbour homologous dsRNA viruses. *Mycological Research* 104:1094–1097

Bing, L. A., and L. C. Lewis. 1992. Endophytic *Beauveria bassiana* (Balsamo) Vuillemin in corn: the influence of the plant growth stage and *Ostrinia nubilalis* (Hübner). *Biocontrol Science and Technology* 2:39–47.

Bissett, J., and P. Widden. 1988. A new species of *Beauveria* isolated from Scottish moorland soil. *Canadian Journal of Botany* 66:361–362.

Brookes, J. F. 1999. The essence of SNPs. *Gene* 234:177–186.

Brumfield, R. T., P. Beerli, D. A. Nickerson, and S. V. Edwards. 2003. The utility of single nucleotide polymorphisms in inferences of population history. *Trends in Ecology and Evolution* 18:249–256.

Cantone, F., and J. D. Vandenberg. 1998. Intraspecific diversity in *Paecilomyces fumosoroseus*. *Mycological Research* 102:209–215.

Castrillo, L. A., and W. M. Brooks. 1998. Differentiation of *Beauveria bassiana* isolates from the darkling beetle *Alphitobius diaperinus*, using isozyme and RAPD analyses. *Journal of Invertebrate Pathology* 72:190–196.

Chase, A. R., L. S. Osborne, and V. M. Ferguson. 1986. Selective isolation of the entomo-pathogenic fungi *Beauveria bassiana* and *Metarhizium anisopliae* from an artificial potting medium. *Florida Entomologist* 69:285–292.

Coates, B. S., R. L. Hellmich, and L. C. Lewis. 2002. *Beauveria bassiana* haplotype deter-mination based on nuclear rDNA internal transcribed spacer PCR-RFLP. *Mycological Research* 106:40–50.

Couteaudier, Y., and M. Viaud. 1997. New insights into population structure of *Beauveria bassiana* with regard to vegetative compatibility groups and telomeric restriction frag-ment length polymorphisms. *FEMS Microbial Ecology* 22:175–182.

Couteaudier, Y., M. Viaud, and G. Riba. 1996. Genetic nature, stability, and improved viru-lence of hybrids from protoplast fusion in *Beauveria*. *Microbial Ecology* 32:1–10.

Cravanzola, F., P. Piatti, P. D. Bridge, and O. Ozino. 1997. Detection of polymorphism by RAPD-PCR in strains of the entomopathogenic fungus *Beauveria brongniartii* isolated from the European cockchafer (*Melolontha* sp.). *Letters in Applied Microbiology* 25:289–294.

de Hoog, G. S. 1972. The genera *Beauveria*, *Isaria*, *Tritirachium* and *Acrodontium* gen. nov. *Studies in Mycology* 1:1–41.

de Hoog, G. S. 1978. Notes on fungicolous hyphomycetes and their relatives. *Persoonia* 10:33–81.

de Hoog G. S., and V. Rao. 1975. Some new Hyphomycetes. *Persoonia* 8:207–212.

Dunn P. H., and B. J. Mechalas. 1963. The potential of *Beauveria bassiana* (Balsamo) Vuillemin as a microbial insecticide. *Journal of Insect Pathology* 5:451–459.

Enkerli J, F. Widmer, C. Gessler, and S. Keller. 2001. Strain-specific microsatellite mark-ers in the entomopathogenic fungus *Beauveria brongniartii*. *Mycological Research* 105:107–109.

Eriksson O. E., H.-O. Baral, R. S. Currah, K. Hansen, C. P. Kurtzman, G. Rambold, and T. Laessøe, eds. 2003. Outline of Ascomycota—2003. *Myconet* 9:1–89

Eyal, J., MD A. Mabud, K. L. Fischbein, J. F. Walter, L. S. Osborne, and Z. Landa. 1994. Assessment of *Beauveria bassiana* Nov. EO-1 strain, which produces a red pigment of microbial control. *Applied Biochemistry and Biotechnology* 44:65–80.

Feng M. G., T. J. Poprawski, and G. G. Khachatourians. 1994. Production, formulation and application of the entomopathogenic fungus *Beauveria bassiana* for insect control: Current status. *Biocontrol Science and Technology* 4:3–34.

Ferron, P. 1978. Biological control of insect pests by entomogenous fungi. *Annual Review of Entomology* 23:409–422.

Ferron, P, J. Fargues, and G. Riba. 1991. Fungi as microbial insecticides against pests. In *Handbook of applied mycology*, ed. D. K. Arora, L. Ajello, and K. G. Mukerji, pp. 665–706. New York: Marcel Dekker.

Field, D., and C. Wills. 1996. Long, polymorphic microsatellites in simple organisms. *Pro-ceedings of the Royal Society of London, Series B* 263:209–215.

Fisher, M. C., G. L. Koenig, T. J. White, and J. W. Taylor. 2000. A test for concordance between the multilocus genealogies of genes and microsatellites in the pathogenic fun-gus *Coccidioides immitis*. *Molecular Biology and Evolution* 17:1164–1174.

Fisher, M. C., G. L. Koenig, T. J. White, and J. W. Taylor. 2002. Molecular and phenotypic description of *Coccidioides posadasii* sp. nov., previously recognized as the non-California population of *Coccidioides immitis*. *Mycologia* 94:73–84.

Gams, W. 1971. *Tolypocladium*. Eine Hyphomycetengattung mit geschwollenen Phialiden. *Persoonia* 6:185–191.

Gillespie, A. T., and E. R. Moorehouse. 1989. The use of fungi to control pests of agricul-tural and horticultural importance. In *Biotechnology of fungi for improving plant growth*, ed. J. M. Whipps and R. D. Lumsden, pp. 55–83. Cambridge: Cambridge University Press.

Glare T. R., and A. J. Inwood. 1998. Morphological characterization of *Beauveria* spp. from New Zealand. *Mycological Research* 102:250–256.

Glass, N. L., and G. A. Kuldau. 1992. Mating type and vegetative compatibility in filamentous ascomycetes. *Annual Review of Phytopathology* 30:201–224.

Glass, N. L., and S. J. Saupe. 2002. Vegetative incompatibility in filamentous fungi. In *Molecular biology of fungal development*, ed. H. D. Osiewacz, pp. 109–131. New York: Marcel Dekker.

Goettel M. S., and G. D. Inglis. 1997. Fungi: Hyphomycetes. In *Manual of techniques in insect pathology,* ed. L. Lacey, pp. 213–249. San Diego, CA: Academic Press.

Goettel, M. S., T. J. Poprawski, J. D. Vandenberg, Z. Li, and D. W. Roberts. 1990. Safety to nontarget invertebrates of fungal biocontrol agents. In *Safety of microbial insecticides*, ed. M. Laird, L. A. Lacey, and E. W. Davidson, pp. 209–232. Boca Raton, FL: CRC Press.

Heckman, D. S., D. M. Geiser, R. E. Brooke, R. L. Stauffer, N. L. Kardos, and S. B. Hedges. 2001. Molecular evidence for the early colonization of land by fungi and plants. *Science* 293:1129–1133.

Hegedus D. D., and G. G. Khachatourians. 1993a. Identification of molecular variants in mitochondrial DNAs of members of the genera *Beauveria, Verticillium, Paecilomyces, Tolypocladium*, and *Metarhizium. Applied and Environmental Microbiology* 59:4283–4288.

Hegedus D. D., and G. G. Khachatourians. 1993b. Construction of cloned DNA probes for the specific detection of the entomopathogenic fungus *Beauveria bassiana* in grasshoppers. *Journal of Invertebrate Pathology* 62:233–240.

Hegedus D. D., and G. G. Khachatourians. 1996. Identification and differentiation of the entomopathogenic fungus *Beauveria bassiana* using polymerase chain reaction and single-strand conformation polymorphism analysis. *Journal of Invertebrate Pathology* 67:289–299.

Huang, B., C.-R. Li, Z.-G. Li, M.-Z. Fan, and Z. Z. Li. 2002. Molecular identification of the teleomorph of *Beauveria bassiana. Mycotaxon* 81:229–236.

Humber, R. A. 2001. Collection of entomopathogenic fungal cultures: Catalog of strains. U. S. Department of Agriculture, Agricultural Research Service. Available: www.ppru.cornell.edu/mycology/Insect_mycology.htm.

Humber, R. A. 1997. Fungi: Identification. In *Manual of techniques in insect pathology*, ed. L. Lacey, pp. 153–185. San Diego, CA: Academic Press.

James, T. Y., J. M. Moncalvo, S. Li, and R. Vilgalys. 2001. Polymorphism at the ribosomal DNA spacers and its relation to breeding structure of the widespread mushroom *Schizophyllum commune. Genetics* 157:149–161.

Johanys, M. 1839. De la muscardine. Des moyens de la developper artifciellement, de modifier les effets de la contagion. *Annales des Sciences Naturelles et Zoologie* 11:65–80.

Kasuga, T., T. J. White, and J.W. Taylor. 2002. Estimation of nucleotide substitution rates in Eurotiomycete fungi. *Molecular Biology and Evolution* 19:2318–2324.

Keller, S., E. Keller, C. Schweitzer, J. A. L. Auden, and A. Smith. 1989. Two large field trials to control the cockchafer *Melolontha melolontha* L. with the fungus *Beauveria brongniartii* (Sacc.) Petch. *British Crop Protection Council Monograph* 43:183–190.

Kirk, P. M., P. F. Cannon, J. C. David, and J. A. Stalpers. 2001. *Dictionary of the fungi.* Wallingford, UK: CAB International.

Kirk, P. M. 2003. *Indexfungorum.* Available: http://www.indexfungorum.org.

Kobayasi, Y., and D. Shimizu. 1982. *Cordyceps* species from Japan: 4. *Bulletin of the National Science Museum, Tokyo, Botany* 8:79–91.

Korolev, N., and G. Gindin. 1999. Vegetative compatibility in the entomopathogen *Verticillium lecanii*. *Mycological Research* 103:833–840.

Kosir, J. M., J. M. MacPherson, and G. G. Khachatourians. 1991. Genomic analysis of a virulent and a less virulent strain of the entomopathogenic fungus *Beauveria bassiana*, using restriction fragment length polymorphisms. *Canadian Journal of Microbiology* 37:534–541.

Koufopanou, V., A. Burt, and J. W. Taylor. 1997. Concordance of gene genealogies reveals reproductive isolation in the pathogenic fungus *Coccidioides immitis*. *Proceedings of the National Academy of Sciences of the USA* 94:5478–5482.

Koufopanou, V., A. Burt, T. Szaro, and J. W. Taylor. 2001. Gene genealogies, cryptic species, and molecular evolution in the human pathogen *Coccidioides immitis* and relatives (Ascomycota, Onygenales). *Molecular Biology and Evolution* 18:1246–1258.

Kroken, S., and J. W. Taylor. 2001. A gene genealogical approach to recognize phylogenetic species boundaries in the lichenized fungus *Letharia*. *Mycologia* 93:38–53.

Kuhner, M. K., P. Beerli, J. Yamato, and J. Felsenstein. 2000. Usefulness of single nucleotide polymorphism data for estimating population parameters. *Genetics* 156:439–447.

Kuldau G. A., H.-F. Tsai, and C. L. Schardl. 1999. Genome sizes of *Epichloë* species and anamorphic hybrids. *Mycologia* 91:776–782.

Leal-Bertioli, S. C. M., T. M. Butt, J. F. Peberdy, and D. J. Bertioli. 2000. Genetic exchange in *Metarhizium anisopliae* strains co-infecting *Phaedon conchleariae* is revealed by molecular markers. *Mycological Research* 104:409–414.

Leslie, J. F. 1993. Fungal vegetative compatibility. *Annual Review of Phytopathology* 31:127–150.

Leslie, J. F. 1996. Fungal vegetative compatibility—promises and prospects. *Phytoparasitica* 24:3–6.

Li, Z, C. Li, B. Huang, and F. Meizhen. 2001. Discovery and demonstration of the teleomorph of *Beauveria bassiana* (Bals.) Vuill., an important entomogenous fungus. *Chinese Science Bulletin* 46:751–753.

MacLeod, D. M. 1954. Investigations on the genera *Beauveria* Vuill. and *Tritirachium* Limber. *Canadian Journal of Botany* 32:818–890.

Major, R. H. 1944. Agostino Bassi and the parasitic theory of disease. *Bulletin of the History of Medicine* 16:97–107.

Maurer P., Y. Couteaudier, P. A. Girard, P. D. Bridge, and G. Riba. 1997. Genetic diversity of *Beauveria bassiana* and relatedness to host insect range. *Mycological Research* 101:159–164.

McEwen, J. G., J. W. Taylor, D. Carter, J. Xu, M. S. Felipe, R. Vilgalys, T. G. Mitchell, T. Kasuga, T. J. White, T. Bui, and C. M. Soares. 2000. Molecular typing of pathogenic fungi. *Medical Mycology* 38:189–197.

Messias, C. L., and J. L. Azevedo. 1980. Parasexuality in the deuteromycete *Metarhizium anisopliae*. *Transactions of the British Mycological Society* 75:473–477.

Mugnai, L, P. D. Bridge, and H. C. Evans. 1989. A chemotaxonomic evaluation of the genus *Beauveria*. *Mycological Research* 92:199–209.

Neuvéglise, C, and Y. Brygoo. 1994. Identification of group-1 introns in the 28s rDNA of the entomopathogenic fungus *Beauveria brongniartii*. *Current Genetics* 27:38–45.

Neuvéglise, C, Y. Brygoo, and R. Riba. 1996. rDNA group-1 introns: A powerful tool for the identification of *Beauveria brongniartii* strains. *Molecular Ecology* 6:373–381.

Neuvéglise, C, Y. Brygoo, B. Vercambre, and G. Riba. 1994. Comparative analysis of molecular and biological characteristics of *Beauveria brongniartii* isolated from insects. *Mycological Research* 98:322–328.

O'Donnell, K., H. C. Kistler, B. K. Tacke, and H. H. Casper. 2002. Gene genealogies reveal global phylogeographic structure and reproductive isolation among lineages of *Fusarium graminearum*, the fungus causing wheat scab. *Proceedings of the National Academy of Science of the USA* 97:7905–7910.

Pacccola-Meirelles, L. D., and J. L. Azevedo. 1991. Parasexuality in *Beauveria bassiana*. *Journal of Invertebrate Pathology* 57:172–176.

Paccola-Meirelles, L. D., and J. L. Azevedo. 1994. Genetic recombination by protoplast fusion in the deuteromycete *Beauveria bassiana*. *Revista Brasileira de Genetica* 17:15–18.

Petch, T. 1926. Studies in entomogenous fungi. *Transactions of the British Mycological Society* 10:244–271.

Pfeifer, T. A., and G. G. Khachatourians. 1993. Electrophoretic karyotype of the entomopathogenic deuteromycete *Beauveria bassiana*. *Journal of Invertebrate Pathology* 61:231–235.

Piatti, P., F. Cravanzola, P. D. Bridge, and O. I. Ozino. 1998. Molecular characterization of *Beauveria brongniartii* isolates obtained from *Melolontha melolontha* in Valle d'Aosta (Italy) by RAPD-PCR. *Letters of Applied Microbiology* 26:317–324.

Poinar, Jr., G. O., and G. M. Thomas. 1984. A fossil entomogenous fungus from Dominican amber. *Experientia* 40:578–579.

Pontecorvo, G. L, J. A. Roper, L. M. Hemmons, K. D. MacDonald, and A.W. J. Bufton. 1953. The genetics of *Aspergillus nidulans*. *Advances in Genetics* 5:141–238.

Pontecorvo, G. 1956. The parasexual cycle in fungi. *Annual Review of Microbiology* 10:393–400.

Poprawski, T. J., G. Riba, W. A. Jones, and A. Aioun. 1988. Variation in isoesterase profiles of geographic populations of *Beauveria bassiana* (Deuteromycotina: Hyphomycetes) isolates from *Sitona* weevils (Coleoptera: Curculionidae). *Environmental Entomology* 17:275–279.

Porter, J. R. 1973. Agostino Bassi bicentennial (1773–1973). *Bacteriological Reviews* 37:284–288.

Rakotonirainy, M. S., M. Dutertre, Y. Brygoo, and G. Riba. 1991. rRNA sequence comparison of *Beauveria bassiana*, *Tolypocladium cylindrosporum* and *Tolypocladium extinguens*. *Journal of Invertebrate Pathology* 57:17–22.

Rehner, S. A., and E. P. Buckley. 2003. Isolation and characterization of microsatellite loci for the entomopathogenic fungus *Beauveria bassiana* (Ascomycota: Hypocreales). *Molecular Ecology Notes* 3:409–411.

Rehner, S. A., and E. P. Buckley. 2004. A *Beauveria* phylogeny inferred from nuclear ITS and EF1α sequences: Evidence for cryptic diversification and links to *Cordyceps* teleomorphs. *Mycologia* in press.

Riba, G., and A. A. M. Ravelojoana. 1984. The parasexual cycle in the entomopathogenous fungus *Paecilomyces fumosoroseus* (Wise) Brown and Smith. *Canadian Journal of Microbiology* 30:922–926.

Riba, G., T. J. Poprawski, and J. Maniania. 1986. Isoesterase variability among geographical populations of *Beauveria bassiana* (fungi imperfecti) isolated from Miridae. In *Fundamental and applied aspects of invertebrate pathology*, ed. R. A. Samson, J. M. Vlak, and D. Peters, pp. 205–209. Wageningen, The Netherlands: Foundation of the Fourth International Colloquium of Invertebrate Pathology.

Samson R. A., and H. C. Evans. 1982. Two new *Beauveria* spp. from South America. *Journal of Invertebrate Pathology* 39:93–97.

Shimazu, M., W. Mitsuhashi, and H. Hashimoto. 1988. *Cordyceps brongniartii* sp. nov., the teleomorph of *Beauveria brongniartii*. *Transactions of the Mycological Society of Japan* 29:323–330.

Shimizu, S., and K. Aizawa. 1988. Serological classification of *Beauveria bassiana*. *Journal of Invertebrate Pathology* 52:348–353.

Shimizu, S., Y. Arai, and T. Matsumoto. 1992. Electrophoretic karyotype of *Metarhizium anisopliae*. *Journal of Invertebrate Pathology* 60:185–187.

Shimizu, S., H. Yoshioka, and T. Matsumoto. 1993. Electrophoretic karyotying of the entomogenous fungus *Paecilomyces fumosoroseus*. *Letters in Applied Microbiology* 16:183–186.

St. Leger R. J., L. L. Allee, R. May, R. C. Staples, and D. W. Roberts. 1992. World-wide distribution of genetic variation among isolates *Beauveria* spp. *Mycological Research* 96:1007–1015.

Steinhaus, E. A. 1949. *Principles of insect pathology*. New York: McGraw-Hill.

Steinhaus, E. A. 1963. *Insect pathology: An advanced treatise*, vol. 2. New York: Academic Press.

Sung, G. -H., J. W. Spatafora, R. Zare, K. Hodge, and W. Gams. 2001. A revision of *Verticillium* sect. *Prostrata*. II. Phylogenetic analysis of SSU and LSU nuclear rDNA sequences from anamorphs and teleomorphs of the Clavicipitaceae. *Nova Hedwigia* 72:311–328.

Sunnucks, P. 2000. Efficient genetic markers for population biology. *Trends in Ecology and Evolution* 15:472–488.

Tanada, Y. and H. K. Kaya. 1993. *Insect pathology*. San Diego, CA: Academic Press.

Taylor, J. W., D. J. Jacobson, S. Kroken, K. Kasuga, D. M. Geiser, D. S. Hibbett, and M. C. Fisher. 2000. Phylogenetic species recognition and species concepts in fungi. *Fungal Genetics and Biology* 31:21–32.

Tinline, R., and C. Noviello. 1971. Heterokaryosis in the entomogenous fungus, *Metarhizium anisopliae*. *Mycologia* 63:701–712.

Urtz, B. E. and W. C. Rice. 1997. RAPD-PCR characterization of *Beauveria bassiana* isolates from the rice water weevil *Lissorhoptorus oryzophilus*. *Letters in Applied Microbiology* 25:405–409.

Viaud, M., Y. Couteaudier, C. Levis, and G. Riba. 1996. Genome organization in *Beauveria bassiana*: Electrophoretic karyotype, gene mapping and telomeric fingerprints. *Fungal Genetics and Biology* 20:175–183.

Vittadini, C. 1852. Della natura del calcino o mal del segno. *Memorie dell'Iimperiale Istituto Veneto di Scienze, Leere, ed Arti* 3:447–512.

Vittadini, C. 1853. Dei mezzi di prevenire il calcino o mal del segno ei bachi do seta. *Memorie dell'Imperiale Istituto Veneto di Scienze, Leere, ed Arti*, Venice 4:241–289.

von Arx, J. A. 1986. *Tolypocladium*, a synonym of *Beauveria*. *Mycotaxon* 25:153–158.

Vuillemin, P. 1912. *Beauveria*, nouveau genre de Verticilliacies. *Bulletin de la Société Botanique de France* 59:34–40.

Wagner, B. L., and L. C. Lewis. 2000. Colonization of corn, *Zea mays*, by the entomopathogenic fungus *Beauveria bassiana*. *Applied and Environmental Microbiology* 66:3468–3473.

Weber, J. L. 1990. Informativeness of human $(dC-dA)n \cdot (dG-dT)_n$ polymorphisms. *Genomics* 7:524–530.

Wu, J., S. J. Saupe, and N. L. Glass. 1998. Evidence for balancing selection operating at the het-c heterokaryon incompatibility locus in a group of filamentous fungi. *Proceedings of the National Academy of Sciences of the USA* 95:12398–12403.

Yarrow, P. J., trans., and Ainsworth, G. C., ed. 1958. Bassi, A. [1835, 1836] Del mal del segno. *Phytopathological Classics 10*. St. Paul, MN: American Phytopathological Society Press.

Zimmerman, G. 1986. The "*Galleria* bait method" for detection of entomopathogenic fungi in soil. *Journal of Applied Entomology* 102:213–215.

2

Phylogeography of *Metarhizium,* an Insect Pathogenic Fungus

Michael J. Bidochka
Cherrie L. Small

Between the years 2000 and 2003, more than 150 publications were devoted to the insect-pathogenic fungus *Metarhizium*. These studies cover molecular biology of pathogenesis, biochemistry, biocontrol, and population genetics. *Metarhizium* species are known to infect more than 200 insect species, many of which are major agricultural and forest insect pests such as spittlebugs, sugar cane borers, termites, scarab grubs, and grasshoppers (St. Leger 1993).

Samson et al. (1988) recognized three species of *Metarhizium*: *M. anisopliae*, *M. flavoviride*, and *M. album*. However, for the purposes of this discussion, these will be described collectively as *Metarhizium* until the taxonomy and phylogeography of the genus is discussed later in this chapter. The topic is important not only because of our academic interest, but also because the fungi have potential economic importance. Several commercial endeavors have registered strains of *Metarhizium* for insect pest management: AGO BIOCONTROL *METARHIZIUM* 50 in Colombia, for garden pests; BioGreen in Australia, for *Adoryphouse couloni* (red-headed cockchafer); GREEN GUARD in Australia for locusts; BIO 1020 in Germany, for vine weevils (*Otiorhynchus sulcatus*); and Green Muscle in South Africa, for *Locustana pardalina* and other locusts and grasshoppers. Despite academic interest, commercial registration, and application of *Metarhizium* for insect pest management, commercial markets for these products have been slow to develop in North America and western Europe. In contrast, South America and southeastern Asia have a burgeoning cottage industry in cultivating and utilizing *Metarhizium* and other insect-pathogenic fungi for biocontrol (Samson et al. 1988).

Several criteria are important in choosing a strain of *Metarhizium* to develop commercially. These include, foremost, a high level of virulence toward a targeted pest insect, because *Metarhizium* strains have a wide range of virulence (Roberts and Hajek

1992). Another consideration is the utilization of strains already existing in the country of use because introduction of exotic strains of fungi could be politically as well as ecologically problematic (Bidochka 2001). Furthermore, after application of a *Metarhizium* formulation, it is essential to be able to distinguish the strain applied in the field against the background populations of this common soil fungus with cosmopolitan distribution. Without genotyping or genetic tagging of the strain, there is no way to ascertain the effectiveness of the applied strain (Bidochka 2001).

Biology of *Metarhizium*

Currently, *Metarhizium* research can be subdivided into three major fields of interest. The first is the infection process, including studies of the biochemistry of enzymes involved in pathogenesis, as well as cloning virulence genes and gene knock-out experiments (St. Leger et al. 1992a). A second area of *Metarhizium* research is the study of population structure. This research includes genotyping strains and correlation of fungal genotype to insect host, geography, or habitat (St. Leger et al. 1992b; Bridge et al. 1993; Fegan et al. 1993; Bidochka et al. 2001a). Finally, other research is focused on the use of *Metarhizium* for biocontrol. While all of these areas are ultimately related to assessing the value of *Metarhizium* as a biological control agent, we emphasize the biology of *Metarhizium* as it relates to the infection process in the following section. The phylogeography of *Metarhizium* is of central importance because there is evidence that some strains show host specificity, and the type and regulation of virulence factors may be related to strain differences and insect host preferences (de Moraes et al. 2003). In this chapter we explore in some depth how virulence to a particular insect relates to the population structure of *Metarhizium*. First, we review the insect infection process.

Insect Infection

The biochemical and molecular mechanisms by which *Metarhizium* infects a host insect have been topics of a large body of research (St. Leger 1993; Hajek and St. Leger 1994; St. Leger and Bidochka 1996). The infection process is similar to that of other fungal pathogens of insects and includes spore recognition and attachment to the host, penetration of the integument through physical and enzymatic mechanisms, evasion of host immune defenses, proliferation, and, finally, reemergence from the host and the next round of spore production.

Spores (more specifically, conidia, which are asexual spores of ascomycetes and basidiomycetes) come into contact with the host and adhere to the cuticle surface via nonspecific hydrophobic mechanisms (Boucias et al. 1988). This step is followed by the production of a mucilaginous coat by the fungus (St. Leger 1993). The attachment of the fungal spores to the host surface is initiated by hydrophobic interactions between conidial hydrophobins (Boucias et al. 1988) and the waxy surface of the insect cuticle (St. Leger 1993). Fungal hydrophobins are a class of hydrophobic-rich proteins that allow the fungi to attach to solid hydrophobic surfaces (Bidochka et al. 2001b).

Even though many university-level entomology textbooks suggest that chitin is the major component of the insect cuticle, it is predominantly composed of protein, with smaller percentages of chitin (Bidochka and Khachatourians 1992). Fungal pathogens produce extracellular enzymes directed at the integument of their hosts. *Metarhizium* produces a potent extracellular subtilisinlike protease called Pr1 that degrades cuticular protein (St. Leger et al. 1992a). A plethora of other extracellular proteases such as trypsin, chymotrypsin, metalloprotease, carboxypeptidases, and aminopeptidases are involved in cuticle degradation (St. Leger and Bidochka 1996). Extracellular chitinases and *N*-acetylglucosaminidases also are produced by the infecting organism.

In the latter stages of infection within the insect hemolymph, certain *Metarhizium* strains produce a cyclodepsipeptide ionophore, destruxin (Samuels et al. 1988; Amiri-Besheli et al. 2000). Other strains do not produce destruxin, but rather ramify (Samuels et al. 1988) in the insect hemocoel and use internal carbohydrates, such as the disaccharide trehalose, via an intracellular fungal trehalase (Xia et al. 2002). Strains that produce destruxins grow sparsely in the insect and kill them relatively quickly, whereas strains that do not produce destruxin grow profusely in the hemolymph and take longer to kill the insect (Samuels et al. 1988). It has been suggested that destruxins facilitate pathogenesis. Finally, *Metarhizium* emerges from the insect cuticle, usually through the less melanized intersegmental folds of the insect, and conidia are produced on the insect surface.

Taxonomy of *Metarhizium*

The classification of *Metarhizium* has been based on morphological characteristics (Tulloch 1976) that discriminated between *M. anisopliae* and *M. flavoviride* and further subdivided *M. anisopliae* based on conidial lengths (i.e., var. *anisopliae* conidia are shorter than those of var. *majus*). Other morphological distinctions include the presence or absence of a subhymenial zone, formation of prismatic columns by laterally adhering conidia, and the color of the conidia and the fungal colony (Humber 1997). A number of additional taxa have been described since Tulloch's classification, including *M. album* (Rombach et al. 1987), *M. brunneum, M. flavoviride* var. *minus* (Rombach et al. 1986), *M. pingshaese, M. cylindrosporae, M. guizhouensis* (Guo et al. 1986), *M. taii* (Liang et al. 1991), and *M. pinsahensis* and *M. biformisporae* (Liu et al. 1989), although each may not be formally recognized as a separate *Metarhizium* species.

DNA and other molecular characters have been used to assess the taxonomy of *Metarhizium*. A population genetic analysis by St. Leger et al. (1992b) using allozyme markers discounted any evidence for a tight association between morphological characteristics (such as conidial shape and color) and genetic groups. More recently, Driver et al. (2000) revised the genus *Metarhizium* based primarily on internal transcribed spacer (ITS) region and 28S ribosomal DNA (rDNA) sequence data and found some discrepancy between the nucleotide data and conidium morphology. Strain FI-152, for example, resembled *M. anisopliae* var. *anisopliae* but clustered within a clade of *M. flavoviride*. Some strains had conidial morphologies that are intermediate between *M. anisopliae* and *M. flavoviride* (Bidochka et al. 1994). Furthermore, Glare et al. (1996) found that phialide morphology could also vary depending on the growth medium.

Three areas of consensus concerning the taxonomy of *Metarhizium* have been reached. First, it is generally agreed that *M. anisopliae* is monophyletic based on phylogenetic analysis of ITS and 28S rDNA sequence data (Driver et al. 2000). However, the designation of varieties (or cryptic species) remains questionable without extensive phylogenetic and vegetative (somatic) compatibility data. Depending on the type of data (morphological or various nucleotide parameters) the investigator uses, the results of *Metarhizium* species or varietal designation vary and morphological classification of varietal designation and nucleotide data may or may not concur. Pipe et al. (1995) showed a clear distinction between *M. anisopliae* var. *majus* and other *Metarhizium* strains, previously identified by morphological characteristics, based on the presence or absence of a *Pst*1 site in the rDNA by restriction fragment length polymorphism (RFLP) analysis. In contrast, random amplification of polymorphic DNA (RAPD) analysis showed that six isolates previously identified morphologically as *M. anisopliae* var. *anisopliae* were more similar to *M. anisopliae* var. *majus* than to other *Metarhizium* strains (Fegan et al. 1993). Of course, there may be discrepancies in taxon sampling, and the issue of varietal designation cannot be resolved until a larger set of strains is assessed for a number of morphological and nucleotide characteristics.

Second, *Metarhizium anisopliae* is related to *M. flavoviride* but can be distinguished using molecular markers (St. Leger et al. 1992b; Bidochka et al. 1994; Tigano-Milani et al. 1995; Driver et al. 2000). Additionally, based on allozyme analysis, *M. album* is related but genetically distinguishable (St. Leger et al. 1992b) from *M. anisopliae* and *M. flavoviride*; however, the phylogenetic resolution is poor, particularly with so few isolates studied to date and almost all of them being from the Philippines (R. Humber, pers. comm.). Third, genetically distinguishable clades (genetic groups or cryptic species) exist within *M. anisopliae* and *M. flavoviride*. Despite these few points of agreement, a taxonomic revision of *Metarhizium* is not possible due to lack of information on all taxa and lack of information on the most informative nucleotide characteristics available. A taxonomic revision of *Metarhizium* is necessary, but only when a larger set of isolates, taxa, and informative nucleotide characteristics become available.

To estimate the number of species of *Metarhizium*, some type of species concept must be used. This agreement is necessary, at least for taxonomists working with entomopathogenic fungi, so that new species are not described without full consensus. Several possibilities have been considered for *Metarhizium* including morphological, biological, and phylogenetic species concepts. The problems of a morphological species concept have already been pointed out due to pleomorphisms of isolates within a clade and the effects of environment (or nutrients) on conidial development and morphology. A biological species concept views speciation primarily as a process of reproductive isolation. A phylogenetic species concept defines a taxon based on the "smallest diagnosable taxonomic units with a clear pattern of parental ancestry" (Geiser et al. 1998, p. 392). Avise (1994) suggested that the biological and phylogenetic species concepts are complementary because the historical transmission of genes that may be used to define a phylogenetic species concept are affected by reproductive isolation. With respect to *Metarhizium*, a biological species concept analysis as well as a phylogenetic species concept analysis can be used. Vegetative compatibility among strains of *Metarhizium* (Bidochka et al. 2000) can support the biological species concept, and molecular data can support the phylogenetic species concept (St. Leger et al. 1992b; Bidochka et al. 1994; Tigano-Milani et al. 1995; Driver et al. 2000). In this light, what is required is a comprehensive evaluation of nucleotide and vegetative compatibility data for a large number of *Metarhizium* isolates and taxa.

In the following sections, we provide evidence that several groups of *M. anisopliae* are reproductively isolated, based on vegetative incompatibility, and they have near fixation of polymorphic loci (Bidochka et al. 2000, 2001a,). Thus, they could be defined as different species. We suggest that as phylogeographical analyses of phylogenetic, morphological, and vegetative incompatibility emerge, a more integrated approach to defining species or varieties of *Metarhizium* will coalesce.

Based on criteria presented in this chapter, we estimate that there are approximately 10–15 different cryptic species of *Metarhizium*, including species of *Metarhizium anisopliae*, *M. flavoviride*, and *M. album*. This number coincides with data presented in studies on *Metarhizium* varieties by Driver et al. (2000) and genetic groups by St. Leger et al. (1992b). We agree with the suggestion of Milner et al. (1994) that the three currently recognized species (*M. anisopliae, flavoviride*, and *album*) should be reduced to a single species, *M. anisopliae*, which can then be further defined as consisting of varieties, genetic groups, or cryptic species, as more phylogenetic and phylogeographic data are obtained. We use the term "cryptic species" to define morphologically indistinguishable yet phylogenetically distinguishable taxonomic units that are reproductively isolated through vegetative incompatibility.

In the following section we review studies on the population genetics of *M. anisopliae* and attempt to integrate information into a phylogeographical assessment of *Metarhizium*.

Phylogeography of *Metarhizium*

To place a phylogeographic perspective on the worldwide population structure of *M. anisopliae*, data from several studies on population genetics of *Metarhizium* were reanalyzed. We assessed phylogenetic trees produced from a number of studies and grouped isolates at the terminus of the tree into genetic groups. The criteria for the genetic distance for which isolates were grouped was based on the genetic distance found to separate genetic groups based on allozyme, RAPD, or RFLP data from the study of Bidochka et al. (2001a). The grouping of isolates according to our analysis is found in appendix 2.1.

The following hypotheses are supported by data from these publications on the population genetics of *Metarhizium* (St. Leger et al. 1992b; Fegan et al. 1993; Rakotonirainy et al. 1994; Pipe et al. 1995; Tigano-Milani et al. 1995; Bridge et al. 1997; Leal et al. 1997; Driver et al. 2000; Mavridou et al. 2000; Bidochka et al.

2001a): (1) An association of *M. anisopliae* genotypes occurs with habitat type in temperate and polar regions. (2) Associations of *Metarhizium* genotypes with certain host insect species probably occur only in tropical and subtropical regions. (3) *M. anisopliae* is actually an assemblage of cryptic species with worldwide distributions. (4) Similar genotypes of *M. anisopliae* traverse large geographical barriers. (5) southeastern Asia is the probable origin of the evolution and diversity of *M. anisopliae*.

Metarhizium anisopliae *Genotypes Are Associated with Habitat Type, Not Insect Species, in Temperate and Near-Arctic Regions*

We collected isolates of insect-pathogenic fungi from various habitats in Ontario, Canada (Bidochka et al. 1998), using the *Galleria* soil-baiting technique (Zimmerman 1986) in which soils obtained from different habitats were sifted and placed into plastic containers with three *Galleria mellonella* (greater wax moth) larvae. The larvae were checked for mortality after 1 week and if dead were washed in a sodium hypochlorite solution to remove opportunistic fungi. The insects were then placed in a plastic vial and incubated with a moistened piece of tissue paper. We next checked for conidial growth on the cuticle. Single-conidium isolates were obtained using a standard microbiological streak-plate method. We obtained primarily *M. anisopliae*, but also another insect-pathogenic fungus, *Beauveria bassiana* (Bidochka et al. 1998).

Our initial analysis showed that *B. bassiana* was preferentially isolated from soils in natural, forested habitats, and *M. anisopliae* was preferentially isolated from agricultural soils. Similar results for *B. bassiana* and *M. anisopliae* have been observed in Europe (Stenzel 1992; Vanninen 1996). Our initial interpretation suggested that the *M. anisopliae* found in forested habitats were part of a statistical anomaly (Bidochka et al. 1998), but these results were reinterpreted after additional study (see below).

We first obtained these strains to investigate host specificity in *M. anisopliae* and to identify genes responsible for host specificity by using a subtractive hybridization technique (Diatchenko et al. 1996). We reasoned that *M. anisopliae* might show insect-host specificity because it has genetic features such as production of the cuticle-degrading subtilisinlike protease, Pr1, related to insect infection (St. Leger 1993) and because the population genetic structure of these fungi were assumed to be influenced primarily by host insect taxa (Riba et al. 1986; St. Leger et al. 1992b; Bridge et al. 1993; Fegan et al. 1993). This haploid, deuteromycetous fungus was assumed to reproduce clonally, and it was also assumed that certain genotypes were related to an insect host, especially if the insect host drives the population structure as has often been suggested (St. Leger et al. 1992b; Bridge et al. 1993; Fegan et al. 1993); in other words, there are fungal isolates or genotypes adapted for pathogenesis toward certain insect taxa.

We used three different types of markers for the *M. anisopliae* isolates: allozyme polymorphisms, RFLPs of the gene encoding Pr1, and RAPDs (Bidochka et al. 2001a). Our initial evaluation of the polymorphisms and the resulting dendrogram showed two major, deeply rooted branches (Bidochka et al. 2001a). To determine insect-host preferences in the two genetic groups, we bioassayed one coleopteran, two lepidopteran, and one orthopteran species against all the *M. anisopliae* isolates, but we did not find any clear associations of virulence with either genetic group (Bidochka et al. 2001a). These results conflicted with the paradigm of host-specific genotypes of *M. anisopliae*. However, A. Kamp, the graduate student working on the study, suggested that an analysis of the habitat might be worthwhile. When the genetic groups were compared to the habitat from which they were isolated, there was a strong statistical correlation of one genetic group with an agricultural habitat and another genetic group with the forested habitat. The previous observation of *M. anisopliae* from a forested habitat was, therefore, not a statistical anomaly, but rather a distinct genetic entity that also had the ability to grow at lower temperatures (10°C). The agricultural habitat group had an ability to grow at higher temperatures (35°C) and was resilient to UV exposure (Bidochka et al. 2001a). The genotypes of the isolates of *M. anisopliae* collected in Ontario were related to habitat preferences, not insect hosts. There are other examples of an absence of genotypic associations between *Metarhizium* isolates from temperate regions and insect host species in the studies by St. Leger et al. (1992b), Pipe et al. (1995), Leal et al. (1997), and Mavridou et al. (2000).

Metarhizium strains have shown wide variation in their abilities to grow at low and high temperatures (Ouedraogo et al. 1997). Evidence of genetic groups associated with low or high temperature growth was presented by Driver et al. (2000), Yip et al. (1992), and McCammon and Rath (1994). Roddam and Rath (1997) isolated cold-active strains of *Metarhizium* from the subantarctic Macquarie Island. Rath et al. (1995) examined carbohydrate utilization and cold-active growth for 134 isolates of *Metarhizium* and distinguished several *Metarhizium* groups based on carbohydrate utilization, as well as a group of cold-active *M. anisopliae* isolates able to germinate at 5°C. The data indicated that carbohydrate utilization clearly distinguished *Metarhizium* strains. The distinct cluster that was seen with cold-active isolates from the remaining strains suggested that the cold-active group forms a new variety of *M. anisopliae*. Based on the morphological, physiological, and biochemical evidence, Rath et al. (1995) named the cold-active strain *M. anisopliae* var. *frigidum*. Driver et al. (2000), however, found that strains from different genetic groups were capable of low temperature growth and suggested that this trait was homoplasious in *Metarhizium*. A taxonomic description based solely on cold-active growth is clearly problematic.

Variation of parameters not directly related to insect pathogenesis has also been observed for *Metarhizium*. Braga et al. (2001) analyzed the tolerance of a worldwide sample of *Metarhizium*, from latitudes from 61°N to 54°S, to UV irradiances of 920 and 1200 mW/m². The *Metarhizium* strains showed a significant quadratic relationship of decreasing UV-B tolerance with increasing latitude after 1- and 2-h exposures. The exposure to two irradiances had a significant effect on the culturability of the 30 *Metarhizium* strains; this further suggests that the environmental persistence of the conidia varies as a function of the normal changes that occur in natural UV regimes. The relationship of UV tolerance to genetic groups was not examined, but data from Bidochka et al. (2001a) suggest that certain genetic groups show a predisposition to UV tolerance.

Evaluation of the population structure of another insect-pathogenic fungus, *Beauveria bassiana*, also showed that genetic groups were related to habitat, growth temperature, and UV resilience (Bidochka et al. 2002), which suggests that other facultative insect-pathogenic fungi may also be genotypically related to habitat preferences rather than to specialization for an insect host.

Associations of M. anisopliae *Genotypes with Certain Host-Insect Species Probably Occur Only in Tropical and Subtropical Regions*

Specificity for certain insects has been reported for three recognized species of *Metarhizium*. Strains of *M. anisopliae* var. *majus*, *M. flavoviride,* and *M. album* are reported to be specific to Coleoptera, Orthoptera, and Hemiptera, respectively. *Metarhizium flavoviride* var. *flavoviride* and var. *minus* also were reported to show specificity to Coleoptera and Homoptera, respectively (Rombach et al. 1986).

A phylogenetic assessment of *Metarhizium* by St. Leger et al. (1992b) based on allozyme polymorphisms found that genotypic class 14 was broadly distributed throughout Brazil and Colombia. Also found in these regions were individuals that occupied rare or singular genotypic classes and were coleopteran pathogens. While single strains representing a genotypic class might simply belong to classes under-represented in the study, another interpretation is possible. The rare isolates may represent distinctive genotypic classes that evolved host specificity. The large genotypic class 14 (St. Leger et al. 1992b) contained homopteran, orthopteran, and lepidopteran pathogens. However, in the rarer classes 21, 37, and 24, each was pathogenic only to Coleoptera. Furthermore, strains isolated from Coleoptera, particularly from scarab beetles, had more fastidious germination requirements in culture, requiring the presence of a crude chitin preparation, suggesting an evolution toward host specificity.

Inglis et al. (1999) analyzed genetic variability of *M. flavoviride* isolates using RFLP analysis of telomeric regions. Telomeric fingerprinting distinguished *M. flavoviride* strains from acridid grasshoppers. Similarly, Leal et al. (1997) found no correlation of *Pr1* RFLP with its insect host, except for strains isolated from grasshoppers in Australia. Since most of Australia has a relatively stable tropical or subtropical environment (as compared to seasonal temperate environments), these isolates might have evolved specificity with acridid hosts. In fact, there are several examples of *M. anisopliae* and *M. flavoviride* that show specificity with acridoid hosts. Bridge et al. (1997) found strains of *M. flavoviride* that showed a close association of genotype with acridids. These genetically uniform strains had a pantropical distribution.

An analysis of rare genotypes from several studies on the population genetics of *Metarhizium* showed that approximately 28% of the genotypes from southeastern Asia are rare (Table 2.1). Again, these single strains may belong to underrepresented genotypic classes but they also may represent distinctive genotypic classes that have evolved host specificity.

Generally, strains of *Metarhizium* that show host specificity have a tropical–subtropical distribution (St. Leger et al. 1992b; Bridge et al. 1997; Leal et al. 1997).

Table 2.1. Geographical distribution of genetic groups of *Metarhizium* from reanalysis of data from seven studies.

Study	N. America	S. America	Europe	Africa	Australia/NZ	SE Asia
			Geographical Distribution of Genetic Groups (*Clusters*)			
St. Leger et al. (1992b)	6, 8, 10, 11	2, 6, 7, 10	4, 6, 9, 10, 12	—	3, 6, 9, 10, 11	1, 6, 7, 9, 10, 14, 15, 16
Driver et al. (2000)	—	2, 7	2, 4, 6	7	2, 3, 6, 7, 8, 9	1, 5, 10
Tigano-Milani et al. (1995)	2, 6	1, 6	1, 5	9	2, 3, 7, 8, 9	4, 6
Bridge et al. (1997)	—	—	11, 13	1, 2, 3, 9, 12	1, 5, 6, 7, 8, 10	10, 11
Rakotonirainy et al. (1994)	2	2	1, 2	1, 2	3, 4	1, 2
Bidochka et al. (1994)	1	3, 6	—	1, 4, 5, 6, 7	1, 6	1, 2, 7

Each number refers to a genetic group. Genetic groups should be compared within each study and not across studies. See appendix 2.1 for the grouping of isolates according to our analysis.

Initially, these strains may have been adapted to certain habitats. However, genetic constraints devoted to habitat adaptation may be relaxed in a relatively stable tropical or subtropical environment that does not have wide seasonal variations in temperature. These strains may have then evolved to specialize as pathogens toward certain insect hosts. In these environments a host insect may have been available year-round for thousands of years, and certain strains of fungi may have specialized on certain insect species. Futuyma (1973) argued that in tropical environments, species diversity, niche adaptation, and specialization result in coevolved species relationships that are not found to the same extent as in temperate environments. By inference, this also may be the case for insect-pathogenic fungi, although this hypothesis has not been fully tested for these fungi. The converse also may be true; *Metarhizium* may have initially been a specialized pathogen of certain insects, but, once they had spread to other, less stable environments, such as temperate and arctic climates, they may have adapted to certain habitats and evolved as facultative pathogens with a wide host range.

M. anisopliae *Is an Assemblage of Cryptic Species Found Worldwide*

Analysis of the population structures of the genetic groups of *M. anisopliae* found in Ontario, Canada, revealed that within a genetic group there was recombination, but between genetic groups there was little recombination and near fixation of alternate polymorphic alleles (Bidochka et al. 2001a). We suggest that these genetic groups fit a description of a cryptic species for *Metarhizium* because of near fixation of alleles. Lack of recombination between cryptic species was further substantiated by a study of double-stranded (ds)RNA viruses found in the Ontario isolates of *M. anisopliae* (Melzer and Bidochka 1998). Although we observed that some of these dsRNA viruses may be of similar sizes when separated by agarose gel electrophoresis, there was little homology (based on Northern hybridization) between dsRNA viruses of two genetic groups, suggesting no horizontal transmission between genetic groups (Bidochka et al. 2000). Furthermore, we produced several *nit* mutants (nitrogen nonutilizing mutants), and complementation between *cnx* and *nia*D mutants was only observed for strains within a genetic group. Mutants from different genetic groups were not complementary, and this effectively acts as a biological barrier to recombination between genetic groups. Based on these analyses, we suggest that the genetic groups of *M. anisopliae* that we found in Ontario are actually cryptic species of *Metarhizium*.

Similar analyses with the asexual fungus *Aspergillus flavus*, from Australia, also showed two reproductively isolated genetic groups that occurred sympatrically and could be considered cryptic species (Geiser et al. 1998). The sexual fungus *Coccidioides immitis* also showed two reproductively isolated taxa (Burt et al. 1996; Koufopanou et al. 1997). Although a more complete data set is lacking for a recombinational component to the population genetics of *Metarhizium*, there is evidence that certain genetic groups may have a near-fixation of alternate alleles (St. Leger et al. 1992b; Bidochka et al. 2001a), suggesting little genetic exchange between genetic groups, or cryptic species, with this genus. Therein, the different genetic groups of *Metarhizium* fulfill the definition of a phylogenetic species concept as well as a biological species concept.

Similar Genotypes of M. anisopliae *Traverse* Large Geographical Barriers

One of the first comprehensive, worldwide studies on the population genetics of *Metarhizium* was performed by St. Leger et al. (1992b). This study grouped a worldwide distribution of *Metarhizium* isolates into 48 genotypic classes based on allozyme polymorphisms. Considerable evidence for geographic clustering in the dendrogram was observed. We reanalyzed data provided in St. Leger et al. (1992b) and, using the genetic distance value of 0.4 as the cutoff value, divided the dendrogram into 16 genetic groups instead of 48 genotypic classes. We chose 0.4 as the cutoff value because in our analyses of *M. anisopliae* in Ontario, we showed that isolates with a genetic similarity value of approximately 0.6 belonged to the same genetic group (Bidochka et al. 2001a). The grouping of isolates according to our analysis is found in appendix 2.1. Table 2.1 shows the phylogeographic distribution of *M. anisopliae* genetic groups using our re-evaluation of St. Leger's et al. (1992b) data, based on genetic distance. The largest group was cluster 6 representing genotypic classes 6–23 of St. Leger et al. (1992b). This cluster spanned eastern North America, South America, Australia, and southeastern Asia. Our analysis also showed cluster 10, representing genotypic classes 31–38 of St. Leger et al. (1992b) spanning all continents.

The St. Leger et al. (1992b) study was a global analysis of *M. anisopliae* genetic variation, while that of Bidochka et al. (2001) concentrated on the population structure in Ontario, Canada. It was fortuitous that both studies included ME1 (=2575) in their analyses. In St. Leger et al. (1992b), ME1 was part of cluster 6 (see appendix 2.1), while the study of Bidochka et al. (2001a) found that ME1 was part of the UV-tolerant and heat-tolerant agricultural genetic group. By inference, this suggests that cluster 6 of St. Leger et al. (1992b) is a widespread genotype that is UV and heat tolerant and occupies a large area of North America (Bidochka et al. 2001a), as well as South America, Australia, and southeastern Asia.

We also observed evidence of long-distance migration of clusters 2 and 6 (table 2.1), using data from Driver et al. (2000). Cluster 2 is found in South America as well as in Australia; cluster 6 is predominant in Europe as well as Australia; and cluster 7 is found in Australia and Africa. Our analysis of the data from Tigano-Milani et al. (1995) found similar genotypes in Brazil and Europe; Bridge et al. (1997) showed similar genotypes in southeastern Asia, Africa, and Australia; Rakotonirainy et al. (1994) showed similar genotypes in South America and Africa; while Bidochka et al. (1994) showed similar genotypes in Brazil, Africa, and Australia, as well as North America, Africa, and the Middle East (table 2.1). Mavridou et al. (2000) identified polymorphisms in group-I introns within the 28S rDNA gene of 30 *M. anisopliae* isolates. Isolates ME1 and ARSEF 703 were 99% identical, and these isolates are from distinctly separated geographic origins and derived from different insect hosts. There is good evidence that certain genotypes of *Metarhizium* have migrated thousands of kilometers, or are now disjunct populations, and occur on different continents.

Southeastern Asia is the Probable Origin of the Evolution and Diversity of M. anisopliae

Our first point of evidence relates to the continental area with largest genetic diversity of *M. anisopliae*, and the second relates to the area where teleomorph–anamorph associations have been recorded for *M. anisopliae*.

First, the largest genotypic diversity of *M. anisopliae* is in southeastern Asia. Culture collections of *M. anisopliae* from some countries or continents may not adequately represent regional population variation to draw inferences about regional diversity. However, when we compiled the data from several studies that have considered genetic variation on a worldwide level, some interesting hypotheses could be postulated. We calculated a genotypic diversity index derived from data from 10 publications of genetic variation of *M. anisopliae* from 6 continental areas: North and South America, Africa, Europe, southeastern Asia, and Australia (table 2.2). Analyses of genetic diversity on a continental level showed that the mean index of genetic variation was highest in southeastern Asia. The premise that large genetic diversity in a geographic region relates to the geographic region of the evolutionary origin of a fungal species was also argued in the case of the potato blight fungus, *Phytophthora infestans* (Goodwin et al. 1994). Mexico was viewed as the area of *Phytophthora* genotypic radiation because the largest genetic diversity was observed there, as well as the largest number of compatible mating types. Similar patterns of genetic variation also have been found for the rice-pathogenic fungus, *Magnaporthe grisea* (Zeigler 1998). In that case populations found in Europe, North America, and South America are composed of a few lineages and often dominated by a single lineage, while levels of genetic variation in Asia, the presumed ancestral area, are high, with up to 45 genotypic lineages within a restricted geographical

Table 2.2. Shannon diversity indices calculated to estimate genetic diversity in *M. anisopliae* from different continental geographical areas.

Location	A	B	C	D	E	F	G	H	I	J	Mean Values
N. America	0.59	ND	0.28	0.48	ND	0.28	ND	ND	0	0.30	0.32
S. America	0.43	0.30	0	0	0.30	0.26	0.60	0	0	0	0.38
Africa	ND	0.30	ND	0	0	0	0.38	ND	0.28	0.71	0.42
Europe	0.61	0.32	0	0.68	0.29	0.30	0.30	0.30	0.55	ND	0.42
SE Asia	0.81	0.56	0.82	0.83	0.47	0.48	0.64	0.67	0.28	0.60	0.62
Australia	0.71	0.22	0.82	0.45	0.43	0.70	0.70	0.65	0.30	0	0.55

Diversity Indices Calculated from Study[a]

Larger values indicate higher indices of genetic diversity. Diversity indices were derived from 10 publications (A–J). Genotypes were placed into genetic groups according to criteria established in the appendix.

[a]Study: A= St. Leger et al. (1992b); B= Leal et al. (1997); C=Pipe et al. (1995); D= Mavridou et al. (2000); E=Driver et al. (2000); F= Tigano-Milani et al. (1995); G= Bridge et al. (1997); H= Curran et al. (1994); I= Rakotonirainy et al. (1994); J= Bidochka et al. (1994).

ND = no data. Zeros were not included in the mean value estimates because these studies had underrepresented samples from those areas.

area (Zeigler 1998). Using a similar premise, southeastern Asia should be considered a likely point of origin of *M. anisopliae*.

Second, southeastern Asia is the region where *Cordyceps* teleomorphs for *Metarhizium* have been found (Liang et al. 1991; Liu et al. 2001), as well as the greatest diversity of *Cordyceps* species worldwide (Nikoh and Fukatsu 2000; Suh et al. 2001). Phylogenetic analysis based on rDNA sequences showed that *Metarhizium* formed a lineage within the Clavicipitaceae (Suh et al. 2001). All the *Metarhizium* anamorphs derived from *Cordyceps* are pathogens of Coleoptera (Liu et al., 2001), and teleomorphs of *Metarhizium* have not been found outside of southeastern Asia. Again, it may be argued that appropriate mating types have not migrated outside of Asia, as was the case for mating types of *Phytophthora* migrating outside of the presumed ancestral population in central Mexico (Goodwin et al. 1994). Similarly, the migration of *M. grisea* rice pathotypes out of Asia resulted in new populations that have low sexual compatability and a predominantly clonal population structure, while in Asia, the population structure is in linkage equilibrium, suggesting high levels of recombination (Zeigler 1998). Again, using a similar premise of the geographic location of anamorph–teleomorph associations, as well as diversity of mating types and potential for recombination, southeastern Asia is likely the place of origin for *Metarhizium*.

Evolutionary Mechanisms

Founder Effects and Genetic Drift

Several random genetic factors may contribute to the population structure of *Metarhizium*. For example, founder populations of *Metarhizium* can be introduced into a new area where no other *Metarhizium* species have established and may be genetically different from the founder population. This could explain differences in population genetic structure of *Metarhizium* from southeastern Asia compared to those from Australia, Africa, or the Americas. Although certain genetic groups are found over a large region, there are small differences in their genotypes. For example, in our analysis (table 2.1) of data collected by St. Leger et al. (1992b), genetic group 6 is composed of a large, genetically related group that could be subdivided (not shown; 6bi and 6bii, for example). In pathogens that have a poorly developed system of recombination, such as in *M. anisopliae*, genetic drift will, however, over time, reduce the number of different clonal lineages present.

Migration

There are several examples of plant-pathogenic fungi that have moved from their endemic biogeographical range to a new environment (Goodwin et al. 1994; Zeigler 1998). Migration can produce profound changes in the resultant population structure and its ability for sexual recombination. This is illustrated by the migration patterns of *Phytophthora infestans* from its presumed ancestral origin in Mexico to North America and Europe (Goodwin et al. 1994). Here, a single major genotype of the A1 mating group has spread worldwide and typifies a genetic bottleneck that defined the clonal population structure of *P. infestans* for over 120 years. This genetic bottleneck reduces an individual's chances of finding a compatible mating partner. In the 1970s, the A2 mating group emerged rapidly in some areas, resulting in recombinant genotypes (Goodwin et al. 1994).

The conidia of *Metarhizium* tend to aggregate into a palisades layer that can be blown about as a unit. The conidia are hydrophobic and tend to readily release from colonies in dry air. Conidia of these fungi have been found in aerial samples and can survive for months at low temperatures and low relative humidities (Clerk and Madelin 1965; Zimmerman 1982). *Metarhizium* could also migrate through infected insects, such as locusts that are known to have crossed the Atlantic Ocean from West Africa to the Caribbean during large outbreaks in the 1980s (Richardson and Nemeth 1991). It is also possible that strains of *Metarhizium* may be transported by humans on fruit, vegetables, or insects infected by *Metarhizium*.

Relationships among populations of *Metarhizium* around the world are consistent with the hypothesis that all strains were derived from southeastern Asia. Figure 2.1 shows a hypothetical path of migration of *Metarhizium* throughout the world. It is, however, difficult to estimate a timeline for the migration, since there is insufficient nucleotide divergence data with *Metarhizium* for such a calculation. In our hypothetical scenario, an ancestral, sexually recombining *Metarhizium*, probably a *Cordyceps* species, arose in southeastern Asia. This insect-pathogenic *Cordyceps* may have been derived from a soil-inhabiting fungus, and Nikoh and Fukatsu (2000) hypothesized that a *Cordyceps* pathogenic on a trufflelike fungi "jumped host" to a soil insect host because both hosts occupy similar habitats.

Conidia of *Metarhizium* anamorphs were dispersed throughout southeastern Asia. Today the *Metarhizium album* lineage is found almost exclusively in the Philippines. The *flavoviride* and *anisopliae* lineages are found in Australia, Asia, Africa, and Europe. *Flavoviride* strains are rare in the Americas, whereas *anisopliae* lineages predominate there. As proposed earlier in this chapter, strains that inhabit tropical or subtropical environments may evolve host-insect specificity, while strains in temperate or arctic habitats become adapted to certain habitats rather than to certain hosts.

Recombination

M. anisopliae is a haploid mitosporic fungus assumed to reproduce clonally. However, recent evidence suggests that, although *M. anisopliae* lacks sexual morphology and is presumed to reproduce clonally, there is substantial recombination within certain genetic groups of this fungus (Bidochka et al. 2001a). Several important mitosporic plant and human pathogens follow this pattern. They also lack sexual morphology and were thought to reproduce clonally. Population genetic analysis, however, showed that recombination had occurred, as evidenced by their present-day population structure. These fungi include the human pathogen *Coccidioides immitis* (Koufopanou et al. 1997), the aflatoxin-producing, opportunistic plant pathogen *Aspergillus flavus* (Geiser et al. 1998), and the plant pathogen *Fusarium oxysporum* (Taylor et al. 1999).

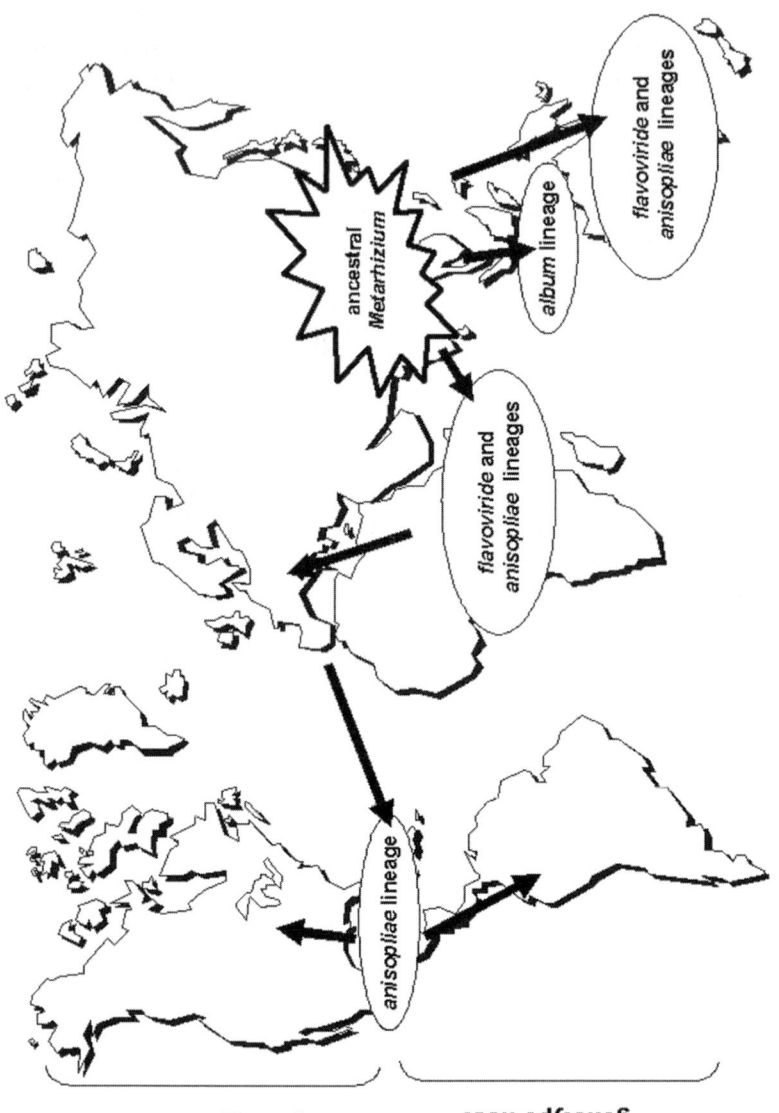

Figure 2.1. Hypothetical paths of the major routes of migration of *Metarhizium* throughout the world starting from a sexually recombining *Metarhizium*, probably a *Cordyceps* species, found in southeastern Asia. Also indicated at the left are temperate regions where fungal genotype-habitat associations predominate and the tropical and subtropical region where fungal genotype-host insect associations predominate. Note that these are the major routes of migration, and this hypothesis does not suggest that other routes of migration are not possible.

Tinline and Noviello (1971) showed that parasexuality exists as a form of mitotic recombination in *Metarhizium*, and a study by Leal-Bertioli et al. (2000) indicated that strains of *M. anisopliae* can recombine when they co-infect the same host insect. The hemolymph of an insect would seem an opportune environment for recombination because there is evidence that, for some insect-pathogenic fungi, the cell wall lacks beta-glucan polymers when the fungus grows in the insect hemolymph (Pendland et al. 1993) and that hydrophobins are lacking in the cell walls when grown in liquid media (Bidochka et al. 1995). The alteration in cell walls under some conditions may provide an opportunity for anastomosis and hyphal fusion and eventually mitotic recombination. Data presented in Bidochka et al. (2001a) indicate a history of recombination within cryptic species of *M. anisopliae*, although we cannot say how or when it occurs. *Metarhizium anisopliae* is a close relative of *Cordyceps* species that are capable of sexual recombination, but it has only been linked in southeastern Asia. In geographic regions outside of southeastern Asia, compatible mating types may not exist, and sexual recombination may not occur. There is, however, still the possibility of parasexual recombination, but it is difficult to judge from available data how frequently recombination (either sexual or parasexual) might occur because very little recombination would be required in each generation to produce the diversity observed in the dendrograms evaluated in this chapter.

It has been argued that the absence of sexual recombination in any population could reduce that population to extinction by the accumulation of deleterious mutations in a phenomenon called Muller's ratchet (Muller 1964). Muller's ratchet is a theoretical process where slightly deleterious mutations will accumulate stochastically with each mitotic generation, as they cannot be eliminated by sexual recombination in asexual organisms, and these mutations can potentially drive a population to extinction. However, a computer model analysis of the lack of sexual recombination and the resulting accumulation of deleterious mutations using parameters derived from *M. anisopliae* showed that under certain realistic conditions, Muller's ratchet may not play a critical role in the fitness of *M. anisopliae*. Population sizes (i.e., production of conidia) are potentially so large that many of the conidia would not carry mutations, so the generational effects of Muller's ratchet may not be realized (Bidochka and de Koning 2001).

Selection and Adaptation

The hypotheses presented in this chapter suggest that selection and adaptation are critical as evolutionary mechanisms in differentiating populations of *Metarhizium*. We suggest that selection operates on at least two levels with regard to the population structure. The first level is genotype-host insect selection and the second is genotype-habitat selection.

There is good evidence that host-specific strains exist in tropical and subtropical habitats. Some isolates of *Metarhizium* are indeed specialized on one host insect (e.g., *M. flavoviride* strains may be specific for acridid infection; Bridge et al. 1997). These strains are generally found in distinctive geographical regions and are not widely distributed. The possibility of selection by host has not been studied in full detail, but the variation in host specialization indicates that no simple generalization can describe all strains of *Metarhizium*. Some strains are specialized on a certain insect host, while other strains show a wide host range.

There also is good evidence of habitat-specific strains in temperate and arctic environments. Cold-active strains are found in temperate regions of North America, Europe, and Australia (McCammon and Rath 1994; Ouedrago et al. 1997; Roddam and Rath 1997; De Croos and Bidochka 1999); these cold-active strains differentially express genes when grown at low temperatures (De Croos and Bidochka 2001). Data by Driver et al. (2000) suggest that cold-active growth could have been derived in phylogenetically diverse lineages of *Metarhizium*. There is also evidence supporting adaptation for high-temperature growth (37°C) and UV resilience (Bidochka et al. 2001a). Habitat-adapted strains are generally widely distributed throughout all continents.

Conclusions

Metarhizium, as a group, has a cosmopolitan distribution and is found in every habitat type, from the arctic to the tropics. We suggest that *Metarhizium* is composed of a number of genetic groups and cryptic species, and that within each genetic group recombination can occur, but little or no recombination occurs between cryptic species. There is good evidence that in temperate environments, genetic groups of *Metarhizium* are adapted to certain environmental conditions and show a wide host range. In tropical and subtropical environments, in contrast, there is some evidence that some genetic groups of *Metarhizium* show host-insect preferences. We also propose that *Metarhizium* may have evolved from a *Cordyceps* in southeastern Asia, perhaps one closely related to a soil-dwelling pathogen of a trufflelike fungus. The availability and use of unambiguous genetic markers, such as allozymes and DNA polymorphisms, should improve our understanding of the paths of migration, relatedness, pathogenicity, and degree of genetic substructuring relationships of *Metarhizium* populations.

Although the population genetics of *Metarhizium* has received some attention, there are several areas of information that are lacking. For example, a comparative pathogenicity analysis has not been done for *Metarhizium*; that is, bioassays of a number of purported wide host-range and narrow host-range species against a wide range of insect species should be carried out. A phylogeographic analysis of *Metarhizium* based on DNA polymorphisms correlated with physiological (e.g., heat and cold activity) and pathological data would add to the description of this species complex. There may be a wealth of opportunity within *Metarhizium* as biocontrol resources for different specific host-insect and habitat preferences that could fit the needs of a variety of insect pest management efforts.

Acknowledgments   This research was supported by an operating grant from the National Sciences and Engineering Research Council of Canada to M.J.B. We thank Barbara MacDonald Buetter for editing this chapter.

Appendix 2.1. Reanalysis of the Geographical Distribution of Genetic Groups from St. Leger et al. (1992a)

The original dendrogram showed 48 genotypic classes of *Metarhizium* species.[1] The dendrogram was revised and divided into 16 clusters, but only 5 clusters were applicable from St. Leger et al. (1992b) in which the branch point similarity matrix value was 0.4. Each number of cluster represents isolates grouped together.

Geographical distribution of genetic groups from Leal et al. (1997) was reanalyzed. The dendrogram was divided into 8 clusters, but only 7 were used from the branch point similarity percentage value of 85. Each number of a cluster represents a genetic group.

Geographical distribution of genetic groups from Pipe et al. (1995) was reanalyzed. The dendrogram was divided into 17 clusters, but only 16 were used with the branch point similarity matrix value of 0.5. Each number of a cluster represents a genetic group.

Mavridou et al. (1998) analyzed intraspecific polymorphisms of *Metarhizium anisopliae* var. *anisopliae* using RFLP analysis of rDNA gene complex and mtDNA. Based on hybridization analysis of total DNA, the 25 isolates were classified in 20 distinct genetic groups.

Driver et al. (2000) redefined *Metarhizium* species using RAPD-polymerase chain reaction (PCR) banding patterns with sequence data from whole ITS region of rDNA to determine a phylogeny. The data revealed 10 distinct clades based on the combined sequence data set.

Tigano-Milani et al. (1995) determined the genetic relationships from isozyme phenotypes among isolates of *Metarhizium*. The dendrogram was divided into 9 clusters, but only 8 were analyzed with a branch point similarity matrix value of 0.5. Each number of a cluster represents a genetic group.

Bridge et al. (1997) analyzed the relationships of 30 Metarhizium strains from locusts and grasshoppers. The dendrogram was divided into 14 clusters with a branch point distance coefficient value of 0.4. Each number of a cluster represents a genetic group.

Curran et al. (1994) reexamined *Metarhizium* isolates to analyze evolutionary relationships using rDNA sequence information. The researchers constructed a phylogenetic tree; therefore, there were no values to represent how the clusters were represented, but the paper gave this information as a representation of a line.

Rakotonirainy et al. (1994) analyzed isoenzyme variability and partial sequences of ribosomal RNA within the genus *Metarhizium*. The researchers constructed a phylogenetic tree based on isoenzyme analysis of *Metarhizium* strains; therefore, there were no values to represent how the clusters were represented, but the paper shows this information by a representation of a line.

Bidochka et al. (1994) assessed the genomic variability among *Metarhizium* isolates by using RAPDs. Coefficients of similarity were based on PCR fragment patterns for strains of *Metarhizium*. Values were calculated from three RAPD primers, and a dendrogram was created with the method of unweighted pair-group arithmetic average with a resultant seven genetic groups.

Note

1. A copy of the strain designations and genetic groups can be obtained by email to bidochka@brocku.ca.

Literature Cited

Amiri-Besheli, B., B. Khambay, S. Cameron, M. L. Deadman, and T. M. Butt. 2000. Inter- and intra-specific variation in destruxin production by insect-pathogenic *Metarhizium* spp., and its significance to pathogenesis. *Mycological Research* 104:447–452.

Avise, J. C. 1994. *Molecular markers, natural history and evolution.* New York: Chapman and Hall.

Bidochka, M. J. 2001. Monitoring the fate of biocontrol fungi. In *Fungi as biocontrol agents: Progress, problems and potential,* ed. T. M. Butt, C. W. Jackson, and N. Magan, pp. 193-218. Wallingford, UK: CABI Publishing.

Bidochka, M. J, A. M. Kamp, T. M. Lavender, J. de Koning, and J. N. A. De Croos. 2001a. Habitat association in two genetic groups of the insect-pathogenic fungus *Metarhizium anisopliae*: Uncovering cryptic species? *Applied and Environmental Microbiology* 67: 1335–1342.

Bidochka, M. J., and G. G. Khachatourians. 1992. Growth of the entomopathogenic fungus *Beauveria bassiana* on cuticular components of the migratory grasshopper, *Melanoplus sanguinipes. Journal of Invertebrate Pathology* 59:165–173.

Bidochka, M. J., J. E. Kasperski, and G. A. M. Wild. 1998. Occurrence of the entomo- pathogenic fungi *Metarhizium anisopliae* and *Beauveria bassiana* in soils from tem- perate and near-northern habitats. *Canadian Journal of Botany* 76:1198–1204.

Bidochka, M. J., and J. de Koning. 2001. Are teleomorphs really necessary? Modeling the potential effects of Muller's ratchet on deuteromycetous entomopathogenic fungi. *Mycological Research* 105:1014–1019.

Bidochka, M. J., J. de Koning, and R. J. St. Leger. 2001b. Analysis of a genomic clone of hydrophobin (*ssg*A) from the entomopathogenic fungus *Metarhizium anisopliae. My- cological Research* 105:360–364.

Bidochka, M. J., M. A. McDonald, R. J. St. Leger, and D. W. Roberts. 1994. Differentia- tion of species and strains of entomopathogenic fungi by random amplification of polymorphic DNA (RAPD). *Current Genetics* 25:107–113.

Bidochka, M. J., M. J. Melzer, T. M. Lavender, and A. M. Kamp. 2000. Genetically related isolates of the entomopathogenic fungus *Metarhizium anisopliae* harbor homologous dsRNA viruses. *Mycological Research* 104:1094–1097.

Bidochka, M. J., F. V. Menzies, and A. M. Kamp. 2002. Genetic groups of the insect- pathogenic fungus *Beauveria bassiana* are associated with habitat and thermal growth preferences. *Archives of Microbiology* 178:531–537.

Bidochka, M. J., R. J. St. Leger, L. Joshi, and D. W. Roberts. 1995. The rodlet layer from aerial and submerged conidia of the entomopathogenic fungus *Beauveria bassiana* contains hydrophobin. *Mycological Research* 99:403–406.

Boucias, D. G., J. C. Pendland, and J. P. Latgé. 1988. Non-specific factors involved in the lattachment of entomopathogenic deuteromycetes to host insect cuticle. *Applied and Environmental Microbiology* 54:1795–1805.

Braga, G. U. L., S. D. Flint, C. D. Miller, A. J. Anderson, and D. W. Roberts. 2001. Vari- ability in response to UV-B among species and strains of *Metarhizium* isolated from sites at latitudes from 61°N to 54°S. *Journal of Invertebrate Pathology* 78:98–108.

Bridge, P. D., C. Prior, J. Sagbohan, C. J. Lomer, M. Carey, and A. Buddie. 1997. Molecular characterization of isolates of *Metarhizium* from locusts and grasshoppers. *Biodiversity and Conservation* 6:177–189.

Bridge, P. D., M. A. J. Williams, C. Prior, and R. R. M. Paterson. 1993. Morphological, biochemical and molecular characteristics of *Metarhizium anisopliae* and *M. flavoviride*. *Journal of General Microbiology* 139:1163–1169.

Burt, A., D. A. Carter, G. L. Koenig, T. J. White, and J. W. Taylor. 1996. Molecular markers reveal cryptic sex in the human pathogen *Coccidioides immitis*. *Proceedings of the National Academy of Sciences of the USA* 93:770–773.

Clerk, G. C., and M. F. Madelin. 1965. The longevity of three insect-parasitizing hyphomycetes. *Transactions of the British Mycological Society* 48:193–209.

Curran, J., F. Driver, J. W. O Ballard, and R. J. Milner. 1994. Phylogeny of *Metarhizium*: Analysis of ribosomal DNA sequence data. *Mycological Research* 98:547–552.

De Croos, A. M. N., and M. J. Bidochka. 1999. Effects of low temperature on growth parameters in the entomopathogenic fungus *Metarhizium anisopliae*. *Canadian Journal of Microbiology* 45:1055–1061.

De Croos, A. M. N., and M. J. Bidochka. 2001. Cold-induced proteins in cold-active isolates of the insect-pathogenic fungus *Metarhizium anisopliae*. *Mycological Research* 105:868–873.

de Moraes, C. K, A. Schrank, and M. H. Vainstein. 2003. Regulation of extracellular chitinases and proteases in the entomopathogen and acaricide *Metarhizium anisopliae*. *Current Microbiology* 46:205–210.

Diatchenko, L., Y.-F. C. Lau,.A.P. Campbell, A. Chenchik, F. Moqadam, B. Huang, S. Lukyanov, K. Lukyanov, N. Gurskaya, E.D. Sverdlov, and P.D. Siebert. 1996. Suppression subtractive hybridization: A method for generating differentially regulated or tissue-specific cDNA probes and libraries. *Proceedings of the National Academy of Sciences of the USA* 93:6025–6030.

Driver, F., R. J. Milner, and J. W. H. Trueman. 2000. A taxonomic revision of *Metarhizium* based on a phylogenetic analysis of rDNA sequence data. *Mycological Research* 104:134–150.

Fegan, M., J. M. Manners, D. J. Maclean, J. A. G. Irwin, K. D. Z. Samuels, D. G. Holdom, and D. P. Li. 1993. Random amplified polymorphic DNA markers reveal a high degree of genetic diversity in the entomopathogenic fungus *Metarhizium anisopliae* var. *anisopliae*. *Journal of General Microbiology* 139:2075–2081.

Futuyma, D. J. 1973. Community structure and stability in constant environments. *American Naturalist* 107:582–583.

Geiser, D. M., J. I. Pitt, and J. W. Taylor. 1998. Cryptic speciation and recombination in the aflatoxin-producing fungus *Aspergillus flavus*. *Proceedings of the National Academy of Sciences of the USA* 95:388–393.

Glare, T. R., R. J. Milne, and C. D. Beaton. 1996. Variation in *Metarhizium*, a genus of fungal pathogens attacking Orthoptera: Is phialide morphology a useful criterion? *Journal of Orthopteran Research* 5:19–27.

Goodwin, S. B., B. A. Cohen, and W. E. Fry. 1994. Panglobal distribution of a single clonal lineage of the Irish potato famine fungus. *Proceedings of the National Academy of Sciences of the USA* 91:11591–11595.

Guo, H. L., B. L. Ye, Y.Y. Yue, Q. T. Chen, and C. S. Fu. 1986. Three new species of *Metarhizium*. *Acta Mycologia Sinica* 5:185–190.

Hajek, A. E., and R. J. St. Leger. 1994. Interactions between fungal pathogens and insect hosts. *Annual Review of Entomology* 39:293–322.

Humber, R. A. 1997. Fungi – Identification. In *Manual of Techniques in Insect Pathology*, ed. L. Lacey, pp. 153–185. London: Academic Press.

Inglis, P. W., B. P. Magalhaes, and M. C. Valadares-Inglis. 1999. Genetic variability in *Metarhizium flavoviride* revealed by telomeric fingerprinting. *FEMS Microbiology Letters* 179:49–52.

Koufopanou, V., A. Burt, and J. W. Taylor. 1997. Concordance of gene genealogies reveals reproductive isolation in pathogenic fungus *Coccidiodes immitis*. *Proceedings of the National Academy of Sciences of the USA* 94:5478–5482.

Leal, S. C. M., D. J. Bertioli, T. M. Butt, J. H. Carder, P. R. Burrows, and J. F. Peberdy. 1997. Amplification and restriction endonuclease digestion of the *Pr1* gene for the detection and characterization of *Metarhizium* strains. *Mycological Research* 101:257–265.

Leal-Bertioli, S. C. M., T. M. Butt, J. F. Peberdy, and D. J. Bertioli. 2000. Genetic exchange in *Metarhizium anisopliae* strains co-infecting *Phaedon cochleariae* is revealed by molecular markers. *Mycological Research* 104:409–414.

Liang, Z. Q., A. Y. Liu, and J. L Liu. 1991. A new species of the genus *Cordyceps* and its *Metarhizium* anamorph. *Acta Mycologia Sinica* 10:257–262.

Liu, A. Y., Z. E. Liang, and L. Cao. 1989. Isolation and identification for *Metarhizium biformisporae* [in Chinese]. *Journal of Guizhou Agricultural College* 2:27–31.

Liu, Z. Y., Z. Q. Liang, A. J. S. Whalley, Y.-J. Yao, and A. Y. Liu. 2001. *Cordyceps brittlebankisoides*, a new pathogen of grubs and its anamorph *Metarhizium anisopliae* var. *majus*. *Journal of Invertebrate Pathology* 78:178–182.

McCammon, S. A., and A. C. Rath. 1994. Separation of *Metarhizium anisopliae* strains by temperature dependent germination rates. *Mycological Research* 98:1253–1257.

Mavridou, A., J. Cannone, and M. A. Typas. 2000. Identification of group-I introns at three different positions within the 28S rDNA gene of the entomopathogenic fungus *Metarhizium anisopliae* var. *anisopliae*. *Fungal Genetics and Biology* 31:79–90.

Melzer, M. J., and M. J. Bidochka. 1998. Diversity of double-stranded RNA viruses within populations of entomopathogenic fungi and potential implications for fungal growth and virulence. *Mycologia* 90:586–594.

Milner, R. J., F. Driver, J. Curran, T. R. Glare, C. Prior, P. D. Bridge, and G. Zimmerman. 1994. Recent problems with the taxonomy of the genus *Metarhizium*, and a possible solution. In Proceedings of the 5th International Colloquium of Invertebrate Pathology and Microbial Control, Montpellier, pp. 109–110.

Muller, H. J. 1964. The relation of recombination to mutational advance. *Mutation Research* 1:2–9.

Nikoh, N., and T. Fukatsu. 2000. Interkingdom host jumping underground: Phylogenetic analysis of entomoparasitic fungi of the genus *Cordyceps. Molecular Biology and Evolution* 17:629–638.

Ouedrago, A., J. Fargues, M. S. Goettel, and C. J. Lomer. 1997. Effect of temperature on vegetative growth among isolates of *Metarhizium anisopliae* and *M. flavoviride*. *Mycopathologia* 137:37–43.

Pendland, J. C., S.-Y. Hungand, and D. G. Boucias. 1993. Evasion of host defense by *in vivo*-produced protoplast-like cells of the insect mycopathogen *Beauveria bassiana*. *Journal of Bacteriology* 175:5962–5969.

Pipe, N. D., D. Chandler, B. W. Bainbridge, and J. B. Heale. 1995. Restriction fragment length polymorphisms in the ribosomal RNA gene complex of isolates of the entomopathogenic fungus *Metarhizium anisopliae*. *Mycological Research* 99:152–161.

Rakotonirainy, M. S., M. L. Cariou, Y. Brygoo, and G. Riba. 1994. Phylogenetic relationships within the genus *Metarhizium* based on 28S rRNA sequences and isozyme comparison. *Mycological Research* 98:225–230.

Rath, A. C., C. J. Carr, and B. R. Graham. 1995. Characterization of *Metarhizium anisopliae* strains carbohydrate utilization (AP150CH). *Journal of Invertebrate Pathology* 65:152–161.

Riba, G., I. Bouvier-Fourcade, and A. Caudal. 1986. Isoenzymes polymorphism in *Metarhizium anisopliae* (Deuteromycotina: Hyphomycetes) entomogenous fungi. *Mycopathologia* 96:161–169.

Richardson, C., and D. J. Nemeth. 1991. Hurricane-borne African locusts (*Schistocerca gregaria*) on the Windward Islands. *GeoJournal* 23:349–357.

Roberts, D. W., and A. E. Hajek. 1992. Entomopathogenic fungi as bioinsecticides. In *Frontiers of industrial mycology*, ed. G. F. Leatham, pp. 144–159. New York: Chapman and Hall.

Roddam, L. F., and A. C. Rath. 1997. Isolation and characterization of *Metarhizium anisopliae* and *Beauveria bassiana* from subantarctic Macquarie Island. *Journal of Invertebrate Pathology* 69:285–288.

Rombach, M. C., R. A. Humber, and H. C. Evans. 1987. *Metarhizium album*, a fungal pathogen of leaf and planthoppers of rice. *Transactions of the British Mycological Society* 88:451–459.

Rombach, M. C., R. A. Humber, and D.W. Roberts. 1986. *Metarhizium flavoviride* var. *minus* var. nov., a pathogen of plant and leafhoppers on rice in the Phillipines and Solomon Islands. *Mycotaxon* 27:87–92.

Samson, R. A., H. C. Evans, and J.-P. Latgé. 1988. *Atlas of entomopathogenic fungi*. Berlin: Springer-Verlag.

Samuels, R. I., A. K. Charnley, and S. E. Reynolds. 1988. The role of destruxins in the pathogenicity of 3 strains of *Metarhizium anisopliae* for the tobacco hornworm *Manduca sexta*. *Mycopathologia* 104:51–58.

St. Leger, R. J. 1993. Biology and mechanisms of insect-cuticle invasion by deuteromycetous fungal pathogens. In *Parasites and pathogens of insects*, vol. 2, eds. N. C. Beckage, S. N. Thompson and B. A. Federici, pp. 211–229. New York: Academic Press.

St. Leger, R. J., and M. J. Bidochka. 1996. Insect-fungal interactions. In *New directions in invertebrate immunology*, eds. K. Soderhall, S. Iwanaga and G. R. Vasta, pp. 443–479. New Jersey: SOS Publications.

St. Leger, R. J., D. C. Frank, D. W. Roberts, and R. C. Staples. 1992a. Molecular cloning and regulatory analysis of the cuticle-degrading protease structural gene from the entomopathogenic fungus *Metarhizium anisopliae*. *European Journal of Biochemistry* 204:991–1001.

St. Leger, R. J., B. May, L. L. Allee, D. C. Frank, R. C. Staples, and D. W. Roberts. 1992b. Genetic differences in allozymes and in formation of infection structures among isolates of the entomopathogenic fungus *Metarhizium anisopliae*. *Journal of Invertebrate Pathology* 60:89–101.

Stenzel, K. 1992. Natural occurrence and population density of *Metarhizium anisopliae* in cultivated and non-cultivated soil in North-Rhine-Westfalia, FRG. In Abstracts of the 25th Annual Meeting of the Society of Invertebrate Pathology, 16–21 August 1992, Heidelberg, Germany, p. 251.

Suh, S.-O., N. Hiroaki, and M. Blackwell. 2001. Insect symbiosis: Derivation of yeast-like endosymbionts within an entomopathogenic filamentous lineage. *Molecular Biology and Evolution* 18:995–1000.

Taylor, J. W., D. J. Jacobsen, and M. C. Fisher. 1999. The evolution of asexual fungi: Reproduction, speciation and classification. *Annual Review of Phytopathology* 37:197–246.

Tigano-Milani, M. S., A. C. M. M. Gomes, and B. W. S. Sobral. 1995. Genetic variability among Brazilian isolates of the entomopathogenic fungus *Metarhizium anisopliae*. *Journal of Invertebrate Pathology* 65:206–210.

Tinline, R. D., and C. Noviello. 1971. Heterokaryosis in the entomogenous fungus *Metarrhizium anisopliae*. *Mycologia* 63:701–712.

Tulloch, M. 1976. The genus *Metarhizium. Transactions of the British Mycological Society* 66:407–411.

Vanninen, I. 1996. Distribution and occurrence of four entomopathogenic fungi in Finland: Effect of geographical location, habitat type and soil type. *Mycological Research* 100:92–101.

Xia, Y., J. M. Clarkson, and K. A. Charnley. 2002. Trehalose-hydrolysing enzymes of *Metarhizium anisopliae* and their role in pathogenesis of the tobacco hornworm, *Manduca sexta. Journal of Invertebrate Pathology* 80:139–147.

Yip, H. Y., A. C. Rath, and T. B. Koen. 1992. Characterization of *Metarhizium anisopliae* isolates from Tasmanian pasture soils and their pathogenicity to redheaded cockchafer (Coleoptera: Scarabaeidae: *Adoryphorus couloni*). *Mycological Research* 96:92–96.

Zeigler, R. S. 1998. Recombination in *Magnaporthe grisea. Annual Review of Phytopathology* 36:249–275.

Zimmerman, G. 1982. Effect of high temperatures and artificial sunlight on the viability of conidia of *Metarhizium anisopliae. Journal of Invertebrate Pathology* 40:36–40.

Zimmerman, G. 1986. The '*Galleria* bait method' for detection of entomopathogenic fungi in soil. *Journal of Applied Entomology* 102:213–215.

3

Interactions between Entomopathogenic Fungi and Arthropod Natural Enemies

Michael J. Furlong
Judith K. Pell

In both natural and agricultural ecosystems, complex multitrophic interactions involving herbivores, predators, parasitoids, and pathogens contribute to arthropod community structure. Studies of the individual relationships among insects and their predators, parasitoids, and pathogenic microorganisms are many and varied, and several represent landmarks in the theory and understanding of population ecology and biological control (Bellows and Hassell 1999). It is common for a herbivore to be exploited by a number of different natural enemies that will interact directly and indirectly not only with each other but also with the herbivore. However, until relatively recently, studies examining the interrelationships between phylogenetically distinct organisms, such as arthropod and pathogen natural enemies utilizing the same insect resource, were uncommon (Hochberg and Lawton 1990; Rosenheim et al. 1995). Only a limited number of these studies have considered interactions between arthropod natural enemies and entomopathogenic fungi (Brooks 1993; Roy and Pell 2000).

Entomopathogenic Fungi

Approximately 750 species of fungi have been documented to infect insects (Hajek 1997). The majority of these are from either the Entomophthorales (Zygomycota) or conidial hyphomycetes of the Ascomycota. Although aspects of their life-history attributes vary, both of these groups of fungi produce conidia or other asexual spores and, in the case of some Entomophthorales, zygospores. The conidium is the infective unit and is acquired by new hosts when they contact sporulating hosts or when dislodged or discharged conidia are encountered in the environment. Both fungal groups are capable of inducing natural epidemics or epizootics within susceptible host populations (Ullyett and Schonken 1940; Smith et al. 1976; Velasco 1983; Thorvilson and Pedigo 1984; Powell et al. 1986), although epizootics of entomophthoralean species are more commonly observed (Pell et al. 2001). The host range of entomopathogenic fungi varies considerably between species and can differ significantly between isolates of the same species.

Parasitoids

Insect parasitoids, principally found in the Hymenoptera and Diptera, are characterized by larval stages completing their entire development by feeding on the body of another arthropod, eventually killing it (Godfray 1994). Free-living adult parasitoids may oviposit at the feeding sites of their hosts and rely on egg ingestion as a means of entering their host or, more commonly, they attack their host directly and deposit eggs either on the external surface of the host cuticle (ectoparasitoids) or directly inside the body of their host (endoparasitoids). Some species are highly host specific, whereas others have broad host ranges (Godfray 1994).

Predators

Insect predators can be polyphagous, consuming a wide range of prey species, or specialists, feeding on a more restricted group of organisms. Adults, larvae, or both stages may actively consume their prey.

Although arthropod natural enemies have received the greatest attention as biological control agents, entomopathogenic fungi have been used successfully in classical, augmentation, and conservation strategies (Goettel et al. 1990; Inglis et al. 2001; Pell et al. 2001). For example, several species of hyphomycetes that are amenable to mass production, formulation, and spray application are now commercially available as inundative pest control products (Shah and Goettel 1999). Entomophthoralean fungi, in contrast, have been largely exploited in inoculative, augmentation, and conservation strategies (Furlong et al. 1995; Pell et al. 2001). Thus, in both natural environments and managed agricultural systems, insect herbivores and their arthropod natural enemies may come into contact with both endemic fungal pathogens (potentially at elevated levels) with which they may have coevolved and shared a close association and also with inundatively applied fungal strains with which they may have no prior association.

Scale of Interactions

When parasitoids, predators, and entomopathogenic fungi utilize the same species of insect host, the spatial and temporal overlap of their populations can result in direct competition for the finite resources of the host. In such situations, pathogen–parasitoid or pathogen–predator interactions may occur at the individual level (e.g., when an insect is simultaneously host to the fungal pathogen and an immature parasitoid or when an infected insect is eaten by a predator) and at the population level (e.g., when the incidence of fungal disease or immature parasitoids within the host population or the number of free-living predators in the system can influence the subsequent population dynamics of each of the competitors). The unique biology of endoparasitoids means that, at the individual level, their interactions with fungal pathogens can occur internal to the host, as when an individual host is both parasitized by a parasitoid and infected by a fungal pathogen, or external to the host, as in direct pathogen infection of free-living adult stages, direct vectoring of fungal conidia to susceptible hosts by adult parasitoids, or altered host behavior in the presence of the parasitoid, which may lead to changes in disease epidemiology. Predators may encounter and consume fungus-infected host tissue as a nutritional resource (Roy and Pell 2000), and it is possible that immature ectoparasitoids will be confronted with fungus-diseased host tissue (Lord 2001); however, the other associations of these groups of natural enemies with fungal pathogens are essentially confined outside of the host. Despite the specific nature of the association, interactions between fungal pathogens and arthropod natural enemies at the individual level underpin and drive interactions at the population level.

Fungal pathogens, parasitoids, and predators represent important resources for the environmentally sound management of many species of insect pest. Each of these natural enemy groups can be important determinants of insect community structure (Waage and Hassell 1982; Anderson and May 1986). Understanding their interactions within a system is imperative to develop effective *de novo* multispecies biological control programs and to prevent disruption of existing biological control by the introduction of anther type of natural enemy. Hochberg et al. (1990) described a general model, based on the assumption of discrete, non-overlapping host generations, for the examination of the consequences of host–parasitoid–pathogen interactions on the population dynamics of each of the organisms involved. Outcomes predicted by the model, which range from coexistence of the parasitoid and pathogen within the host population to total exclusion of one of the natural enemies by the other, depend on the relative rates of pathogen transmission and parasitoid searching efficiency, the finite rate of increase of the host population, the distribution of parasitoid attacks, and the degree of interference between natural enemies developing within the common host (fig. 3.1). The relative abilities of a parasitoid and a pathogen developing within a common host to outcompete each other is measured by interference competition, whereas relative abilities of the organisms to locate and parasitize or infect susceptible hosts is measured by exploitation competition. The bias of interference competition and the relative rates of exploitation competition between the parasitoid and the pathogen in a given host–parasitoid–pathogen association will determine the outcome of the competitive interaction (Hochberg and Lawton 1990; Hochberg et al. 1990; fig. 3.1).

Interactions at the Individual Level

The safety of using fungal pathogens as biological control agents has often been of paramount concern when associations between entomopathogenic fungi and

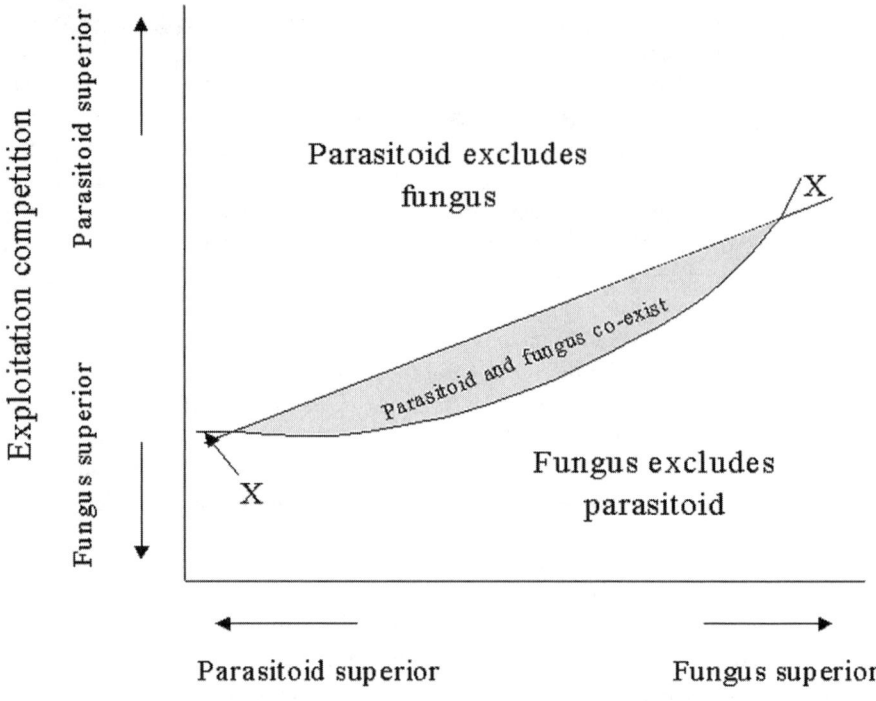

Figure 3.1. Conceptual model of population–level interactions between a parasitoid and an entomopathogenic fungus attacking a common insect host, adapted from Hochberg and Lawton (1990) and Hochberg et al. (1990). The vertical axis represents the relative abilities of the parasitoid and the fungus to locate and parasitize or locate and infect susceptible hosts. Active mechanisms for host searching and location used by parasitoids confer an advantage over fungal pathogens, and they tend to be superior exploitative competitors. The horizontal axis represents the relative abilities of the parasitoid and the fungus to directly compete for resources within a common individual host. The shorter intra-host development times of fungi confer an advantage over parasitoids, and they tend to be superior interference competitors. As the differences between interference and exploitation capacities of the parasitoid and the fungal pathogen become smaller, the probability that the two natural enemies will co-exist within the host population increases (gray shaded area of graph). When both the interference and exploitation capacities of one of the natural enemies far exceed those of the other, the superior competitor will prevail and the inferior competitor will be excluded. In regions "X," the outcome of the interaction depends on the initial population densities.

arthropod natural enemies of insects have been considered (Goettel et al. 1990; Vinson 1990), and the only review to consider the reciprocal relationships between insect parasitoids and fungal pathogens specifically appeared more than a decade ago (Brooks 1993).

    In the intervening period the development of entomopathogenic fungi as biological control agents has continued apace and, to appreciate the fundamentals of the ecology of insect–fungus associations and to ensure the compatibility of candidate biological control agents in integrated pest management programs, interactions between a large number of fungal pathogens and arthropod natural enemies have been investigated (Roy and Pell 2000). Although further research is required, several detailed studies of specific associations at the intra- and extra-host levels have increased our knowledge considerably.

Intra-host Interactions

Direct Competition for Resources

The potential intra-host interactions that can occur between entomopathogenic fungi and insect parasitoids are summarized in figure 3.2. An insect that is the potential host for both an entomopathogenic fungus and a parasitoid is unlikely to be attacked by both natural enemies at exactly the same time. When joint attack does occur, the order of attack and the physiological time that elapses between attacks by the different natural enemies have profound consequences for the interaction and can determine its outcome.

As the intra-host development time for fungal pathogens is generally shorter than that of parasitoids (Brooks 1993), if a fungal pathogen infects a host insect before a parasitoid, then the pathogen will usually exclude the parasitoid because it has a temporal advantage (i.e., interference competition is skewed in favor of the fungus). In most cases, the temporal nature of fungal infection versus parasitism is of paramount importance in the interaction, and only if a host insect within which a parasitoid is developing becomes infected when the parasitoid larva is at a relatively advanced stage of development can a viable adult parasitoid emerge.

Powell et al. (1986) found that the fungal pathogen *Pandora* (=*Erynia*) *neoaphidis* (Zygomycota: Entomophthorales) took 3–4 days to complete its development and kill its aphid host, whereas the parasitoid *Aphidius rhopalosiphi* (Hymenoptera: Braconidae) took 8 or more days after oviposition to pupate. When parasitized aphids were infected by *P. neoaphidis* fewer than 4 days after parasitism, the fungus outcompeted the parasitoid for the host resources and caused host death before the parasitoid could complete its development. Infection of parasitized host aphids 4 or more days after parasitoid oviposition allowed parasitoids to complete their development, and in some cases both *P. neoaphidis* and *A. rhopalosiphi* developed successfully within the same host. In a similar study of the interactions among the fungus *Zoophthora radicans* (Zygomycota: Entomophthorales), the diamondback moth, *Plutella xylostella* (Lepidoptera: Yponomeutidae), and two of its larval parasitoids, *Diadegma semiclausum* (Hymenoptera: Ichneumonidae) and *Cotesia plutellae* (Hymenoptera: Braconidae), the fungal pathogen was also the superior competitor for host resources (Furlong and Pell 2000). Interference competition between the fungus and each of the parasitoids was skewed in favor of the fungus, and both parasitoids required a significant temporal advantage over the pathogen to develop successfully. The premature death of parasitized insects that were infected by *Z. radicans* 3 days after parasitoid oviposition resulted in the mortality of immature parasitoids. Infection of parasitized hosts 5 or more days after parasitoid oviposition reduced, but did not

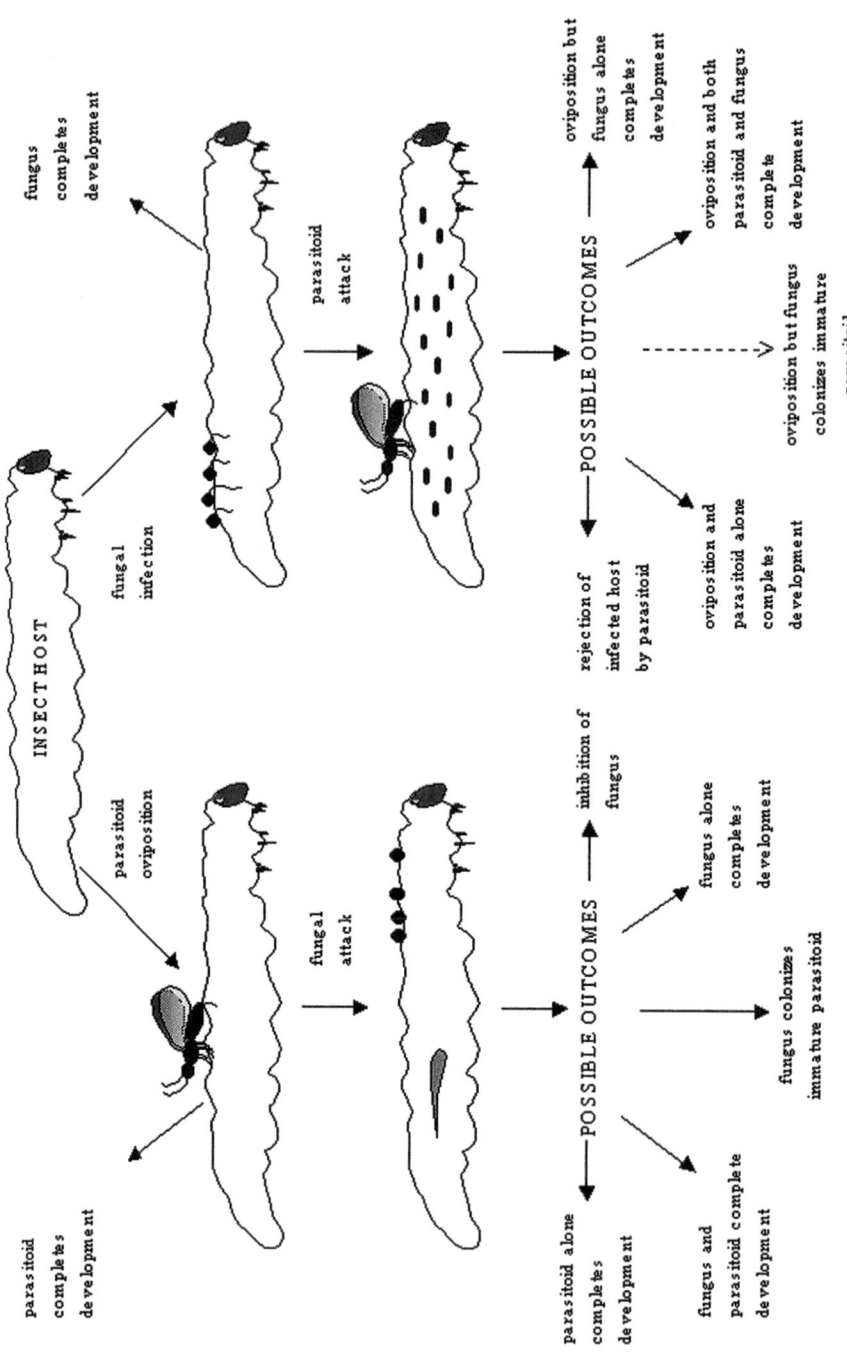

Figure 3.2. Diagrammatic representation of the potential interactions between a parasitoid and a fungal pathogen attacking the same insect host. Solid arrows represent known outcomes that are recorded in the literature; the dotted arrow represents a hypothetical outcome for which there is no record.

preclude, successful parasitoid development, and the negative competitive effects of *Z. radicans* on parasitoid survival decreased as the interval between parasitoid oviposition and initiation of infection increased (fig. 3.3). In some host individuals both competitors developed successfully; parasitoids formed viable pupae, and the fungus produced conidia (fig. 3.4a). However, the number of conidia produced and fitness parameters of the emerging parasitoid, both of which could have been negatively affected by interference competition, were not quantified.

Fransen and van Lenteren (1994), studying the intra-host interactions among the greenhouse whitefly, *Trialeurodes vaporariorum* (Homoptera: Aleurodidae), its fungal pathogen *Aschersonia aleyrodis* (Ascomycota: Hypocreales), and the parasitoid *Encarsia formosa* (Hymenoptera: Aphelinidae), also found that increasing the time between parasitoid oviposition and fungal infection decreased the competitive advantage of the pathogen over the parasitoid. If parasitized hosts were treated with fungus 4 or more days after parasitoid oviposition, the parasitoid had absolute competitive advantage over the pathogen, and parasitoid development was completed with no detrimental effect on the reproductive potential of emerging adults. In a study of the interactions among the Russian wheat aphid, *Diuraphis noxia* (Homoptera: Aphididae), its parasitoid *Aphelinus asychis* (Hymenoptera: Aphelinidae), and the pathogen *Paecilomyces fumosoroseus* (Ascomycota: Hypocreales), Mesquita and Lacey (2001) showed that treatment with the pathogen 24–96 h after parasitoid oviposition did not affect cocoon production, but adult parasitoid emergence from cocoons was depressed following pathogen application 24 h after parasitism. When aphid hosts were treated with the fungus before exposure to parasitoids, the number of cocoons produced was significantly affected by the degree of temporal separation between fungal attack and parasitoid oviposition. When the fungus and the parasitoid attacked host aphids at the same time, the number of cocoons produced was not reduced, but if the delay in parasitoid attack increased by 24 h or more the number of cocoons produced declined, approaching zero when the delay was 72 h. In this system, due to the similar development time of the two natural enemies, the fungus did not have a complete competitive advantage over the parasitoid, providing a rare example of successful parasitoid development in hosts previously infected by a fungal pathogen.

In situations where the temporal separation of parasitoid oviposition and fungus infection is such that neither natural enemy has an absolute competitive advantage, host-plant effects may significantly influence the marginal outcome of the intra-host interaction. Fuentes-Contreras et al. (1998) demonstrated that the increased development time of the aphid parasitoid *A. rhopalosiphi* within *Sitobion avenae* (Homoptera: Aphididae) feeding on resistant wheat cultivars increased the competitive advantage of the aphid pathogen *P. neoaphidis* in hosts attacked by both natural enemies and led to significantly greater immature parasitoid mortality. Host plants can significantly affect the susceptibility of herbivorous insects to fungal pathogens (Hare and Andreadis 1983), and the toxins they produce can have deleterious effects on pathogen survival (Costa and Gaugler 1989). Recently, Rostás and Hilker (2003) showed that phytopathogen-mediated changes in the host plant increased the susceptibility of herbivorous larvae to a fungal entomopathogen. Properties of host plants that increase herbivore susceptibility to the fungus are likely to

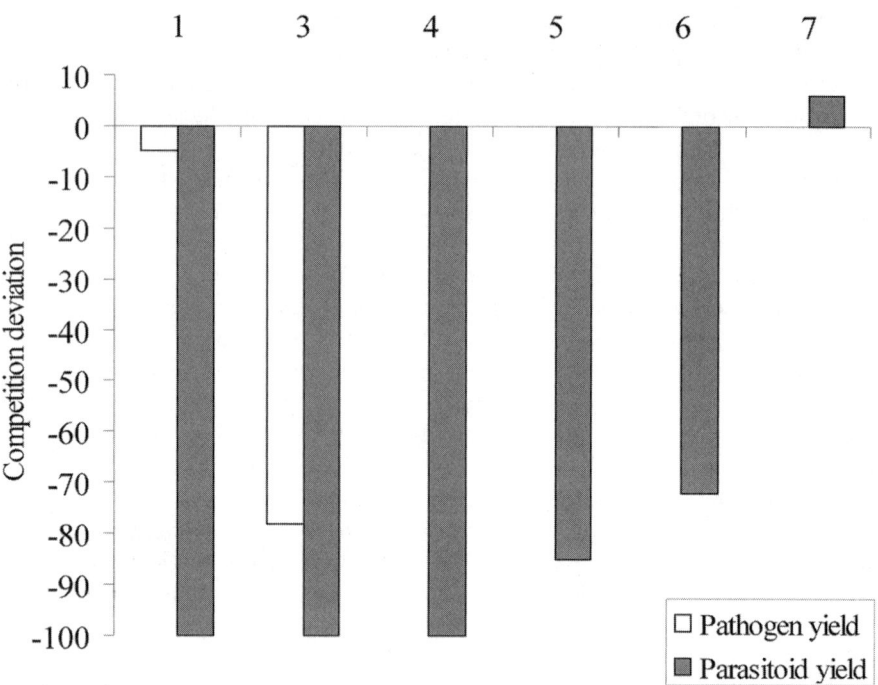

Figure 3.3. Interference between *Cotesia plutellae* and *Zoophthora radicans* within larvae of their common host, the diamondback moth. Competition deviations for *Z. radicans* (conidium production per cadaver in the presence of *C. plutellae* minus conidium production per cadaver in the absence of *C. plutellae*) and *C. plutellae* (number of cocoons produced in the presence of *Z. radicans* minus number of cocoons produced in the absence of *Z. radicans*) when *Z. radicans* infected parasitized hosts on given days after parasitism. Conidia production was only measured for individuals infected by *Z. radicans* 1 and 3 days after parasitism. When fungal infection occurred 1 day after parasitoid oviposition, the presence of the parasitoid had a minimal impact on the pathogen, but as the period between parasitoid oviposition and fungal infection increased, interference by the developing parasitoid reduced pathogen yield. Fungal infection of the host 4 days or fewer following parasitoid oviposition completely excluded the parasitoid, but as the period between parasitoid oviposition and fungal infection increased beyond this point, the negative interference effects of the pathogen on parasitoid yield were reduced (adapted from Furlong and Pell 2000).

be detrimental to co-occurring parasitoids as they skew interference competition further in favor of the pathogen, whereas host-plant properties that cause a decrease in susceptibility are likely to favor the parasitoid. The complexities of these multitrophic interactions render case-specific predictions difficult, and additional studies investigating such subtle effects in explicit host–parasitoid–fungal pathogen systems are clearly required.

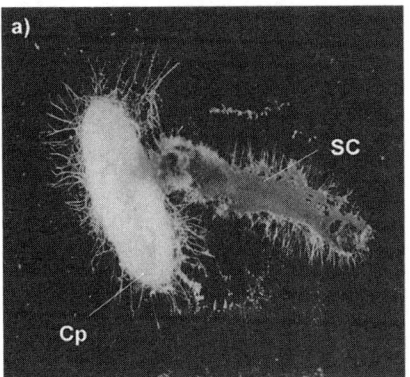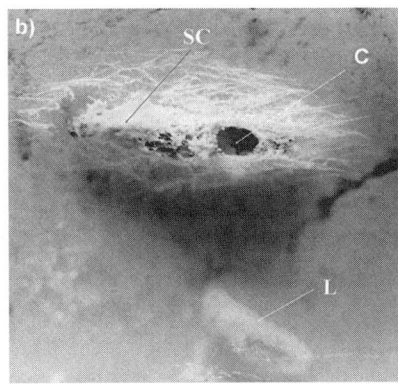

Figure 3.4. Outcomes of intra-host competitive interactions between *Zoophthora radicans* and two parasitoids of their common host, the diamondback moth. (a) Host insect infected by *Z. radicans* 5 days after oviposition by *Cotesia plutellae*. Both the pathogen and the parasitoid developed within the common host, the parasitoid formed a viable pupa within a cocoon (Cp) after emerging from the *Z. radicans*-infected final instar larva, the remains of which subsequently sporulated (SC). (b) Host insect infected by *Z. radicans* 3 days after oviposition by *Diadegma semiclausum*. Both the pathogen and the parasitoid developed within the common host that was killed by the pathogen before the larval parasitoid had completed its development. The noninfected, immature parasitoid larva (L) emerged from the sporulating cadaver of its infected host (SC), exposing a large cavity (C).

Intra-host Susceptibility of Parasitoids to Infection

In the previous examples, the detrimental effects of infection on developing parasitoids can be attributed to the ability of fungal pathogens to outcompete immature, co-occurring parasitoids. In these studies, which used several different fungal entomopathogens, no reports of intra-host infection of immature parasitoids were reported. Furlong and Pell (2000) found that *D. semiclausum* larvae developing within host insects, which subsequently became infected by *Z. radicans*, continued their development until the fungus eventually used all remaining host resources and sporulated. At this point noninfected but immature parasitoid larvae still emerged from the sporulating host (fig. 3.4b shows the large cavity in a sporulating *P. xylostella* cadaver from which an immature *D. semiclausum* larva emerged). In the first definitive report of its kind, Askary and Brodeur (1999) demonstrated intra-host invasion of the tissues of immature *Aphidius nigripes* (Hymenoptera: Braconidae) larvae by *Lecanicillium* (=*Verticillium*) *lecanii* (Ascomycota: Hypocreales) hyphal bodies developing within the same heavily infected *Macrosiphum euphorbiae* (Homoptera: Aphididae) host. Subsequent studies on interactions between the diamondback moth, *D. semiclausum*, and the fungus *Beauveria bassiana* (Ascomycota: Hypocreales) support the concept of intra-host invasion of the parasitoid by the fungus (Furlong 2004). Treatment of parasitized larvae with a high dose of the fungal pathogen 2 days after parasitoid oviposition precluded *D. semiclausum* development, as all host larvae were killed by the pathogen.

Treatment of parasitized larvae with a low dose of the fungus at this time allowed some *D. semiclausum* pupae to develop; adult parasitoids, however, emerged from only a small proportion of these. Some pupae were subsequently the sites of spore formation, indicating that parasitoids became infected by the fungus while developing within *B. bassiana*-infected hosts. Treatment of parasitized larvae with a high dose of the pathogen 4 days after parasitoid oviposition resulted in the development of some *D. semiclausum* pupae, but almost 90% of these were subsequently found to be infected by the fungus. Although the formation of pupae was not affected by treatment with the low dose of *B. bassiana* 4 days after parasitoid oviposition, adult emergence was reduced, and a significant number of pupae were found to be infected based on fungal spore production. Thus, the intra-host interaction between developing parasitoids and entomopathogenic fungi may not be a simple case of direct competition for host resources as often supposed. The developing parasitoid also may have to contend with fungal invasion, increasing its competitive disadvantage and skewing the interaction further in favor of the co-occurring fungal pathogen.

Changes in Host Susceptibility to Infection after Parasitism

Immature parasitoids and fungal pathogens can compete directly for the resources of their common host only if the host is colonized by both of the natural enemies. A change in the susceptibility of host insects to fungal pathogens after parasitoid oviposition has not been widely studied. The larval cuticle of *Pieris brassicae* (Lepidoptera: Pieridae) and *Cydia pomonella* (Lepidoptera: Tortricidae) is chemically altered and less stable when parasitized by braconid wasps (El-Sufty and Führer 1981a). Although the induced changes to cuticles facilitated the penetration of *B. bassiana* into the hemolymph (El-Sufty and Führer 1981b, 1985), subsequent fungus development was severely checked in both hosts, and immature parasitoids were afforded a degree of protection from the detrimental effects of fungal invasion. The mechanism underlying this effect is not clear in the *C. pomonella*–braconid system, but in the *P. brassicae*–braconid system, a fungistatic compound produced by teratocytes in the host hemolymph inhibited fungal colonization (El-Sufty and Führer 1981b). Fransen and van Lenteren (1994) noted a similar decrease in the susceptibility of immature greenhouse whiteflies to infection by *A. aleyrodis* when parasitized by *E. formosa*. Whitefly nymphs treated with *A. aleyrodis* up to 3 days after parasitoid oviposition, the period before parasitoid eggs hatched, showed no change in susceptibility to the fungus, but older parasitized nymphs were significantly less susceptible to *A. aleyrodis* infection than were nonparasitized nymphs. Fransen and van Lenteren (1994) speculated that the hatching of the parasitoid larva induced changes in the host cuticle that inhibited *A. aleyrodis* penetration or that fungistatic or fungicidal substances induced or produced by the immature parasitoid were present in the host hemolymph. Clearly, this is an area that requires further research.

King and Bell (1978) reported that parasitism of *Helicoverpa* (=*Heliothis*) *zea* (Lepidoptera: Noctuidae) larvae by *Microplitis croceipes* (Hymenoptera: Braconidae) increased the susceptibility of hosts to *Nomuraea rileyi* (Ascomycota: Hypocreales) when applied to larvae immediately after parasitoid attack. Powell et al. (1986) showed that parasitized individuals of *Metopolophium dirhodum* (Homoptera: Aphididae) were more susceptible than nonparasitized conspecifics to infection by *P. neoaphidis* 2 days after parasitism, but increased susceptibility of parasitized individuals disappeared as the interval between parasitoid oviposition and challenge by the fungal pathogen increased.

Parasitoid Avoidance of Infected Hosts

Parasitoids may evade direct intra-host competition with fungal pathogens by avoiding oviposition in infected hosts. Brobyn et al. (1988) showed that the parasitoid *A. rhopalosiphi* avoided *M. dirhodum* hosts that had been infected with *P. neoaphidis* 3 days previously but did not avoid infected hosts in which mycosis was less advanced. In more comprehensive experiments investigating the interaction among the Russian wheat aphid, *D. noxia*, its parasitoid *A. asychis*, and the fungal pathogen *P. fumosoroseus*, Mesquita and Lacey (2001) demonstrated that parasitoids discriminated between infected and healthy aphids only after probing potential hosts with their ovipositors. The number of eggs laid in probed hosts was not measured, but based on analysis of probing behavior (Bai and Mackauer 1990), it was concluded that parasitoids were less likely to accept aphids in which *P. fumosoroseus* infection was well advanced. In similar experiments, Fransen and van Lenteren (1993) dissected healthy and *A. aleyrodis*-infected whitefly nymphs after exposure to *E. formosa,* and they demonstrated that following internal examination of hosts with its ovipositor, the parasitoid laid fewer eggs in infected individuals (including those at an early stage of infection) than in noninfected individuals. Lord (2001) studied the effect of *B. bassiana* infection in grain beetles (*Oryzaephilus surinamensis*; Coleoptera: Silvanidae) on oviposition by the ectoparasitoid *Cephalonomia tarsalis* (Hymenoptera: Bethylidae) and concluded that the parasitoid's inability to avoid attacking infected hosts (leading to rapid infection and death of immature parasitoids) was due to the exotic nature of the interaction. As *C. tarsalis* is unlikely to encounter hosts naturally infected by *B. bassiana*, it had never experienced selection pressures to avoid oviposition on individuals upon which its progeny would be unable to survive (Lord 2001). The lack of studies on relationships among ectoparasitoids, their hosts, and entomopathogenic fungi with which they have coevolved makes it impossible to determine whether the ectoparasitoids are less able to avoid unsuitable hosts than endoparasitoids, but the intimate ovipositor examination of potential hosts by endoparasitoids may offer an advantage in this regard.

Sublethal Effects

Few studies have analyzed sublethal effects on parasitoids surviving development within a fungus-infected host. In the *T. vaporariorum–E. formosa–A. aleyrodis* and *D. noxia–A. asychis–P. fumosoroseus* associations discussed previously (Fransen and van Lenteren 1994; Mesquita and Lacey 2001), no detrimental effects on the fecundity or longevity of adult parasitoids emerging from fungus infected hosts were reported. However, this is another area of research that requires further exploration.

Even less effort has been directed toward the possible detrimental consequences of fungus and parasitoid co-occurrence on the fungus. *Plutella xylostella* larvae treated with *Z. radicans* 1 day after parasitism by *C. plutellae* did not show a reduction in *Z. radicans* conidium yield when compared to nonparasitized larvae of the same age. However, parasitized larvae treated with *Z. radicans* 3 days after parasitism, while always killed by the fungus, resulted in a decrease in almost 80% fewer conidia than in nonparasitized controls (Furlong and Pell 2000; fig. 3.3). Whether this was due simply to the larger size of the parasitoid larvae (which could not be colonized by the fungus) or whether it was due to a more subtle effect of the parasitoid on the fungus remains unclear.

Extra-host Interactions

Direct Infection of Insect Natural Enemies

Adult parasitoids and arthropod predators that forage in an environment where they encounter the free conidia of entomopathogenic fungi may become infected by the fungus. The host ranges of entomopathogenic fungi vary enormously and can be significantly affected by the conditions under which pathogenicity is tested, often leading to very different estimates of physiological and ecological host ranges for a given pathotype of a single species of fungus (Hajek and Butler 1999). Asexual ascomycetes (hyphomycetes) generally are considered to have broader host ranges than species of Entomophthorales. However, there is a diverse assemblage of genotypes within groups such as the conidial ascomycetes *Beauveria* (chapter 1) and *Metarhizium* (chapter 2), and individual isolates can have very restricted host ranges (Goettel et al. 2000). Similarly, although the relatively cosmopolitan entomophthoralean fungus *Z. radicans* has been isolated from insects belonging to seven orders, individual strains of the pathogen appear to have some host specificity, infecting taxonomically related insect hosts (Papierok et al. 1984; Pell et al. 1993). Studies on other species of entomophthoralean fungi have found little phylogenetic relatedness between susceptible hosts and suggest that the ability of a fungal pathogen to infect an insect host is determined more by the properties of the host surface than by host phylogenetic relatedness to the insect from which the fungus was originally isolated (Steinkraus and Kramer 1987; Hajek et al. 1995).

Many studies designed to test the safety of a candidate fungal biological control agent for a given pest species have been laboratory based and investigated the exposure of known predators and adult parasitoids of the pest to elevated doses of the fungus. The majority of these studies have ended with the determination of physiological susceptibility to infection (Goettel et al. 1990). For example, Ignoffo (1981) exposed three species of predatory insects and four species of parasitoids to extremely high doses of *N. rileyi* in the laboratory but found that the fungus killed none of the natural enemies. In another laboratory study, exposure of two hymenopteran parasitoids of the diamondback moth, *P. xylostella*, to extremely high concentrations of conidia of a strain of *Z. radicans* isolated from the diamondback moth indicated that one species, *D. semiclausum,* was susceptible but that the other, *C. plutellae*, was not (Furlong and Pell 1996).

Magalhaes et al. (1988) demonstrated that although two species of predatory coccinellid were susceptible to infection by topically applied *B. bassiana* in the laboratory, infection rates were considerably reduced if the predators received the inoculum only when they walked over contaminated leaf surfaces (the most likely mode of inoculum acquisition in the field). Such studies provide useful information, as species that are not physiologically susceptible to infection under optimal laboratory conditions for the fungus, where environmental and behavioral constraints to infection have been removed, are unlikely to be ecologically susceptible to infection in the field. However, physiological susceptibility in the laboratory does not necessarily translate into field ecological susceptibility, where the impact of a given fungus on a physiologically susceptible predator or parasitoid may be minimal (Jaronski et al. 1998). Pell and Vandenberg (2002) showed that under optimal conditions for the fungus, stressed individuals of *Hippodamia convergens* (Coleoptera: Coccinellidae) were killed by high doses of *P. fumosoroseus* after 72 h at 100% relative humidity in the laboratory, but they speculated that as these conditions are unlikely to be replicated in the field, this important aphid natural enemy would be unlikely to be affected by field application rates of the fungus. In similar studies, James and Lighthart (1994) showed that *H. convergens* was physiologically susceptible to *B. bassiana, Metarhizium anisopliae* (Ascomycota: Hypocreales), and *P. fumosoroseus* in laboratory tests. However, in field tests only *B. bassiana* caused significant mortality in *H. convergens* populations and then only early in the season when environmental conditions were favorable. Applications later in the season, when temperatures were higher than the optimum for infection, had little impact on the insect (James et al. 1995, 1998). There are no reports of predatory species becoming infected by entomophthoralean fungi in laboratory based susceptibility tests.

Poprawski et al. (1998) found that feeding the predator *Serangium parcesetosum* (Coleoptera: Coccinellidae) whitefly prey that had been contaminated with *B. bassiana* 24, 48, 72, or 96 h before significantly increased predator mortality, and they speculated that, in addition to the possibility of direct infection by the pathogen, *S. parcestosum* may have been adversely affected by toxins such as oosporein produced by the fungus within its whitefly hosts. Such detrimental interactions have previously been suggested as a likely consequence of intra-host fungus–parasitoid interactions (Brooks 1993) but require further study within the context of predatory and scavenging arthropods that may consume infected prey.

Sublethal and Premortality Effects

Few studies have considered the impact of sublethal fungal infections or the premortality effects of lethal fungal infections on the ecological fitness of adult parasitoids, and no studies have investigated such effects in predatory species that may be susceptible to fungal infection. Furlong and Pell (1996) demonstrated that the parasitoid *D. semiclausum* was susceptible to infection by *Z. radicans*, although

>100-fold less susceptible than adult or larval stages of its host *P. xylostella*. In addition, the reproductive capacity of infected females was significantly depressed within 24 h of infection (Furlong and Pell 2000). Lacey et al. (1997) provided evidence that *A. asychis* infected by *P. fumosoroseus* was less mobile than noninfected individuals, but daily per capita cocoon production by infected individuals was not affected before death.

Predator and Parasitoid Feeding on Infected Prey

Studies on the interactions between arthropod natural enemies and entomopathogenic fungi usually have considered the interaction from the perspective of the potential detrimental impact of the fungus on the arthropod natural enemy population (Roy and Pell 2000). However, arthropod predators can ingest entomopathogenic fungi by consuming infected host insects (Rosenheim et al. 1995). For example, in nonchoice tests, the beetle *Pterostichus madidus* (Coleoptera: Carabidae) consumed more *P. neoaphidis*-infected and sporulating aphid cadavers than uninfected cadavers (Roy et al. 1998). However, the consumption of *P. neoaphidis*-infected aphids in the field is unlikely to affect the epidemiology of the disease because the beetle forages on the ground where it can only contact dislodged cadavers already removed from the environment of transmission on the plant. Both adults and larvae of *Coccinella septempunctata* (Coleoptera: Coccinellidae) can also feed on *P. neoaphidis*-infected sporulating cadavers, occasionally consuming them entirely (Roy et al. 1998, 2003). Although partial consumption of cadavers significantly reduced conidium production, it did not impact subsequent transmission rates in laboratory experiments, again an indication of the limitation of a detrimental impact in this interaction (Roy et al. 1998). Other common arthropod predators of aphids in the United Kingdom, the hoverfly *Episyrphus balteatus* (Diptera: Syrphidae) and the lacewing *Chrysoperla carnea* (Neuroptera: Chrysopidae), never fed on *P. neoaphidis*-infected sporulating aphid cadavers (Roy et al. 1998). In these studies predators foraging on *P. neoaphidis*-infected aphids never became infected by the pathogen; however, the potentially more subtle sublethal effects of consuming diseased prey were not examined (Roy et al. 1998, 2003). In this system, the most detrimental impact on the fungus is likely to occur when predators consume infected live aphids, preventing sporulation and any further contribution to disease transmission (Roy et al. 1998).

Encounters with sporulating cadavers on the soil surface and the subsequent acquisition of *B. bassiana* conidia by *Leptinotarsa decemlineata* (Coleoptera: Chrysomelidae) pre-pupae searching for pupation sites in the soil of the potato agroecosystem is essential to the secondary cycling of the disease in the pest population (Long et al. 2000). *Harpalus rufipes* (Coleoptera: Carabidae), which commonly preys on and caches seeds in the potato agroecosystem of northern Maine (Hartke et al. 1998), was suspected of being responsible for the rapid removal of tethered mummified and sporulating *B. bassiana*-infected *L. decemlineata* cadavers from the soil surface and their subterranean storage (S. Fernandez, per. comm.). Such changes in the distribution of sporulating cadavers have the potential to seriously reduce the secondary transmission of the fungus to the pest population.

In addition to using host insects as nutritional sources for the development of their progeny, some parasitoids use host insects as adult food sources (Jervis and Kidd 1986). Adult parasitoids that also prey on host insects can interact with fungal pathogens that attack the host at the intra-host level, as described above, or in essentially the same way as predators. Fransen and van Lenteren (1993) showed that adult *E. formosa* females fed on whitefly hosts infected with *A. aleyrodis* but to a lesser extent than on noninfected hosts. In nonchoice tests, Mesquita and Lacey (2001) showed that *A. asychis* females consumed fewer aphid hosts already killed by *P. fumososroseus* than infected living aphids or healthy aphids. In these studies that appear to be the only investigations to explore the effect of host fungal infection on parasitoid host feeding explicitly, the implications of host feeding were not explored further. Such interactions are likely to be case specific and will require testing if host feeding parasitoids and entomopathogenic fungi are to be used together in pest management strategies.

Increased Transmission and Mechanical Vectoring of Infection

Predators and parasitoids may act as mechanical vectors and transfer infective conidia to susceptible individuals. Their presence also may alter the behavior of their host/prey so that the dynamics of the interaction with the pathogen are significantly affected. In laboratory and field experiments, foraging *C. septempunctata* increased dispersal of *P. neophidis* to susceptible aphid hosts by direct vectoring of conidia with which they had become contaminated during foraging in infected aphid populations (Pell et al. 1997; Roy et al. 1998, 2001). The presence of foraging *C. septempunctata* in aphid populations also significantly increased local transmission of *P. neoaphidis* (Roy et al. 1998). Increased escape movements of aphids brought them into contact with more infective conidia, hence encouraging transmission. Parasitoids foraging in infected aphid populations had the same effect (Fuentes-Contreras et al. 1998).

There is little evidence in the literature indicating that foraging parasitoids act as direct vectors of entomopathogenic fungi by transferring conidia to hosts. However, Fransen and van Lenteren (1993) reported that a small proportion of parasitoids that probed whitefly nymphs infected by *A. aleyrodis* infected healthy nymphs via fungal elements attached to their ovipositors. In laboratory studies, *D. semiclausum* and *C. plutellae* foraging for host larvae in the presence of sporulating cadavers killed by *Z. radicans* did not vector conidia to their hosts (Furlong and Pell 1996). However, in a manner similar to the aphid system, foraging *D. semiclausum* induced behavioral changes and greater movement in host larvae, which caused them to encounter more conidia and increased the number that became infected.

Interactions at the Population Level

Interactions between fungal pathogens and arthropod natural enemies at the population level are difficult and time consuming to study. Therefore, our understanding of population-level effects lags significantly behind studies conducted at the individual level, most of which have been performed under laboratory conditions. In a laboratory experiment designed to examine interactions at the population level, the complementary action of the parasitoid *A. rhopalosiphi* and the fungal pathogen *P. neoaphidis* was demonstrated to reduce aphid populations (Fuentes-Contreras and Niemeyer 2000). In manipulated experiments under field conditions, Mesquita et al. (1997) demonstrated the potential of *A. asychis* and *P. fumosoroseus* used in combination to control the Russian wheat aphid, *D. noxia*, although the authors stressed that any detrimental effect of the fungus on the parasitoid may not have been manifest due to the unfavorable environmental conditions at the time of experimentation. Such studies provide useful information on the short-term compatibility between various natural enemies and are invaluable when a fungal pathogen is used as a biopesticide with the aim of rapid suppression of pest populations complemented by the action of endemic or released parasitoids or predators. However, these studies do not yield information regarding the long-term ecological effects of such introductions, for which prolonged field studies are required. As endemic entomopathogenic fungi and arthropod natural enemies have coevolved with their hosts and as populations of parasitoids and predators survive epizootics of such fungi within their host populations, it is generally assumed that long-term ecological effects of artificially augmenting levels of endemic fungal inoculum are unlikely to be detrimental. In situations where the introduction of a fungal pathogen or a parasitoid or predator into the ecosystem results in new associations, the long-term ecological effects are less predictable and can be assessed robustly only by extended field experiments.

Fungal Pathogen–Parasitoid Interactions in the Field

Although the co-occurrence of endemic parasitoids and pathogens attacking common hosts has been recorded frequently in the literature (e.g., Powell et al. 1986; Feng et al. 1991), comprehensive studies of their long-term interactions under field conditions are often lacking. Ullyett and Schonken (1940) recorded that *Z. radicans* epizootics in *P. xylostella* populations in South Africa killed such a large proportion of the host population that survival of the larval parasitoid *D. semiclausum* was severely reduced, resulting in major outbreaks of *P. xylostella* the next growing season. A similar negative interaction has been reported between *Z. radicans* and the braconid parasitoid *C. plutellae* in *P. xylostella* populations in the Philippines (Velasco 1983). Perhaps the most studied and best understood associations are those among the alfalfa weevil, *Hypera postica* (Coleoptera: Curculionidae), two of its larval parasitoids, *Bathyplectes curculionis* (Hymenoptera: Ichneumonidae) and *B. anurus* (Hymenoptera: Ichneumonidae), and the fungal pathogen *Zoophthora phytonomi* (Zygomycota: Entomphthorales) in eastern North America.

    Bathyplectes anurus and *B. curculionis* were introduced into the eastern United States in the late 1950s as part of a classical biological control program against *H. postica* (Brunson and Coles 1968), an accidental introduction to North America from Europe decades before (Titus 1910; Poos and Bisell 1953). *Z. phytonomi* was discovered attacking *H. postica* in Ontario in 1973 (Harcourt et al. 1977). Los and

Allen (1983) showed that there was a negative correlation between rates of *B. anurus* parasitism and *Z. phytonomi* infection in field populations of *H. postica* in Virginia and speculated that fungal epizootics could be as potentially detrimental to the alfalfa ecosystem as insecticidal sprays. Goh et al. (1989) presented similar findings after studying the *H. postica–B. curculionis–Z. phyonomi* interaction in the years immediately after the first occurrence of the fungus in Oklahoma in 1983 and speculated on the possible long-term negative effects the fungus on the abundance of *B. curculionis* in the local alfalfa agroecosystem. Harcourt et al. (1984) credited a severe epizootic of *Z. phytonomi* in the late 1970s with the local elimination of the introduced parasitoid *Tetrastichus incertus* (Hymenoptera: Eulophidae) from areas of Ontario.

Harcourt (1990) implicated *Z. phytonomi* in the rapid displacement of *B. curculionis* by *B. anurus* in eastern Ontario. *Bathyplectes anurus* has several attributes that confer an intrinsic superiority over *B. curculionis* (e.g., greater reproductive potential, faster host-handling time, aggressive larval stages), but its preference for oviposition in older hosts (second and third instars) compared with *B. curculionis* (which prefers first and second instars) reduces the probability that its host will be killed by the fungal disease (Harcourt 1990). Similar conclusions have been drawn from studies on the same pest–fungus–parasitoid complex in Illinois (Oloumi-Sadaghi et al. 1993) and Iowa (Giles et al. 1994), but studies in Oklahoma (Berberet and Bisges 1998) suggested that *B. anurus* was becoming more abundant than *B. curculionis*, rather than replacing it, as a consequence of the greater susceptibility of *B. curculionis* to mortality during *Z. phytonomi* epizootics. These studies illustrate that the introduction of a fungal pathogen to an ecosystem may cause significant destabilization and have long-term consequences that affect the parasitoid complex associated with its host by reducing the relative competitiveness of a member of the complex (in this case reduced abundance and possible local exclusion of *B. curculionis*). Epizootics of *Z. phytonomi* in *H. postica* populations following periods of high rainfall frequently disrupted the stability of *H. postica*–parasitoid systems (Harcourt 1990), leading to pest outbreaks in dry weather that could only be managed economically by widespread application of insecticide (Giles et al. 1994). Clearly, dependable and predictive pest management strategies based on the *H. postica* parasitoid complex and *Z. phytonomi* can only be established if the interactions among the fungus and parasitoids are clearly understood and underpinned with a detailed appreciation of the effect of climatic factors on the potential for *Z. phytonomi* epizootics.

Fungal Pathogen–Predator Interactions in the Field

No studies have examined the long-term effects of the application of fungal pathogens for pest management on predator populations, although the occurrence of *B. bassiana* epizootics in hibernating populations of coccinellids (Mills 1981) suggests that such studies would be opportune when these natural enemies are proposed for integrated pest management. In similar studies, Baltensweiler and Cerutti (1986) and Parker et al. (1997) applied *Beauveria brongniartii* (Ascomycota: Hypocreales) and *B. bassiana*, respectively, to forest environments and monitored the effects on nontarget predatory arthropods. Both studies recorded a very low incidence of fungal disease in arthropods sampled (1.1–11.6%) and showed that viable inoculum was rapidly lost from the system. It was concluded in each case that the applications were unlikely to seriously impact nontarget predatory arthropod fauna.

Conclusions

There is now a considerable body of theoretical and empirical research considering interactions between insect natural enemies from different kingdoms. A vast number of studies on arthropod–fungus interactions have focused on natural enemies that are candidate biological control agents in agricultural ecosystems. However, some of these studies relate to organisms that have coevolved, and others relate to new associations between natural enemies. These studies provide a basis with which to begin to predict potential risks that are likely to occur when a new natural enemy is introduced to an existing coevolved community or when populations of one natural enemy are enhanced above natural levels. The studies also provide insight into the dynamic and ongoing evolution of communities in particular ecosystems and present valuable information to examine the coevolutionary role of fungal pathogens and arthropod natural enemies in determining herbivore–host-plant relationships (Elliot et al. 2000). The outcome of any intraguild interaction is coexistence of the two species or exclusion of one species. However, it is important to remember that this is within the context of spatial and temporal scale, biotic and abiotic environmental variables, and the behavior of the species involved. For example, in the alfalfa weevil system, although the natural enemies were coevolved with each other and with their host, changing the geographic region, and therefore the environmental conditions, affected the outcome of competitive interactions among the natural enemies. In some of the coevolved systems, it is also clear that competition has led to the selection of mechanisms by which competition can be avoided (e.g., parasitoids that can detect the presence of a superior fungal competitor within a host and thereby avoid oviposition in that host). Over evolutionary time it is also possible that competing natural enemies undergo niche differentiation so that their populations become spatially or temporally segregated.

Although it is possible that the interactions within a given system will need to be considered on a case-by-case basis, interactions should be examined both at the level of the individual and at the population scale. Predictions made at the level of the individual under controlled laboratory conditions can indicate potential interactions at the population scale in the field, but this will not always be the case. Continued research in this area will contribute greatly to our understanding of insect community structure in natural and agricultural systems and help us develop effective biological control strategies with minimal environmental impact.

Acknowledgments   M.J.F. is funded by the Australian Centre for Agricultural Research. J.K.P. was funded by the Department for Environment, Food and Rural Affairs, UK. Rothamsted

Research receives grant-aided support from the Biotechnology and Biological Sciences Research Council, UK.

Literature Cited

Anderson, R. M., and R. M. May. 1986. The invasion, persistence, and spread of infectious diseases within animal and plant communities. *Philosophical Transactions of the Royal Society of London, Series B* 314:533–570.

Askary, H., and J. Brodeur. 1999. Susceptibility of larval stages of the aphid parasitoid *Aphidius nigripes* to the entomopathogenic fungus *Verticillium lecanii*. *Journal of Invertebrate Pathology* 73:129–132.

Bai, B., and M. Mackauer. 1990. Host discrimination by the aphid parasitoid *Aphelinus asychis* (Hymenoptera: Aphelinidae): When superparasitism is not adaptive. *Canadian Entomologist* 122:363–372.

Baltensweiler, W., and F. Cerutti. 1986. Bericht uber die Nebenwirkungen einer Bekampfung des Maikafers (*Melolontha melolontha* L.) mit dem Pilz *Beauveria brongniartii* (Sacc.) Petch auf die Arthropodenfauna des Waldrandes. *Miteilungen der Schweizerischen Entomologischen Gesellschaft* 59:267–274.

Bellows, T. S., and M. P. Hassell. 1999. Theories and mechanisms of natural population regulation. In *Handbook of biological control*, ed. T. S. Bellows and T. W. Fisher, pp. 17–44. San Diego, CA: Academic Press.

Berberet, R. C., and A. D. Bisges. 1998. Potential for competition among natural enemies of larvae of *Hypera postica* (Coleoptera: Curculionidae) in the southern plains. *Environmental Entomology* 27:743–751.

Brobyn, P. J., S. J. Clark, and N. Wilding. 1988. The effect of fungus infection of *Metopolophium dirhodum* (Hom.: Aphididae) on the oviposition behaviour of the aphid parasitoid *Aphidius rhopalosiphi* (Hym.: Aphidiidae). *Entomophaga* 33:333–338.

Brooks, W. M. 1993. Host-parasitoid-pathogen interactions. In *Pathogens*, vol. 2, *Parasites and pathogens of insects*, ed. N. E. Beckage, S. N. Thompson, and B. A. Frederici, pp. 231–272. San Diego, CA: Academic Press.

Brunson, M. H., and L. W. Coles. 1968. The introduction, release, and recovery of parasites of the alfalfa weevil in eastern United States. Production Research Report 101. Washington, DC: U.S. Department of Agriculture.

Costa, S. D., and R. Gaugler. 1989. Sensitivity of *Beauveria bassiana* to solanine and tomatine: Plant defensive chemicals inhibit an insect pathogen. *Journal of Chemical Ecology* 15:697–706.

Elliot, S. L., M. W. Sabelis, A. Jansen, L. P. S. van der Geest, E. A. M Beerling, and J. Fransen. 2000. Can plants use entomopathogens as bodyguards? *Ecology Letters* 3:228–235.

El-Sufty, R., and E. Führer. 1981a. Parasitäre veränderungen der wirtskutikula bei *Pieris brassicae* und *Cydia pomonella* durch entomophage endoparasiten. *Entomologia Experimentalis et Applicata* 30:134–139.

El-Sufty, R., and E. Führer. 1981b. Wechselbeziehungen zwischen *Pieris brassicae* L. (Lep., Pieridae), *Apanteles glomeratus* L. (Hym., Braconidae) und dem pilz *Beauveria bassiana* (Bals.) Vuill. *Zeitschrift fur Angewandte Entomologie* 92:321–329.

El-Sufty, R., and E. Führer. 1985. Wechselbeziehungen zwischen *Cydia pomonella* L. (Lep., Totricidae), *Ascogaster quadridentatus* Wesm. (Hym., Braconidae) und dem pilz *Beauveria bassiana* (Bals.) Vuill. *Zeitschrift fur Angewandte Entomologie* 99:504–511.

Feng, M. G., J. B. Johnson, and S. E. Halbert. 1991. Natural control of cereal aphids

(Homoptera: Aphididae) by entomopathogenic fungi (Zygomycetes: Entomophthorales) and parasitoids (Hymenoptera: Braconidae and Encyrtidae) on irrigated spring wheat in southwestern Idaho. *Environmental Entomology* 20:1669–1710.

Fransen, J. J., and J. C. van Lenteren. 1993. Host selection and survival of the parasitoid *Encarsia formosa* on greenhouse whitefly, *Trialeurodes vaporariorium*, in the presence of hosts infected with the fungus *Aschersonia aleyrodis*. *Entomologia Experimentalis et Applicata* 69:239–249.

Fransen, J. J., and J. C. van Lenteren. 1994. Survival of the parasitoid *Encarsia formosa* after treatment of parasitized greenhouse whitefly larvae with fungal spores of *Aschersonia aleyrodis*. *Entomologia Experimentalis et Applicata* 71:235–243.

Fuentes-Contreras, E., and H. M. Niemeyer. 2000. Effect of wheat resistance, the parasitoid *Aphidius rhopalosiphi*, and the entomopathogenic fungus *Pandora neoaphidis*, on population dynamics of the cereal aphid *Sitobion avenae*. *Entomologia Experimentalis et Applicata* 97:109–114.

Fuentes-Contreras, E., J. K. Pell, and H. M. Niemeyer. 1998. Influence of plant resistance at the third trophic level: interactions between parasitoids and entomopathogenic fungi of cereal aphids. *Oecologia* 117:426–432.

Furlong, M. J. 2004. Infection of the immature stages of *Diadegma semiclausum*, an endolarval parasitoid of the diamondback moth, by *Beauveria bassiana*. *Journal of Invertebrate Pathology* 86:52–55.

Furlong, M. J., and J. K. Pell. 1996. Interactions between the fungal entomopathogen *Zoophthora radicans* Brefeld (Entomophthorales) and two hymenopteran parasitoids attacking the diamondback moth, *Plutella xylostella*. *Journal of Invertebrate Pathology* 68:15–21.

Furlong, M. J., and J. K. Pell. 2000. Conflicts between a fungal entomopathogen, *Zoophthora radicans*, and two larval parasitoids of the diamondback moth. *Journal of Invertebrate Pathology* 76:85–94.

Furlong, M. J., J. K. Pell, P. C. Ong, and A. R. Syed. 1995. Field and laboratory evaluation of a sex pheromone trap for the autodissemination of the fungal entomopathogen *Zoophthora radicans* (Entomophthorales) by the diamondback moth, *Plutella xylostella* (Lepidoptera: Yponomeutidae). *Bulletin of Entomological Research* 85:331–337.

Giles, K. L., J. J. Obrycki, T. A. Degooyer, and C. J. Orr. 1994. Seasonal occurrence and impact of natural enemies of *Hypera postica* (Coleoptera: Curculionidae) larvae in Iowa. *Environmental Entomology* 23:167–176.

Godfray, H. C. J. 1994. *Parasitoids: Behavioural and evolutionary ecology*. Princeton, NJ: Princeton University Press.

Goettel, M. S., G. D. Inglis, and S. P. Wraight. 2000. Fungi. In *Field manual of techniques in invertebrate pathology*, ed. L. A. Lacey and H. K. Kaya, pp. 255–282. Dordrecht: Kluwer Academic.

Goettel, M. S., T. J. Poprawski, J. D. Vandenburg, Z. Li, and D. W. Roberts. 1990. Safety of nontarget invertebrtaes to fungal biocontrol agents. In *Safety of microbial insecticides*, eds. M. Laird, L. A. Lacey and E. W. Davidson, pp. 209–231. Boca Raton, FL: CRC Press.

Goh, K. S., R. C. Berberet, L. J. Young, and K. E. Conway. 1989. Mortality of the parasite *Bathyplectes curculionis* (Hymenoptera: Ichneumonidae) during epizootics of *Erynia phytonomi* (Zygomycetes: Entomophthorales) in the alfalfa weevil. *Environmental Entomology* 18:1131–1135.

Hajek, A. E. 1997. Ecology of terrestrial fungal entomopathogens. *Advances in Microbial Ecology* 15:193–249.

Hajek, A. E., and L. Butler. 1999. Predicting the host range of entomopathogenic fungi. In *Nontarget effects of biological control*, ed. P. A. Follett and J. J. Duan, pp. 263–276. Dortrecht: Kluwer Academic.

Hajek, A. E., L. Butler, and M. M.Wheeler. 1995. Laboratory bioassays testing the host range of the gypsy moth fungal pathogen *Entomophaga maimaiga. Biological Control* 5:530–544.

Harcourt, D. G. 1990. Displacement of *Bathyplectes curculionis* (Thoms.) (Hymenoptera: Ichneumonidae) by *B. anurus* (Thoms.) in eastern Ontario populations of the alfalfa weevil, *Hypera postica* (Gyll.) (Coleoptera: Curculionidae). *Canadian Entomologist* 122:641–645.

Harcourt, D. G., J. C. Guppy, and M. R. Binns. 1977. The analysis of intrageneration change in eastern Ontario populations of the alfalfa weevil, *Hypera postica* (Coleoptera: Curculionidae). *Canadian Entomologist* 10:1521–1534.

Harcourt, D. G., J. C. Guppy, and M. R. Binns. 1984. Analysis of numerical change in subeconomic populations of the alfalfa weevil, *Hypera postica* (Coleoptera: Curculionidae), in eastern Ontario. *Environmental Entomology*. 13:1627–1633.

Hare, D. J., and T. G. Andreadis. 1983. Variation in the susceptibility of *Leptinotarsa decemlineata* (Coleoptera: Chrysomelidae) when reared on different plants to the fungal pathogen, *Beauveria bassiana* in the field and the laboratory. *Environmental Entomology* 12:1892–1897.

Hartke, A., F. A. Drummond, and M. Liebman. 1998. Seed feeding, seed caching and burrowing behaviors of *Harpalus rufipes* De Geer larvae (Coleoptera: Carabidae) in the Maine potato agroecosystem. *Biological Control* 13:91–100.

Hochberg, M. E., M. P. Hassell, and R. M. May. 1990. The dynamics of host-parasitoid-pathogen interactions. *American Naturalist* 135:74–94.

Hochberg, M. E., and J. H. Lawton. 1990. Competition between kingdoms. *Trends in Ecology and Evolution* 5:367–371.

Ignoffo, C. M. 1981. The fungus *Nomuraea rileyi* as a microbial insecticide. In *Microbial control of pests and plant diseases: 1970–1980*, ed. H. D. Burges, pp. 513–538. New York: Academic Press.

Inglis, G. D., M. S. Goettel, T. M. Butt, and H. Strasser. 2001. Use of hyphomycetous fungi for managing insect pests. In *Fungi as biocontrol agents. Progress, problems and potential*, ed. T. M. Butt, C. W. Jackson, and N. Magan, pp. 23–69. Wallingford, UK: CABI Publishing.

James, R. R., and B. Lighthart. 1994. Susceptibility of the convergent lady beetle (Coleoptera: Coccinellidae) to four entomogenous fungi. *Environmental Entomology* 23:190–192.

James, R. R., B. T. Shaffer, B. Croft, and B. Lighthart. 1995. Field evaluation of *Beauveria bassiana*: its persistence and effects on the pea aphid and a non-target coccinellid in alfalfa. *Biocontrol Science and Technology* 5:425–437.

James, R. R., B. A. Croft, B. T. Shaffer, and B. Lighthart. 1998. Impact of temperature and humidity on host-pathogen interactions between *Beauveria bassiana* and a coccinellid. *Environmental Entomology* 27:1506–1513.

Jaronski, S. T., J. Lord, J. Rosinka, C. Bradley, K. Hoelmer, G. Simmons, R. Osterlind, C. Brown, R. Staten, and L. Antilla. 1998. Effect of *Beauveria bassiana*-based mycoinsecticide on beneficial insects under field conditions. In *The 1998 Brighton Conference—pests and diseases*, pp. 651–656. Brighton: British Crop Protection Council.

Jervis, M. A., and N. A. C. Kidd. 1986. Host-feeding strategies in hymenopteran parasitoids. *Biological Reviews* 61:395–434.

King, E. G., and J. V. Bell. 1978. Interactions between a braconid, *Microplitis croceipes*, and a fungus *Nomuraea rileyi*, in laboratory-reared bollworm larvae. *Journal of Invertebrate Pathology* 31:337–340.

Lacey, L. A., A. L. M. Mesquita, G. Mercardier, R. Debire, D. J. Kazmer, and F. Leclant. 1997. Acute and sublethal activity of the entomopathogenic fungus *Paecilomyces fumosoroseus* (Deuteromycotina: Hyphomycetes) on adult *Aphelinus asychis* (Hymenoptera: Aphelinidae). *Environmental Entomology* 26:1452–1460.

Long, D. W., F. A. Drummond, E. Groden, and D. W. Donahue. 2000. Modelling *Beauveria bassiana* horizontal transmission. *Agricultural and Forest Entomology* 2:19–34.

Lord, J. C. 2001. Response of the wasp *Cephalonomia tarsalis* (Hymenoptera: Bethylidae) to *Beauveria bassiana* (Hyphomycetes: Moniliales) as free conidia or infection in its host, the sawtoothed grain beetle, *Oryzaephilus surinamensis* (Coleoptera: Silvanidae). *Biological Control* 21:300–304.

Los, L. M., and W. A. Allen. 1983. Incidence of *Zoophthora phytonomi* (Zygomycetes: Entomophthorales) in *Hypera postica* (Coleoptera: Curculionidae) larvae in Virginia. *Environmental Entomology* 12:1318–1321.

Magalhaes, B. P., J. C. Lord, S. P. Wraight, R. A. Daoust, and D. W. Roberts. 1988. Pathogenicity of *Beauveria bassiana* and *Zoophthora radicans* to the coccinellid predators *Coleomegilla maculata* and *Eriopis connexa*. *Journal of Invertebrate Pathology* 52:471–473.

Mesquita, A. L. M., and L. A. Lacey. 2001. Interactions among the entomopathogenic fungus, *Paecilomyces fumosoroseus* (Deuteromycotina: Hyphomycetes), the parasitoid, *Aphelinus asychis* (Hymenoptera: Aphelinidae) and their aphid host. *Biological Control* 22:51–59.

Mesquita, A. L. M., L. A. Lacey, and F. Leclant. 1997. Individual and combined effects of the fungus, *Paecilomyces fumososroseus* and parasitoid, *Aphelinus asychis* Walker (Hym., Aphelinidae) on combined populations of Russian wheat aphid, *Diuraphis noxia* (Mordvilko) (Hom., Aphididae) under field conditions. *Journal of Applied Entomology* 121:155–163.

Mills, N. J. 1981. The mortality and fat content of *Adalia bipunctata* during hibernation. *Entomologia Experimentalis et Applicata* 30:265–268.

Oloumi-Sadaghi, H., K. L. Steffey, S. J. Roberts, J. V. Maddox, and E. J. Armrust. 1993. Distribution and abundance of two alfalfa weevil (Coleoptera: Curculionidae) larval parasitoids in Illinois. *Environmental Entomology* 22:220–225.

Papierok, B., B. V. L. Torres, and M. Arnault. 1984. Contribution a l'étude de la specificité parasitaire du champignon entomopathogène *Zoophthora radicans* (Zygomycetes, Entomophthorales). *Entomophaga* 29:109–119.

Parker, B. L., M. Skinner, V. Gouli, and M. Brownbridge. 1997. Impact of soil applications of *Beauveria bassiana* and *Mariannaea* sp. on nontarget forest arthropods. *Biological Control* 8:203–206.

Pell, J. K., J. Eilenberg, A. E. Hajek, and D. C. Steinkraus. 2001. Biology, ecology and pest management potential of Entomophthorales. In *Fungi as biocontrol agents. Progress, problems and potential*, ed. T. M. Butt, C. W. Jackson and N. Magan, pp. 71–153. Wallingford, UK: CABI Publishing.

Pell, J. K., R. Pluke, S. J. Clark, M. G. Kenward, and P. G. Alderson. 1997. Interactions between two aphid natural enemies, the entomopathogenic fungus *Erynia neoaphidis* Remaudiere and Hennebert (Zygomycetes: Entomophthorales) and the predatory beetle *Coccinella septempunctata* L. (Coleoptera: Coccinellidae). *Journal of Invertebrate Pathology* 69:261–268.

Pell, J. K., and J. D. Vandenberg. 2002. Interactions among the aphid *Diuraphis noxia*, the entomopathogenic fungus *Paecilomyces fumosoroseus* and the coccinellid *Hippodamia convergens*. *Biocontrol Science and Technology* 12:217–214.

Pell, J. K., N. Wilding, A. L. Player, and S. J. Clark. 1993. Selection of an isolate of *Zoophthora radicans* (Zygomycetes: Entomophthorales) for biocontrol of the diamondback moth *Plutella xylostella* (Lepidoptera: Yponomeutidae). *Journal of Invertebrate Pathology* 61:75–80.

Poos, F. W., and T. L. Bisell. 1953. The alfalfa weevil in Maryland. *Journal of Economic Entomology* 46:178–179.

Poprawski, T. J., J. C. Legaspi, and P. E. Parker. 1998. Influence of entomopathogenic fungi on *Serangium parcesetosum* (Coleoptera: Coccinellidae), an important predator of whiteflies (Homoptera: Aleyrodidae). *Environmental Entomology* 27:785–795.

Powell, W., N. Wilding, P. J. Brobyn, and S. J. Clark. 1986. Interference between parasitoids (Hym.: Aphidiidae) and fungi (Entomophthorales) attacking cereal aphids. *Entomophaga* 31:293–302.

Rosenheim, J. A., H. K. Kaya, L. E. Ehler, J. J. Marois, and B. A. Jaffee. 1995. Intraguild predation among biological-control agents: theory and evidence. *Biological Control* 5:303–335.

Rostás, M., and M. Hilker. 2003. Indirect interactions between a phytopathogenic and an entomopathogenic fungus. *Naturwissenscaften* 90:63–67.

Roy, H. E., and J. K. Pell. 2000. Interactions between entomopathogenic fungi and other natural enemies: Implications for biological control. *Biocontrol Science and Technology* 10:737–752.

Roy, H. E., P. G. Alderson, and J. K. Pell. 2003. Effect of spatial heterogeneity on the role of *Coccinella septempunctata* as an intra-guild predator of the aphid pathogen *Pandora neoaphidis*. *Journal of Invertebrate Pathology* 82:85–95.

Roy, H. E., J. K. Pell, and P. G. Alderson. 2001. Targeted dispersal of the aphid pathogenic fungus *Erynia neoaphidis* by the aphid predator *Coccinella semptumpunctata*. *Biocontrol Science and Technology* 11:99–110.

Roy, H. E., J. K. Pell, S. J. Clark, and P. G. Alderson. 1998. Implications of predator foraging on aphid pathogen dynamics. *Journal of Invertebrate Pathology* 71:236–247.

Shah, P., and M. S. Goettel. 1999. Directory of microbial control products and services. Gainesville, FL: Microbial Control Division, Society for Invertebrate Pathology.

Smith, J. W, E. J. King, and J. V. Bell. 1976. Parasites and pathogens among *Heliothis* species in the central Mississippi delta. *Environmental Entomology* 5:224–226.

Steinkraus, D. C., and J. P. Kramer. 1987. Susceptibility of sixteen species of Diptera to the fungal pathogen *Entomophthora muscae* (Zygomycetes: Entomophthoraceae). *Mycopathologia* 100:55–63.

Thorvilson, H. G., and L. P. Pedigo. 1984. Epidemiology of *Nomuraea rileyi* (Fungi: Deuteromycotina) in *Plathypena scabra* (Lepidoptera: Noctuidae) populations from Iowa soybeans. *Environmental Entomology* 13:1491–1497.

Titus, E. G. 1910. The alfalfa leaf-weevil. *Utah Agricultural Experiment Station Bulletin* 110:17–72.

Ullyett, G. C., and D. B. Schonken. 1940. A fungus disease of *Plutella maculipennis*, Curt., in South Africa, with notes on the use of entomogenous fungi in insect control. *Scientific Bulletin of the Department of Agriculture and Forestry Union of South Africa* 218:539–553.

Velasco, L. R. I. 1983. Field parasitism of *Apanteles plutellae* Kurdj. (Braconidae, Hymenoptera) on the diamond-back moth of cabbage. *Philippines Entomologist* 6:539–553.

Vinson, S. B. 1990. Potential impact of microbial insecticides on beneficial arthropods in the terrestrial environment. In *Safety of microbial insecticides*, ed. M. Laird, L. A. Lacey and E. W. Davidson, pp. 43–64. Boca Raton, FL: CRC Press.

Waage, J. K., and M. P. Hassell. 1982. Parasitoids as biological control agents—a fundamental approach. *Parasitology* 84:241–268.

4

Ecology and Evolution of Fungal Endophytes and Their Roles against Insects

A. Elizabeth Arnold

Leslie C. Lewis

S ymbiotic associations between fungi and photosynthetic organisms are both ancient and ubiquitous (Alexopoulos et al. 1996; Berbee 2001; Heckman et al. 2001). Comprising interactions spanning mutualism to antagonism, fungi associated with living plants shape both the diversity and species composition of a wide array of terrestrial communities (Clay and Holah 1999; Clay 2001; Wilson et al. 2001; Castelli and Casper 2003; Gilbert 2002; Gehring 2003; Packer and Clay 2003). Yet, ecological interactions have been catalogued for only an extreme minority (<5%) of the 1.5 million species of fungi thought to exist (Hawksworth 1991, 2001) with most research focusing on above- and below-ground plant pathogens (Agrios 1997) and on rhizosphere symbionts such as mycorrhizal fungi (Rygiewicz and Andersen 1994; Husband et al. 2002). In contrast, the diversity, species composition, ecological relevance, and evolutionary importance of diverse and abundant fungi occurring in the phyllosphere have not been established for most plant–fungus associations. Such is the case for fungal endophytes—those fungi that colonize and form unapparent infections in healthy plant tissues (Petrini 1991)—which are known from photosynthetic tissues in all major lineages of plants studied thus far, but which have been studied extensively in only a few focal plant clades.

Fungal endophytes associated with foliage comprise a diverse group, primarily of ascomycetous fungi (Fröhlich and Hyde 1999; Arnold 2002), which are considered ubiquitous, having been recovered from nonvascular plants, ferns, conifers, and both monocotyledonous and dicotyledonous angiosperms (Petrini et al. 1982; Clay 1988; Legault et al. 1989; Schulz et al. 1993; Rodrigues 1994; Carroll 1995; Fisher 1996; Faeth and Hammon 1997; Arnold et al. 2001; Cannon and Simmons 2002). Endophyte records date at least from de Bary's (1863) description of the endophytic habit of *Epichlöe typhina* and Sampson's (1933) characterization of endophytes associated with *Festuca* (see Clay 1986), but research regarding the diversity, ecology, and evolution of endophytes has expanded rapidly in the last two decades. Recent studies indicate that endophytes are a cryptic component of every terrestrial plant community (and some marine plant and algal communities; Cubit 1974; Stanley 1992), thereby gaining the interest of ecologists, evolutionary biologists, plant biologists, mycologists, bioprospectors, and agriculturalists (Lewis and Cossentine 1986; Fox 1993; Arnold et al. 2000; Schulz et al. 2002). As records of host associations, transmission patterns, spatial structure, and host affinity of endophytes have increased (see Stone et al. 2000), similarities and differences in endophytic symbioses with regard to different clades of plants and fungi have begun to emerge, allowing researchers to use some endophyte–host associations as model systems for framing questions of ecological and evolutionary significance (Clay and Schardl 2002).

Because endophytes are obligate heterotrophs that subsist on nutrition gained from host plants (apoplastic carbon; Clay 2001), the potential for endophytic fungi to serve as mild parasites or latent pathogens of hosts has been recognized for some time (Petrini 1986). More recently, compelling research assessing interactions between grass endophytes and their hosts has led to a general expectation that endophytes compensate plants for the cost of infection by serving as defensive mutualists that confer protection against biotic and abiotic stressors (Clay 1990; Hoveland 1993; Malinowski and Belesky 1999; Cheplick et al. 2000). The generality of mutualism among endophyte–host symbioses has been actively debated through recent work in natural systems both with feral grasses (Faeth 2002) and woody plants (Carroll 1986, 1995), with the only general rule appearing to be a continuum of diverse interactions between endophytes and hosts (Saikkonen et al. 1998) that must be viewed through the varied lenses of plant physiology, plant reproductive biology, evolutionary theory, and community ecology.

The expected prevalence of undiscovered biodiversity among endophytes (Dreyfuss and Chapela 1994; Fröhlich and Hyde 1999) suggests that general knowledge regarding both direct and indirect interactions of endophytes and their host plants is truly in its infancy. Equally challenging are the uncertainties associated with endophyte sampling methods, which limit assessments of the diversity and species composition of fungi associated with focal hosts (Petrini 1986; Carroll 1995; Guo et al. 2001; Arnold 2002; Gamboa et al. 2003). Similarly, the possibility that some costs and benefits of endophytism may be subtle, and evident only via indirect measures in the context of multispecies interactions, has not been addressed fully. In part because of the model afforded by grass endophytes, however, research regarding roles of endophytic fungi in mediating plant responses to agents of selection has become an active area of investigation (Saikkonen et al. 1998; Pinto et al. 2000; Arnold 2002; Clay and Schardl 2002).

Biotic stressors, including herbivores, are thought to have played an important role in plant evolution, as demonstrated by the diversity and variation in chemical and structural plant defenses (Coley and Barone 1996). Herbivorous insects, which represent both a strong selective force in natural communities (Marquis and

Alexander 1992) and an economically important component of agricultural systems (Rao et al. 2000), have the potential to interact closely with foliar endophytes through their association with host plants. For some endophyte–host associations, including both temperate grasses and conifers, effects of endophytes on plant–insect interactions have been surveyed in relative detail. However, these cases represent only a subset of endophytic associations and differ in fundamental ways from other endophyte symbioses. In this chapter, we seek to provide a general vision of the ecology and evolution of endophytic fungi with regard to insects, focusing on herbivorous insects attacking above-ground tissues of asymptomatic plants. Following an overview of endophytic symbioses, we focus on the horizontally transmitted endophytes of angiosperms for understanding endophyte–plant–insect interactions in an ecological and evolutionary context.

An Introduction to Endophytes

When first introduced, "endophyte" was used broadly to refer to any organism found within tissues of living autotrophs (de Bary 1866, cited in Carroll 1986). The term has undergone various redefinitions, leading to some confusion regarding the breadth of its application and its implicit statements about host interactions. For example, some definitions have included only the clavicipitaceous endophytes of grasses, or only those fungi within plant tissues that confer benefits upon hosts (Stone et al. 2000). In 1991 Petrini provided a working definition for endophytes that has since been widely accepted and that we apply for the purposes of this chapter: endophytes comprise "organisms inhabiting plant organs that at some time in their life . . . colonize internal plant tissues without causing apparent harm to their host" (p. 179). So defined, endophytes represent diverse microorganisms, including bacteria, cyanobacteria, and fungi, which inhabit the rhizosphere, phyllosphere, vascular tissues, and interior parts of reproductive structures of living plants (Chanway 1998; Arnold 2002). Furthermore, endophytes may exhibit more than one type of life history at distinct life stages, include pathogenic organisms that exhibit latency between infection and manifestation of symptoms, and encompass microorganisms that exist simultaneously, or over time, both upon and within plant tissues (Petrini 1986). Microbial endophytes thus comprise a diverse, polyphyletic group with the potential for myriad interactions with host plants.

The majority of endophyte research has focused on endophytic fungi, which appear to be ubiquitous among terrestrial plants: every plant species examined to date harbors at least one endophyte species (Petrini 1991; Arnold 2001), and many plant species may harbor tens to hundreds of species of endophytes within apparently healthy tissues (Stone et al. 2000). Within asymptomatic hosts, fungal endophytes may inhabit all available tissues, including leaves, petioles, stems, twigs, branches, bark, xylem, roots, fruit, flowers, and seeds (Boddy et al. 1987; Chapela and Boddy 1988; Sieber 1989; Clay 1990; Bills and Polishook 1991; Bohn 1993; Lodge et al. 1996; Lupo et al. 2001; Meyer et al. 2001). Accordingly, endophyte biologists with an eye to species interactions might agree that plants represent complex habitats filled with diverse, if unapparent, inhabitants (Arnold 2002).

Fungal Endophytes Associated with Foliage

For the purposes of this chapter, we focus on endophytic fungi inhabiting aerial plant tissues with particular attention to fungi living within asymptomatic foliage (hereafter, "endophytes"). Two groups of endophytes associated with leaves have received particular attention in terms of species interactions: the vertically transmitted endophytes associated with temperate grasses (Clay 1991) and the horizontally transmitted endophytes occurring in leaves of temperate conifers (Legault et al. 1989; Carroll 1995). Current knowledge regarding the taxonomic distributions, infection processes, ecological roles, and evolutionary significance of these fungi has been reviewed elsewhere (Schardl 1996; Saikkonen et al. 1998; Stone et al. 2000). Here, we briefly address these endophytic symbioses as a means to contrast them with the horizontally transmitted endophytes of woody and herbaceous angiosperms, on which we focus the majority of our discussion.

The vertically transmitted endophytes associated with temperate grasses (Clavicipitaceae, Ascomycota) represent the most thoroughly studied group of endophytic fungi. Known to infect at least 80 genera and 300 species of grasses (Clay and Schardl 2002), these systemic endophytes are well known for their ability to produce secondary compounds *in planta* with direct and antagonistic consequences for foraging herbivores (Clay 1986). Many studies have documented negative effects of these mycotoxins, particularly of alkaloids, for an array of herbivorous insects (Clay et al. 1985; Siegel et al. 1990; Schardl 1996; Wilkinson et al. 2000; Brem and Leuchtmann 2001) and have recorded concomitant decreases in herbivore damage to infected versus uninfected plants. Most data in such studies come from two agronomically important grass species (tall fescue, *Festuca arundinaceae*, and perennial ryegrass, *Lolium perenne*) and herbivores with broad diet ranges (e.g., the fall armyworm, *Spodoptera frugiperda*; Lepidoptera: Noctuidae). However, the generality of mutualistic interactions between nonagricultural grasses and specialist herbivores is not certain. Faeth and Sullivan (2003) found that infection of Arizona fescue, a native grass, by the endophyte *Neotyphodium* led to decreased growth and reproduction of the host. However, Brem and Leuchtmann (2001) found evidence for a beneficial effect of *Epichloë* infection in the woodland grass *Brachypodium sylvaticum*, suggesting a benefit for at least some nonagricultural species. Further, the apparent benefits of endophytism in well-studied agricultural systems may mask more complex interactions. For example, Kunkel and Grewal (2003) used field and laboratory trials to show that consumption of *L. perenne* by the black cutworm (*Agrotis ipsilon*; Lepidoptera: Noctuidae) resulted in lower levels of infection of cutworms by a lethal entomopathogenic nematode (*Steinernema carpocapsae*) when the endophyte *Neotyphodium lolii* was present in grass tissues. However, a similar study assessing interactions among tall fescue and its endophyte, a scarabeid herbivore, and a heterorhabdiid nematode found no effect in field surveys (Koppenhofer and Fuzy 2003).

These and similar cases lend support to the suggestion by Saikkonen et al. (1998) that there is a continuum of interactions between endophytes and hosts, particularly with regard to interactions with insect herbivores. However, the weight of evidence argues that clavicipitaceous, vertically transmitted endophytes of grasses and some sedges frequently do provide benefits to their hosts, ranging from physiological enhancements to defense against herbivores and pathogens (Clay and Schardl 2002). These symbioses, characterized by relatively species-poor assemblages of endophytes in individual hosts, host specificity, maternal transmission, systemic growth, and host benefits, are consistent with tenets of evolutionary theory (Bull 1994) and thus represent a useful model system for understanding evolution of species interactions. Following Carroll (1986), the clavicipitaceous endophytes of grasses are generally thought to comprise constitutive mutualists that confer direct benefits on host plants, with an important component of that benefit occurring in terms of plant–insect interactions.

Endophytes also are common in leaves of conifers, and in certain hosts, these fungi have been studied extensively (*Pseudotsuga menziesii, Taxus* spp*., Picea* spp., *Abies* spp., *Juniperus* spp., *Pinus* spp., *Larix* spp., and others; Bernstein and Carroll 1977; Carroll and Carroll 1978; Petrini and Muller 1979; Petrini and Carroll 1981; Sherwood-Pike et al. 1986; Stone 1987; Legault et al. 1989; Petrini 1991; Johnson and Whitney 1992; Rollinger and Langenheim 1993; Carroll 1995; Dobranic et al. 1995; Stanosz et al. 1997; Kriel et al. 2000; Deckert et al. 2001; Muller et al. 2001). Unlike the vertically transmitted, systemic endophytes of grasses, these symbionts generally are transmitted horizontally among hosts, whereby they infect tissues via contagious spread of propagules. Conifer endophytes such as *Meria parkeri* (teleomorph: *Rhabdocline parkeri*) grow within single epidermal cells, whereas others (e.g., *Phyllosticta* sp.) persist intercellularly and at low density in asymptomatic needles (Stone 1987, 1988). Endophytes of conifers are characterized by numerous localized infections that increase in number and density as leaves age; however, the localized nature of infection restricts the total biomass resulting from any single infection (Stone 1985, 1988; Carroll 1986). Reproduction occurs via spores released from senescent and abscised leaves, such that reproduction of host plants and endophytes is strongly decoupled.

Although there is some evidence for toxicity of endophyte-derived compounds for herbivorous insects (e.g., spruce budworm, *Choristoneura fumiferana*, Lepidoptera: Tortricidae; Findlay et al. 2003), it is generally thought that these infections serve important roles not only as a direct deterrent of insects, but as a source of inoculum for use when tissues are damaged by folivores (Carroll 1986). Evidence for this hypothesis can be found in Carroll's (1986) demonstration that infection of galls on Douglas fir by the endophyte *Rhabdocline parkeri* led to increased larval mortality among gall midges (*Contarinia* sp., Diptera: Cecidomyiidae). Carroll (1986) further noted that dense infections in older trees might provide inoculum for younger trees in nearby sites. However, in some cases, the presence of endophyte-derived secondary compounds has been shown to directly deter herbivores from feeding on already infected tissues and to influence growth rates of insect pests (Miller et al. 2002), such that the distinction between endophytes' roles as inoculum sources versus direct deterrents may be blurred. Based on available studies, it appears that the low-biomass, horizontally transmitted endophytes of conifers generally fit under Carroll's (1986) definition of inducible mutualists, which are characterized by benefits conferred though a relatively loose association with host plants, and that at least a subset of endophytes found consistently in conifer needles have positive effects for their host plants.

A third major group of endophytes has received significantly less attention in terms of species interactions, yet their ubiquity underscores the need to assess their roles with respect to both their host plants and the natural enemies that attack them. These are the horizontally transmitted endophytes associated with nonconiferous plants. For the remainder of our discussion, we focus on these fungi, with special attention to horizontally transmitted endophytes associated with foliage of angiosperms.

Horizontally Transmitted Endophytes of Angiosperms

Horizontally transmitted endophytes have been found within asymptomatic tissues of diverse angiosperms, including grasses, bamboo, palms, and other monocots as well as broad-leaved species (Petrini et al. 1982; Stone et al. 2000; Arnold et al. 2001; Cannon and Simmons 2002). These fungi are similar in habit to the conifer-needle endophytes described above, comprising both localized and small-scale infections that accumulate in density as tissues age (Arnold 2002). Similarly, endophytes of woody and herbaceous angiosperms accumulate via horizontal transmission (Faeth and Hammon 1997; Bayman et al. 1998; Arnold and Herre 2003), with propagules from senescent or dead tissues colonizing living leaves. Infection by particular endophyte species may be spatially heterogeneous over the range of the host taxon, as demonstrated by surveys of focal host species at geographic scales (e.g., *Theobroma cacao*, Malvaceae; Arnold et al. 2003), or at sites where the host is native versus introduced (Fisher et al. 1993; Bayman et al. 1998). Like conifer endophytes, many horizontally transmitted endophytes of angiosperms appear to be closely related to known plant pathogens or to represent known pathogens of other host taxa. For example, species known to be pathogenic to some hosts, including *Colletotrichum gloeosporioides* and *Botryosphaeria dothidea*, are routinely isolated as endophytes from asymptomatic tropical foliage (Arnold 2002). Unlike conifer endophytes, however, the horizontally transmitted endophytes of angiosperms generally do not appear to include so high a proportion of species in the Leotiomycetes (e.g., *Rhabdocline* sp. in Douglas fir, Stone 1985; or *Lophodermium* sp. in *Pinus* sp., Eells et al., unpublished data), instead comprising a diversity of Sordariomycetes, Dothideomycetes, some Leotiomycetes, and other groups of filamentous Ascomycota (including taxa traditionally delimited as conidial ascomycetes, as well as Chaetothyriomycetes and Eurotiomycetes; Arnold 2002). Basidiomycota occur as endophytes of angiosperms but are rare compared to Ascomycota (Petrini 1986; Fröhlich and Hyde 1999). In general, horizontally transmitted endophytes associated with angiosperms represent a diverse guild: individual species of plants may harbor tens to hundreds of endophyte species in their asymptomatic aerial tissues (Lodge et al. 1996; Fröhlich and Hyde 1999; Arnold et al. 2000; Stone et al. 2000).

A review of the recent literature demonstrates that horizontally transmitted endophytes are ubiquitous among angiosperms, having been recorded from foliage of Acanthaceae, Chenopodiaceae, and Aizoaceae (Kumaresan and Suryanarayanan 2001), Aceraceae (Sieber and Dorworth 1994), Anacardiaceae, Flacourtiaceae,

Lecythidaceae, Melastomataceae, and Sterculiaceae (Arnold et al. 2001), Araceae, Bromeliaceae, and Orchidaceae (Petrini and Dreyfuss 1981), Araliaceae (Laessøe and Lodge 1994), Ericaceae (Petrini et al. 1982; Petrini 1985), Meliaceae (Gamboa and Bayman 2001; Arnold et al. 2001), Musaceae (Pinto et al. 2000), Myrtaceae, Annonaceae, Sapindaceae, Burseraceae, Simaroubaceae, Chrysobalanaceae, Connaraceae, Violaceae, Clusiaceae, Moraceae, Melastomataceae, and Lecythidaceae (Arnold 2002; Fisher et al. 1993), Ochnaceae and Olacaceae (Arnold et al. 2000), Piperaceae and Crassulaceae (Dreyfuss and Petrini 1984), Poaceae (Meijer and Leuchtmann 1999), Rubiaceae (Arnold 2002; Posada et al. 2003), Rutaceae (Meyer et al. 2001), and Sapotaceae (Lodge et al. 1996; Bayman et al. 1998), among many others (see also Stone et al. 2000). Although surveys traditionally have been biased toward northern temperate regions (e.g., northern Europe: Petrini 1984; northern United States: Carroll and Carroll 1978), recent studies have included taxa in Panamá (Arnold et al. 2000), Puerto Rico (Lodge et al. 1996; Gamboa and Bayman 2001), French Guyana (Cannon and Simmons 2002), Brunei and Australia (Fröhlich and Hyde 1999), India (Suryanarayanan et al. 2002), Hong Kong (Guo et al. 2001), and other tropical sites (Posada et al. 2003).

In general, horizontally transmitted endophytes associated with foliage are more abundant in tropical than in temperate regions (Arnold 2002), but even in the temperate zone they may densely infect foliage. Petrini et al. (1982) encountered endophytes in 78% of leaves sampled among numerous evergreen shrubs in western Oregon, while Faeth and Hammon (1997) reported up to 100% infection in *Quercus emoryi* leaves in Arizona, USA. In lowland tropical forests, infection of 100% of mature leaves has been shown for more than 30 species of trees and shrubs representing 24 families and 14 orders of angiosperms (Arnold 2002), consistent with surveys of other species in other tropical sites. In contrast, proportions of leaf area colonized by endophytes appear to vary widely among sites and host species. Rodrigues (1994) documented that an average of 30% of leaf pieces sampled per leaf were colonized in fronds of Amazonian palms. Lodge et al. (1996) found a greater infection rate in Puerto Rican *Manilkara bidentata* (90–95%), as did Gamboa and Bayman (2001) for leaves of *Guarea guidonia* in Puerto Rico (>95%). Arnold et al. (2001) reported a similarly high mean colonization rate (98.7%) for three species studied at Barro Colorado Island, Panamá, and recent surveys suggest that such infection rates are consistent for members of 14 orders of angiosperms in that area (Arnold 2002). In general, these values appear to exceed those typical of temperate-zone species; for example, Arnold and Lutzoni (unpublished data) found that 2-year-old leaves of *Magnolia grandiflora* (Magnoliaceae) in North Carolina, USA, typically bear endophytes in about 20% of sampled leaf area. Concomitant measures of richness suggest that endophytes are more diverse in tropical versus temperate host species, but even in temperate sites, diversity may be quite high. In *M. grandiflora*, for example, >30 species of endophytes were encountered from only 9 leaves sampled in midwinter (Arnold and Lutzoni, unpublished data). In contrast, Arnold et al. (2002) encountered >400 morphotypes of endophytes in association with two host species in a lowland forest in Panama, with individual leaves often containing up to 20 species of endophytic fungi (see also Lodge et al. 1996; Fröhlich and Hyde 1999).

Horizontally Transmitted Endophytes of Angiosperms: Known Interactions with Insects

In accordance with evolutionary theory, horizontal transmission and high diversity within hosts are generally consistent with antagonistic interactions between symbionts and hosts (Bull 1994; Herre 1995; Leigh 1999). Further, Boyle et al. (2001) argued that endophytes that form numerous, localized infections are less likely to confer mutualistic benefits upon their host plants than are endophytes with systemic growth. Several authors (Fröhlich and Hyde 1999; Arnold et al. 2003) have documented high turnover in endophyte species composition among conspecific hosts at a landscape scale, indicating inconsistent infection of given hosts by particular endophyte taxa, and thereby violating one of Carroll's (1986) criteria for mutualistic endophyte–host interactions. These observations, coupled with a general lack of evidence for direct antiherbivore defense by horizontally transmitted endophytes of angiosperms (Faeth and Hammon 1997), have led some authors to suggest that these fungi are unlikely to play protective or mutualistic roles with the host plants they inhabit (see Faeth 2002). At present, functional roles of these ubiquitous and obligately heterotrophic symbionts remain generally unknown and are the subject of active research (Pinto et al. 2000; Arnold et al. 2003), which has recently been augmented by phylogenetic methods and a concomitantly evolutionary perspective.

Evolution of Endophytism with Regard to Insect Antagonism

The ubiquity of horizontally transmitted endophytic fungi among plants and within plant tissues and the observation that fungi have been associated with plants since the first colonization of land (Heckman et al. 2001) suggest that plants and endophytes share a long and intimate history. However, general patterns regarding endophyte–host interactions in ecological timeframes have not been established, such that general conclusions regarding the evolution of endophytism and the evolutionary import of endophytes for plant–insect interactions are difficult to draw. Such a lack of clarity extends beyond insect-related interactions and represents an important suite of general questions in endophyte biology. For example, whether the asymptomatic nature of endophyte infection represents plant-mediated control of the interaction (Redman et al. 2001), competition among endophytes within plant tissues (Herre et al. unpublished data), or a general pattern of evolution from pathogenicity to a less virulent state (see Kuldau et al. 1997) is not yet known. Similarly, whether horizontally transmitted endophytes of angiosperms are descended from plant pathogenic ancestors (as is thought for clavicipitaceous grass endophytes; Clay 1991, but see Kuldau et al. 1997) is not yet clear. Addressing such questions requires that experimental data and concomitant phylogenetic approaches be used to infer general patterns of endophyte–host interactions and that these data be used as a basis for establishing ancestral character states. At present, however, resolution of endophyte–host interactions in ecological time is further clouded because distinctions between endophytes and pathogens are difficult to establish: many foliar endophytes may act pathogenically under some conditions of host stress (e.g., *Sphaeropsis sapinea* in *Pinus* spp.; Stanosz et al. 1997, 2001). Further, Freeman and Rodriguez (1993) demonstrated that for one species of *Colletotrichum*, the transition from pathogenicity to endophytism may require as little as a single mutation at one locus. These data, coupled with the observation that many endophytes are closely related to known pathogens (Carroll 1995), suggest that the evolutionary transition from symptomatic pathogenicity to the endophytic habit may have occurred relatively easily, and many times, across the fungal kingdom and may represent a labile interaction dependent on abiotic factors and the species or genotype of endophytes and hosts in question.

In the lack of explicit data regarding evolutionary history and implications of endophytes, ecological observations may provide some general patterns for understanding the nature of extant endophyte symbioses. For example, it has been proposed that horizontally transmitted endophytes generally represent neutral or mildly parasitic inhabitants of their host plants (see Saikkonen et al. 1998 for a general review). Neutrality may reflect either a lack of effects on hosts or a relative balance of positive and antagonistic effects. However, it seems unlikely that neutrality per se describes most endophyte–host interactions. The observation that endophytes fruit from senescent tissues suggests that neutrality would be accompanied by a delay in reproduction for endophytic fungi, which in turn would prove costly to endophytes and could therefore lead to selection against the endophyte habit. This scenario is most likely under conditions of strong host control of the endophyte–plant interaction, but such effective plant-mediated control seems unlikely given the diversity of endophytes that may be encountered in host tissues (e.g., 19 species in individual leaves of tropical trees; see Lodge et al. 1996).

In turn, the hypothesis that endophytes are mild parasites of their hosts is well supported by the observation that fungal endophytes are obligate heterotrophs and must therefore subsist on carbon and other nutrients from host tissues (Clay 2001). As for neutrality, however, parasitism may take several forms; indirect effects between endophytes and hosts could, at the endophytes' benefit, prove costly to the plants they inhabit. For example, evidence from plants in lowland Panama suggests that cuticular wounding via folivory by hesperiid larvae increases the local density of infection by endophytic fungi by nearly twofold in leaves in a focal tropical tree (*Gustavia superba*, Lecythidaceae; Arnold, unpublished data). Similarly, Faeth and Hammon (1997) found that mining activity by *Cameraria*, a microlepidopteran leaf-miner of oak, was associated with increased endophytic fungal infections, reflecting either an increase in successful leaf colonization by fungi or a change in host physiology. These examples are not exclusive of the inducible mutualism posited for conifer endophytes (Faeth and Hammon 1997; Carroll 1986), as infections following herbivory could serve a defensive role against subsequent damage. However, herbivore damage also increases the probability of successful symptom development by leaf pathogens (Garcia-Guzman and Dirzo 2001). Further, Monk and Samuels (1990) noted that endophytic fungi can pass successfully through the gut of orthopteran herbivores in a palaeotropical rainforest, suggesting that herbivores may play an important role in dispersal of endophytic fungi in natural systems. As

Saikkonen et al. (1998) note, a positive association between folivory and the spread and infectivity of endophytes could lead to tolerance or even promotion of herbivory by leaf-inhabiting endophytes. Evidence for this hypothesis has been reported from some plant pathogen–insect–host interactions (Johnson et al. 2003), wherein infection by a pathogen is positively associated with subsequent herbivore damage, which in some cases may lead to further spread of the disease. The general potential for such indirect but antagonistic effects in the vast majority of plant–endophyte associations has not been assessed.

A third and nonexclusive hypothesis is that horizontally transmitted endophytes may compensate for the cost of heterotrophism by benefiting their hosts under some conditions. Such positive associations could lead to the evolution of endophytism in diverse fungal lineages (see Carroll 1986). To date, however, documented benefits of horizontally transmitted endophytes have been relatively few among diverse angiosperm hosts. Although Arnold et al. (in review) showed that horizontally transmitted endophytes of an economically important tropical tree (*Theobroma cacao*) conferred local resistance against a foliar pathogen, data indicating a strong role of horizontally transmitted endophytes in protecting foliage from herbivores such as insects are generally lacking. It is possible, however, that unlike the directly antagonistic effects of grass endophytes, effects of horizontally transmitted endophytes on insects may be markedly subtler. Given that plants have developed a tremendous array of diverse chemical defenses against insects (Coley and Barone 1996), many of which may be metabolically costly, it is safe to conclude that insects play a major selective role with regard to plant evolution (Marquis and Alexander 1992; Sagers and Coley 1995; Siemens et al. 2002). Could herbivore pressure in angiosperms be mitigated via endophytism as a cryptic mode of defense?

This hypothesis has been presented in various forms by numerous authors (e.g., Carroll 1991), yielding three generalizations regarding modes by which endophytes may augment host defense against herbivorous insects: direct chemical defense, resulting from the production of secondary compounds that are toxic to or that deter insects (Saikkonen et al. 1998); defense via a mosaic effect, whereby diverse endophytes provide a biochemical mosaic among otherwise similar leaves of individual hosts, thereby allowing parts of a genetically uniform plant to differ unpredictably in terms of palatability or quality for herbivores (Carroll 1991; Saikkonen et al. 1998); and the potential for plants to harbor entomopathogens as endophytic fungi (Lewis and Bing 1991; Elliott et al. 2000). As noted above, direct antagonism of herbivores is well known among the vertically transmitted endophytes of grasses and has been observed among endophytes of conifers, but it is poorly known among horizontally transmitted endophytes of angiosperm foliage. In turn, mosaic-type defenses are compelling but are difficult to assess experimentally. Although several studies have indicated heterogeneity in endophyte assemblages among leaves of individual trees (Arnold 2002), thus corroborating the potential for biochemical diversity among tissues of individual plants, experimental assessment of the costs and benefits of such mosaics have not been conducted and may prove difficult to perform. For the purposes of this chapter, we focus on the potential for endophytic entomopathogens to play an important role in host defense against insects.

Entomopathogens as Endophytes

Much like known endophytes, the majority of insect pathogens outside of the Entomophthorales (Zygomycota; Humber 1984) also occur within the Pezizomycotina or are mitosporic fungi with phylogenetic affinities in the filamentous Ascomycota (e.g., *Metarhizium* spp. and *Beauveria* spp., Clavicipitaceae; Cooke 1977). The degree to which entomopathogenic and endophytic guilds of fungi overlap and/or share life-history traits is not generally known. Penetration of host cuticles, be they of insects or plants, requires physical and enzymatic activity (Cooke 1977), and it is likely that enzymatic agents are specialized to major host groups. For example, fungi infecting plants cope with cellulose, lignin, and other structural components of plant tissues. Accordingly, enzymes with cellulolytic activity are known in many fungi that penetrate plant organs (Hart et al. 2002; Khalil 2002). In contrast, the integuments of insects contain a layer composed of protein and chitin. As expected, diverse lipases, proteases, and chitinases are known among insect pathogens (Askary et al. 1999; Gimenez-Pecci et al. 2002). For example, *Cordyceps* spp. produce chitinolytic enzymes (see Cooke 1977 for an overview), which are important for successful colonization of insect hosts.

Together, these observations would suggest that entomopathogenic fungi and endophytes likely represent mutually exclusive guilds or, at best, groups with little taxonomic overlap. Further, infection of insects by entomopathogens generally occurs via cuticular penetration by germinating propagules, rather than via consumption of infected plant tissues. For example, propagules of *Cordyceps* have not been recovered from the gut of infected hosts, and thus infection is generally thought to occur via passage of germ tubes through the cuticle (Cooke 1977). Similarly, ingested spores of *Aspergillus* fail to germinate in caterpillars, instead initiating growth after contact with insect cuticles (Rawlins 1984). Thus, the hyphal state of endophytes *in planta* would not appear to provide an important source of entomopathogenic infections. However, evidence suggests that several major pathogens of insects, including *Beauveria bassiana* (Lewis and Bing 1991; Posada et al. 2003), *Aspergillus* (Southcott and Johnson 1997; Cao et al. 2002), and *Paecilomyces* sp. (Arnold 2002), may occur as endophytes within living tissues of plants. By inference, plants harboring these fungi as endophytes will benefit by high inoculum volume of fungal propagules produced on senescent tissues. Although Cooke (1977) noted that at least some insect pathogens likely persist as saprobes, the potential for entomopathogens to occur as endophytes represents a relatively new and exciting area of research. Here, we focus on endophytic *B. bassiana* as a case study, highlighting recent research to characterize the endophytic symbiosis of this entomopathogenic fungus with maize.

Beauveria bassiana *in* Zea mays: *A Case Study*

Beauveria bassiana is a ubiquitous entomopathogen that is bioactive against numerous species of both pest and nonpest insects (Bruck and Lewis 2002a; Shah and Pell 2003). As the first microorganism documented as an insect pathogen (chapter 1), it has been used as a microbial insecticide to protect ornamentals, row crops, and orchards from insect pests (Labatte et al. 1996; Legaspi et al. 2000; Todorova et al. 2002). In North America, *B. bassiana* is used as a biological control agent for the European corn borer (*Ostrinia nubilalis*, Lepidoptera: Pyralidae), a major pest of maize (Lewis et al. 2002).

Although *B. bassiana* is well known as a soilborne fungus, several studies have shown that it also forms an endophytic symbiosis with maize (Lewis and Cossentine 1986; Lewis and Bing 1991). As a horizontally transmitted endophyte, *B. bassiana* infects hosts via germination of dry conidia following hydration on the leaf surface (Wagner and Lewis 2000). The majority of epiphyllous conidia do not germinate successfully: fewer than 3% of conidia may germinate after application to plant surfaces, and less than 1% of these may penetrate the leaf surface directly. At germination, germ tubes gradually elongate into hyphae and randomly spread across leaf epidermal cells. As for many phytopathogens and other endophytes, hyphae penetrate the epidermal cell layer to infect host tissues. The typical method of invasion is directly through the epidermal cell wall and into the leaf interior, most often at the junction of two epidermal cells. Examination of hyphae inside the leaf shows that they grow through the air spaces between parenchyma cells and that no haustoria are formed. Hyphae of *B. bassiana* also have been observed in the xylem vessels of maize, and by following this continuous path throughout the plant, the fungus may invade plant tissues far from the primary inoculation point. In contrast to other horizontally transmitted endophytes of angiosperms, *B. bassiana* thus can become systemic within host tissues. However, inoculation of reproductive-stage maize has shown that systemic infection of corn occurs very quickly due to spore transport within plant tissues, rather than due to hyphal growth throughout the plant. After this initial dispersal, spores may continue to disperse in the stem and leaves. Subsequent hyphal growth within plant tissues completes the establishment of the endophytic association. Infections by *B. bassiana* remain asymptomatic, with no evidence for an effect of infection on seed germination, plant growth, or dry matter accumulation.

Lewis and Cossentine (1986) assessed the intraplant epizoology of *B. bassiana* in maize with regard to damage by the European corn borer. In agricultural settings, the life history of *O. nubilalis* is tightly coupled to *B. bassiana* infections. In the major corn-producing regions of the United States, the European corn borer emerges from its overwintering hibernaculum as a moth in mid-May to mid-June and preferentially oviposits on maize in the mid-whorl stage. Emergent larvae feed within the whorl and bore into the plant as late fourth- to early fifth instars. After pupation, second-generation moths emerge in late July to early August, ovipositing on maize in the early reproductive stage. Young larvae feed in leaf midribs, bore into the plant as late fourth to early fifth instars, enter diapause, and overwinter as larvae in crop residue before emerging in spring as first-generation moths.

Through an experimental approach, Lewis and Cossentine (1986) tested the hypothesis that *B. bassiana* propagules, generated from infected first-generation European corn borers in whorl-stage maize, remain on the plant and are available to infect conspecific larvae from the second generation within the same plant. Insect damage in control plants was compared with that on plants exposed to *B. bassiana*-infected cadavers of European corn borers after a prolonged incubation time (40 days and 60 days in paired experiments with maize in different developmental stages). Plants on which cadavers were placed had significantly less tunneling by European corn borers than did control plants. Reduction in damage in whorl-stage corn was credited to the immediate effect of infection of European corn borers by *B. bassiana* from cadavers, whereas tunnel reduction in mature plants was due to colony forming units remaining on the plant from either the original cadavers or cadavers generated during development of larvae from the first generation. However, a later review of the data by Lewis and colleagues suggested that during whorl stage, treated plants were being colonized endophytically by *B. bassiana*, which then provided larval suppression in mature plants. Further research has since indicated successful reisolation of *B. bassiana* from internal tissues of maize after application of the fungus as a commercial formulation to whorl-stage plants (Lewis and Bing 1991), as well as the ability of whorl-stage applications to suppress the European corn borer throughout the growing season.

Development of a reliable protocol for establishing endophytic symbioses between *B. bassiana* and its plant host suggests a model system for exploring the ecology and evolution of entomopathogenic endophytes. Field isolates may be cultured on selective media, identified using molecular and morphological techniques, catalogued in culture collections, and formulated by scraping dried plates into an aqueous suspension. The resulting inoculum can be introduced via injection or applied topically after formulation on corn kernel grits.

Using these methods, Lewis and Bing (1991) found evidence for a strong effect of abiotic factors in shaping colonization patterns by *B. bassiana*. In the first year of a 2-year study, during which rainfall was scarce and humidity low, the entomopathogen was isolated rarely from inoculated plants: the fungus was recovered from the injection site in 14.3% of treated plants, but less frequently from other nodes of those individuals. In the second year, when both humidity and rainfall were greater, the fungus was recovered at harvest from all sections of injected plants, with most frequent isolation occurring in tissues relatively distant from the original inoculation site (node below the primary ear: 65.0% of plants). The sensitivity of infection by *B. bassiana* to factors such as humidity may reflect an important component of natural insect epizootics: ambient humidity is an important factor in penetration of plant tissues by endophytic fungi (Arnold 2002) and of insect cuticles by entomopathogenic fungi (Rawlins 1984). Bruck and Lewis (2002a) also showed that rainfall plays an important role in transferring *B. bassiana* propagules from soil to living plant tissues. Consistent with other horizontally transmitted endophytes of angiosperms (Faeth and Hammon 1997), insect vectors also appear to play a potentially important role as dispersers of this fungus. Several species of Nitidulidae (Coleoptera) have been encountered in the tunnels made by European corn borer larvae, and at least one species is an effective vector of *B. bassiana* infections via both mechanical means and fecal transmission (Bruck and Lewis 2002b).

B. bassiana in maize also represents a useful model for examining the geographical distribution of endophyte symbioses. To assess whether *B. bassiana* occurs as an endophyte only where corn is intensively cultivated and where soil is rich in organic matter, maize stalks from various parts of the U.S. Corn Belt were evaluated for infection by the entomopathogen. Over a 4-year period, plants were assayed from 16 states with different cultivation methods and pest pressure. From these data, it is evident that *B. bassiana* as an endophyte of maize is widespread (table 4.1). Subsequent investigations were conducted to determine if plant genetics specifically controlled this endophyte association. Assays using foliar application of *B. bassiana* showed that inbred lines of corn ($N = 10$) differed significantly in formation of endophyte associations (table 4.2). Further, diallel crosses using those inbred lines differed significantly in endophytism, suggesting some degree of genetic control of the symbiosis. Interestingly, further work has shown that *B. bassiana* readily forms an endophytic symbiosis with both transgenic and non-transgenic maize (Lewis et al. 2001).

Beyond Z. mays

The above studies provide a comprehensive view of an entomopathogenic endophyte associated with an economically important species while generating an array of questions of agricultural, ecological, and evolutionary importance. For example, what is the host breadth of endophytic *B. bassiana*, and does its virulence against insects differ among plant hosts? *B. bassiana* has been successfully isolated from a variety of weedy species in Iowa, USA, including velvet leaf, orchard grass, brome grass, red clover, and yellow clover. However, the virulence of these isolates toward insects has not been established. Further work in North Carolina indicated

Table 4.1. Geographical distribution of *Beauveria bassiana* as an endophyte of *Zea mays* based on cultural studies of field-collected maize stalks in 16 U.S. states with different latitudes, weather regimes, and cultivation practices, 1997-2000 (from Lewis and Gunnarson, unpublished ms.).

	1997		*1998*		*1999*		*2000*	
State	No. of Plants	% With Endophytic *B. bassiana*	No. of Plants	% With Endophytic *B. bassiana*	No. of Plants	% With Endophytic *B. bassiana*	No. of Plants	% With Endophytic *B. bassiana*
New York	20	40.0	50	4.0	50	8.0	50	18.0
Kansas	20	55.0			50	0.0	30	26.7
Michigan	20	0.0	40	10.0				
Delaware	20	5.0	40	2.5	50	0.0	50	26.0
South Dakota	50	60.0	49	8.2				
North Dakota	40	17.5	50	6.0			30	13.3
Iowa	50	22.0	115	27.8	160	26.9	90	61.1
Illinois	40	42.5	50	12.0			50	42.0
Minnesotta	60	50.0	20	0.0				
Maryland	10	20.0						
Montana	40	5.0						
Wisconsin			49	0.0	50	2.0	50	30.0
Kentucky					50	10.0		
Mississippi					50	0.0		
Ohio					50	16.0		
Vermont							48	8.3

Table 4.2. Endophytism with regard to 10 inbred lines of *Zea mays* following a foliar application of *Beauveria bassiana*.

Inbred Lines of Maize	No. of Plants	% With Endophytic *B. bassiana*	No. of Nodes	% With Endophytic *B. bassiana*
W153R	53	43.4	292	11.6
B52	60	90.0	360	42.8
MO17	60	71.7	359	24.5
B37	60	80.0	352	35.5
B86	60	40.0	351	13.7
B96	60	71.7	359	23.1
A619	60	56.7	349	23.2
B73	54	48.2	323	13.9
W182BN	60	26.7	355	8.2
A632	60	45.0	358	16.2

Data indicate the number of plants sampled per line, the prevalence of *B. bassiana* infections among plants, the number of nodes examined, and the percentage of nodes containing *B. bassiana* as an endophyte (from Lewis and Gunnarson, unpublished ms.).

that *B. bassiana* occurs as an endophyte in cotton, corn, and jimsonweed (Jones 1994), but not all plants sampled were positive for the fungus. Efforts are underway throughout the tropics to assess whether *B. bassiana* occurs as an endophyte of *Coffea* (Posada et al. 2003).

The occurrence of *B. bassiana* in both monocotyledonous and dicotyledonous hosts suggests a wide host range and the facility for diverse hosts to form endophytic symbioses with an entomopathogenic fungus. Could these plant lineages harbor other important entomopathogens among their endophytic fungal communities? The frequent isolation of *Paecilomyces* sp. from tropical trees (Arnold 2002), of *Aspergillus* from temperate and tropical hosts (Posada et al. 2003), and of various *Verticillium* spp. from an angiosperm species in North Carolina (Arnold and Lutzoni, unpublished data) suggests that entomopathogens may occur with some frequency as endophytes. However, the entomopathogenic activity of these and other endophytic isolates has not been established. Further, several taxonomic issues remain to be resolved; for example, uncertainty exists with regard to the taxonomic congruence of endophytic *Verticillium* and entomopathogenic species previously considered members of the genus (now *Lecanicillium*; Zare and Gams 2001). Given the paucity of data regarding the species composition of fungal endophytes in most plants, the efficacy of traditional culture-based studies in accurately capturing species composition, the scale of undiscovered endophyte diversity in tropical forests, and the causes of mortality in herbivorous insects in natural systems, it would appear that endophytic fungi would be useful for exploring known and new entomopathogenic species.

Assessing the validity of this prediction will benefit from a synthesis of methods. First, identifying causal agents of mortality among insect folivores will be important for understanding the diversity of insect pathogens. Such surveys could focus on agroecosystems but also could seek epizootics in natural systems wherein endophyte communities may be less disturbed. Concurrently, molecular characterization of insect-pathogenic isolates will allow development of species-specific probes that could be used to search for focal fungi within plant tissues. Concomitant isolations using specific media could elucidate the presence of entomopathogens in plant tissues as well. Development of rapid assays for screening endophytes in existing culture collections for entomopathogenic activity could determine the efficacy of focal strains or species as control agents for particular insect pests and could take advantage of growing collections of diverse tropical endophytes. Finally, each of these approaches will gain from an understanding of phylogenetic relationships among known entomopathogens and plant-associated fungi, allowing explicit hypothesis testing with regard to the evolution of insect-pathogenic and endophytic fungi.

Conclusions

Fungal endophytes represent an important but cryptic component of the earth's fungal biodiversity and comprise myriad but poorly known interactions with other organisms. Through the hosts they inhabit, endophytes have the opportunity to interact closely with herbivorous insects, against which some may act antagonistically via direct antagonism, mosaic-type defenses, or as entomopathogens. Work with *B. bassiana* has shown that this entomopathogen can be harbored as an endophyte in a variety of hosts, including both agronomic and weedy species. In contrast to the constitutive mutualism embodied by other grass endophytes in the Clavicipitaceae, *B. bassiana* is transmitted among hosts by infected herbivores and by liberation of propagules from senescent tissues by rain and other disturbances. Moreover, it persists as an infective reservoir within living plant tissues. The endophytic symbiosis of *B. bassiana* with *Z. mays* blurs some of the general boundaries among major types of endophytic symbioses and thus represents a model system for understanding general aspects of the ecology and evolution of endophytism and the roles of endophytic fungi with regard to insects. We suggest that the especially high diversity of horizontally transmitted endophytes in tropical forests represents a particularly useful resource for seeking novel entomopathogens among plant symbionts, and we anticipate that such research could generate new and interesting insect pathogens for systematics, agriculture, and biological control research.

Literature Cited

Agrios, G. N. 1997. *Plant pathology*, 4th ed. San Diego, CA: Academic Press.

Alexopoulos, C. J., C. W. Mims, and M. Blackwell. 1996. *Introductory mycology*. New York: John Wiley and Sons.

Arnold, A. E. 2001. Fungal endophytes in neotropical trees: Abundance, diversity, and ecological interactions. In *Tropical ecosystems: Structure, diversity, and human welfare*, ed. K. N. Ganeshaiah, R. Uma Shaanker and K. S. Bawa, pp. 739–745. New Delhi: Oxford and IBH Publishing.

Arnold, A. E. 2002. Neotropical fungal endophytes: diversity and ecology. Ph.D. dissertation, University of Arizona, Tucson.

Arnold, A. E. and E. A. Herre. 2003. Canopy cover and leaf age affect colonization by tropical fungal endophytes: Ecological pattern and process in *Theobroma cacao* (Malvaceae). *Mycologia* 95:388–398.

Arnold, A. E., L. C. Mejía, D. A. Kyllo, E. I. Rojas, Z. Maynard, N. Robbins, and E. A. Herre. 2003. Fungal endophytes limit pathogen damage in a tropical tree. *Proceedings of the National Academy of Sciences of the USA* 100:15649–15654.

Arnold, A. E., Z. Maynard, and G. S. Gilbert. 2001. Fungal endophytes in dicotyledonous neotropical trees: Patterns of abundance and diversity. *Mycological Research* 105:1502–1507.

Arnold, A. E., Z. Maynard, G. S. Gilbert, P. D. Coley, and T. A. Kursar. 2000. Are tropical fungal endophytes hyperdiverse? *Ecology Letters* 3:267–274.

Askary, H., N. Benhamou, and J. Brodeur. 1999. Ultrastructural and cytochemical characterization of aphid invasion by the hyphomycete *Verticillium lecanii*. *Journal of Invertebrate Pathology* 74:1–13.

Bayman, P., P. Angulo-Sandoval, Z. Baez-Ortiz, and D. J. Lodge. 1998. Distribution and dispersal of *Xylaria* endophytes in two tree species in Puerto Rico. *Mycological Research* 102:944–948.

Berbee, M. L. 2001. The phylogeny of plant and animal pathogens in the Ascomycota. *Physiological and Molecular Plant Pathology* 59:165–187.

Bernstein, M. E., and G. C. Carroll. 1977. Internal fungi in old-growth Douglas fir foliage. *Canadian Journal of Botany* 55:644–653.

Bills, G. F., and J. D. Polishook. 1991. Microfungi from *Carpinus caroliniana*. *Canadian Journal of Botany* 69:1477–1482.

Boddy, L., D. W. Bardsley, and O. M. Gibson. 1987. Fungal communities in attached ash branches. *New Phytologist* 107:143–154.

Bohn, M. 1993. *Myrothecium groenlandicum sp. nov.*, a presumed endophytic fungus of *Betula nana* (Greenland). *Mycotaxon* 46:335–341.

Boyle, C., M. Götz, U. Dammann-Tugend, and B. Schulz. 2001. Endophyte-host interactions III. Local vs. systemic colonization. *Symbiosis* 31:259–281.

Brem, D., and A. Leuchtmann. 2001. *Epichloë* grass endophytes increase herbivore resistance in the woodland grass *Brachypodium sylvaticum*. *Oecologia* 126:522–530.

Bruck, D. J., and L. C. Lewis. 2002a. Rainfall and crop residue effects on soil dispersion and *Beauveria bassiana* spread to corn. *Applied Soil Ecology* 20:183–190.

Bruck, D. J., and L. C. Lewis. 2002b. *Carpophilus freemani* (Coleoptera: Nitidulidae) as a vector of *Beauveria bassiana*. *Journal of Invertebrate Pathology* 80:188–190.

Bull, J. J. 1994. Perspective: Virulence. *Evolution* 48:1423–1437.

Cannon, P. F., and C. M. Simmons. 2002. Diversity and host preference of leaf endophytic fungi in the Iwokrama Forest Reserve, Guyana. *Mycologia* 94:210–220.

Cao, L. X., J. L. You, and S. N. Zhou. 2002. Endophytic fungi from *Musa acuminata* leaves and roots in South China. *World Journal of Microbiology and Biotechnology* 18:169–171.

Carroll, G. C. 1986. The biology of endophytism in plants with particular reference to woody perennials. In *Microbiology of the phyllosphere*, ed. N.J. Fokkema and J. van den Huevel, pp. 205–222. Cambridge: Cambridge University Press.

Carroll, G. C. 1991. Beyond pest deterrence. Alternative strategies and hidden costs of endophytic mutualisms in vascular plants. In *Microbial ecology of leaves*, ed. J. H. Andrews and S. S. Hirano, pp. 358–375. New York: Springer-Verlag.

Carroll, G. C. 1995. Forest endophytes: pattern and process. *Canadian Journal of Botany* 73:S1316–S1324.

Carroll, G. C., and F. E. Carroll. 1978. Studies on the incidence of coniferous needle endophytes in the Pacific Northwest. *Canadian Journal of Botany* 56:3034–3043.

Castelli, J. P., and B. B. Casper. 2003. Intraspecific AM fungal variation contributes to plant-fungal feedback in a serpentine grassland. *Ecology* 84:323–336.

Chanway, C. P. 1998. Bacterial endophytes: ecological and practical implications. *Sydowia* 50:149–170.

Chapela, I. H., and L. Boddy. 1988. Fungal colonization of attached beech branches. 1. Early stages of development of fungal communities. *New Phytologist* 110:39–45.

Cheplick, G. P., A. Pereira, and K. Koulouris. 2000. Effect of drought on the growth of *Lolium perenne* genotypes with and without fungal endophytes. *Functional Ecology* 14:657–667.

Clay, K. 1986. Grass endophytes. In *Microbiology of the phyllosphere*, ed. N. J. Fokkema and J. van den Huevel, pp. 188–204. Cambridge: Cambridge University Press.

Clay, K. 1988. Fungal endophytes of grasses—a defensive mutualism between plants and fungi. *Ecology* 69:10–16.

Clay, K. 1990. Fungal endophytes of grasses. *Annual Review of Ecology and Systematics* 21:275–297.

Clay, K. 1991. Endophytes as antagonists of plant pests. In *Microbial ecology of leaves*, ed. J. H. Andrews and S. S. Hirano, pp. 331–357. New York: Springer-Verlag.

Clay, K. 2001. Symbiosis and the regulation of communities. *American Zoologist* 41: 810–824.

Clay, K., T. N. Hardy, and A. M. Hammond, Jr. 1985. Fungal endophytes of grasses and their effects on an insect herbivore. *Oecologia* 66:1–6.

Clay, K., and J. Holah. 1999. Fungal endophyte symbiosis and plant diversity in successional fields. *Science* 285:1742–1744.

Clay, K., and C. Schardl. 2002. Evolutionary origins and ecological consequences of endophyte symbiosis with grasses. *American Naturalist* 160:S99–S127.

Coley, P. D., and J. A. Barone. 1996. Herbivory and plant defenses in tropical forests. *Annual Review of Ecology and Systematics* 27:305–335.

Cooke, R. 1977. *The biology of symbiotic fungi*. London: John Wiley and Sons.

Cubit, J. D. 1974. Interactions of seasonally changing physical factors and grazing affecting high intertidal communities on a rocky shore. PhD dissertation, University of Oregon.

de Bary, A. 1863. Ueber die Entwickelung der *Sphaeria typhina* Pers. und Bail's "Mycologische Studien". *Flora* 46:401–401.

de Bary, A. 1866. *Morphologie und Physiologie der Pilze, Flechten und Myxomyceten*. Leipzig: Engelmann.

Deckert, R. J., L. H. Melville, and R. L. Peterson. 2001. Structural features of a *Lophodermium* endophyte during the cryptic life-cycle phase in the foliage of *Pinus strobus*. *Mycological Research* 105:991–997.

Dobranic, J. K., J. A. Johnson, and Q. R. Alikhan 1995. Isolation of endophytic fungi from eastern larch (*Larix laricina*) leaves from New Brunswick, Canada. *Canadian Journal of Microbiology* 41:194–198.

Dreyfuss, M., and I. H. Chapela. 1994. Potential of fungi in the discovery of novel, low-molecular weight pharmaceuticals. In *The discovery of natural products with therapeutic potential*, ed. V.P. Gull, pp. 49–80. London: Butterworth-Heinemann.

Dreyfuss, M., and O. Petrini. 1984. Further investigations on the occurrence and distribution of endophytic fungi in tropical plants. *Botanica Helvetica* 94:33–40.

Elliott, S. L, M. W. Sabelis, L. P. S. van der Geest, E. A. M. Beerling, and J. Fransen. 2000. Can plants use entomopathogens as bodyguards? *Ecology Letters* 3:228–235.

Faeth, S. H. 2002. Are endophytic fungi defensive plant mutualists? *Oikos* 98:25–36.

Faeth, S. H., and K. E. Hammon. 1997. Fungal endophytes in oak trees: Long-term patterns of abundance and associations with leafminers. *Ecology* 78:810–819.

Faeth, S. H., and T. J. Sullivan. 2003. Mutualistic asexual endophytes in a native grass are usually parasitic. *American Naturalist* 161:310–325.

Findlay, J. A., G. Q. Li, J. D. Miller, and T. O. Womiloju. 2003. Insect toxins from spruce endophytes. *Canadian Journal of Chemistry* 81:284–292.

Fisher, P. J. 1996. Survival and spread of the endophyte *Stagonospora pteridiicola* in *Pteridium aquilinum*, other ferns and some flowering plants. *New Phytologist* 132:119–122.

Fisher, P. J., O. Petrini, and B. C. Sutton. 1993. A comparative study of fungal endophytes in leaves, xylem and bark of *Eucalyptus* in Australia and England. *Sydowia* 45:338–345.

Fox, F. M. 1993. Tropical fungi: their commercial potential. In *Aspects of tropical mycology*, ed. S. Isaac, J. C. Frankland, R. Watling, and A. J. S. Whalley, pp. 253–264. Cambridge: Cambridge University Press.

Freeman, S., and R. J. Rodriguez. 1993. Genetic conversion of a fungal plant pathogen to a nonpathogenic, endophytic mutualist. *Science* 260:75–78.

Fröhlich, J., and K. D. Hyde. 1999. Biodiversity of palm fungi in the tropics: Are global fungal diversity estimates realistic? *Biodiversity and Conservation* 8:977–1004.

Gamboa, M. A., and P. Bayman. 2001. Communities of endophytic fungi in leaves of a tropical timber tree (*Guarea guidonia*: Meliaceae). *Biotropica* 33:352–360.

Gamboa, M. A., P. Laureano, and P. Bayman. 2003. Measuring diversity of endophytic fungi in leaf fragments: Does size matter? *Mycopathologia* 156:41–45.

Garcia-Guzman, G., and R. Dirzo. 2001. Patterns of leaf-pathogen infection in the understory of a Mexican rain forest: Incidence, spatiotemporal variation, and mechanisms of infection. *American Journal of Botany* 88:634–645.

Gehring, C. A. 2003. Growth responses to arbuscular mycorrhizae by rain forest seedlings vary with light intensity and tree species. *Plant Ecology* 167:127–139.

Gilbert, G. S. 2002. Evolutionary ecology of plant diseases in natural ecosystems. *Annual Review of Phytopathology* 40:13–43.

Gimenez-Pecci, M. D., M. R. Bogo, L. Santi, C. K. de Moraes, C. T. Correa, M. H. Vainstein, and A. Shrank. 2002. Characterization of mycoviruses and analyses of chitinase secretion in the biocontrol fungus *Metarhizium anisopliae*. *Current Microbiology* 45:334–339.

Guo, L. D., K. D. Hyde., and E. C. Y. Liew. 2001. Detection and taxonomic placement of endophytic fungi within frond tissues of *Livistona chinensis* based on rDNA sequences. *Molecular Phylogenetics and Evolution* 20:1–13.

Hart, T. D., F. A. A. M.de Leij, G. Kinsey, J. Kelly, and J. M. Lynch. 2002. Strategies for the isolation of cellulolytic fungi for composting of wheat straw. *World Journal of Microbiology and Biotechnology* 18:471–480.

Hawksworth, D. L. 1991. The fungal dimension of biodiversity: Magnitude, significance, and conservation. *Mycological Research* 95:641–655.

Hawksworth, D. L. 2001. The magnitude of fungal diversity: the 1.5 million species estimate revisited. *Mycological Research* 105:1422–1432.

Heckman, D. S., D. M. Geiser, D. B. Eidell, R. L. Stauffer, N. L. Kardos, and B. Hedges. 2001. Molecular evidence for the early colonization of land by fungi and plants. *Science* 293:1129–1133.

Herre, E. A. 1995. Factors affecting the evolution of virulence: Nematode parasites of fig wasps as a case study. *Parasitology* 111:S179–S191.

Hoveland, C. S. 1993. Importance and economic significance of the *Acremonium* endophytes to performance of animals and grass plants. *Agriculture, Ecosystems and Environment* 44:3–12.

Humber, R. A. 1984. Foundations for an evolutionary classification of the Entomophthorales

(Zygomycetes). In *Fungus-insect relationships: Perspectives in ecology and evolution*, ed. Q. Wheeler and M. Blackwell, pp. 166–183. New York: Columbia University Press.

Husband, R., E. A. Herre, and J. P. W. Young. 2002. Temporal variation in the arbuscular mycorrhizal communities colonising seedlings in a tropical forest. *FEMS Microbiology Ecology* 42:131–136.

Johnson, J. A., and N. J. Whitney. 1992. Isolation of fungal endophytes from black spruce (*Picea mariana*) dormant buds and needles from New Brunswick, Canada. *Canadian Journal of Botany* 70:1754–1757.

Johnson, S. N., A. E. Douglas, S. Woodward, and S. E. Hartley. 2003. Microbial impacts on plant-herbivore interactions: The indirect effects of a birch pathogen on a birch aphid. *Oecologia* 134:388–396.

Jones, K. D. 1994. Aspects of the biology and biological control of the European corn borer in North Carolina. PhD dissertation, North Carolina State University.

Khalil, A. I. 2002. Production and characterization of cellulolytic and xylanolytic enzymes from the ligninolytic white-rot fungus *Phaerochaete chrysosporium* grown on sugarcane bagasse. *World Journal of Microbiology and Biotechnology* 18:753–759.

Koppenhofer, A. M., and E. M. Fuzy. 2003. Effects of turfgrass endophytes (Clavicipitaceae: Ascomycetes) on white grub (Coleoptera: Scarabaeidae) control by the entomopathogenic nematode *Heterorhabditis bacteriophora* (Rhabditida: Heterorhabditidae). *Environmental Entomology* 32:392–396.

Kriel, W. M., W. J. Swart, and P. W. Crous. 2000. Foliar endophytes and their interactions with host plants, with specific reference to the Gymnospermae. *Advances in Botanical Research* 33:1–34.

Kuldau, G. A., J.-S. Liu, J. F. White, Jr., M. R. Siegel, and C. L. Schardl. 1997. Molecular systematics of Clavicipitaceae supporting monophyly of genus *Epichloë* and form genus *Ephelis*. *Mycologia* 89:431–441.

Kumaresan, V., and T. S. Suryanarayanan. 2001. Occurrence and distribution of endophytic fungi in a mangrove community. *Mycological Research* 105:1388–1391.

Kunkel, B. A., and P. S. Grewal. 2003. Endophyte infection in perennial ryegrass reduces the susceptibility of black cutworm to an entomopathogenic nematode. *Entomologia Experimentalis et Applicata* 107:95–104.

Labatte, J. M., S. Meusnier, A. Migeon, J. Chuafaux, Y. Couteaudier, G. Riba, and B. Got. 1996. Field evaluation of and modeling the impact of three control methods on the larval dynamics of *Ostrinia nubilalis* (Lepidoptera: Pyralidae). *Journal of Economic Entomology* 89:852–862.

Laessøe, T., and D. J. Lodge. 1994. Three host-specific *Xylaria* species. *Mycologia* 86:436–446.

Legaspi, J. C., T. J. Poprawski, and B. C. Legaspi. 2000. Laboratory and field evaluation of *Beauveria bassiana* against sugarcane stalkborers (Lepidoptera: Pyralidae) in the Lower Rio Grande Valley of Texas. *Journal of Economic Entomology* 93:54–59.

Legault, D., M. Dessureault, and G. Laflamme. 1989. Mycoflora of the needles of *Pinus banksiana* and *Pinus resinosa*. 1. Endophytic fungi. *Canadian Journal of Botany* 67:2052–2060.

Leigh, E. G., Jr. 1999. *Tropical forest ecology*. Oxford: Oxford University Press.

Lewis, L. C., and L. A. Bing. 1991. *Bacillus thuringiensis* Berliner and *Beauveria bassiana* (Balsamo) Vuillemin for European corn borer control: Program for immediate and season-long suppression. *Canadian Entomologist* 123:387–393.

Lewis, L. C., D. J. Bruck, and R. D. Gunnarson, and K. G. Bidne. 2001. Assessment of plant pathogenicity of endophytic *Beauveria bassiana* in Bt transgeneic and non-transgenic corn. *Crop Science* 41:1395–1400.

Lewis, L. C., D. J. Bruck, and R. D. Gunnarson. 2002. On-farm evaluation of *Beauveria bassiana* for control of *Ostrinia nubilalis* in Iowa, USA. *Biocontrol* 47:167–176.

Lewis, L. C., and J. E. Cossentine. 1986. Season long intraplant epizootics of entomopathogens, *Beauveria bassiana* and *Nosema pyrausta*, in a corn agroecosystem. *Entomophaga* 31:363–369.

Lodge, D. J., P. J. Fisher, and B. C. Sutton. 1996. Endophytic fungi of *Manilkara bidentata* leaves in Puerto Rico. *Mycologia* 88:733–738.

Lupo, S., S. Tiscornia, and L. Bettucci. 2001. Endophytic fungi from flowers, capsules and seeds of *Eucalyptus globulus*. *Revista Iberoamericana de Micología* 18:38–41.

Malinowski, D. P., and D. P. Belesky. 1999. Tall fescue aluminum tolerance is affected by *Neotyphodium coenophialum* endophyte. *Journal of Plant Nutrition* 22:1335–1349.

Marquis, R. J., and H. M. Alexander. 1992. Evolution of resistance and virulence in plant herbivore and plant pathogen interactions. *Trends in Ecology and Evolution* 7:126–129.

Meijer, G., and A. Leuchtmann. 1999. Multistrain infection of the grass *Brachypodium sylvaticum* by its fungal endophyte *Epichloë sylvatica*. *New Phytologist* 141:355–368.

Meyer, L., B. Slippers, L. Korsten, J. M. Kotze, and M. Wingfield. 2001. Two distinct *Guignardia* species associated with citrus in South Africa. *South African Journal of Science* 97:191–194.

Miller, J. D., S. Mackenzie, M. Foto, G. W. Adams, and J. A. Findlay. 2002. Needles of white spruce inoculated with rugulosin-producing endophytes contain rugulosin reducing spruce budworm growth rate. *Mycological Research* 106:471–479.

Monk, K. A., and G. J. Samuels. 1990. Mycophagy in grasshoppers (Orthoptera, Acrididae) in Indo-Malayan rain forests. *Biotropica* 22:16–21.

Müller, M. M., R. Valjakka, A. Suokko, and J. Hantula. 2001. Diversity of endophytic fungi of single Norway spruce needles and their role as pioneer decomposers. *Molecular Ecology* 10:1801–1810.

Packer, A., and K. Clay. 2003. Soil pathogens and *Prunus serotina* seedling and sapling growth near conspecific trees. *Ecology* 84:108–119.

Petrini, O. 1984. Endophytic fungi in British Ericaceae: a preliminary study. *Transactions of the British Mycological Society* 83:510–512.

Petrini, O. 1985. Wirtsspezifität endophytischer Pilze bei enheimischen Ericaceae. *Botanica Helvetica* 95:213–218.

Petrini, O. 1986. Taxonomy of endophytic fungi of aerial plant tissues. In *Microbiology of the phyllosphere*, ed. N.J. Fokkema and J. van den Huevel, pp. 175–187. Cambridge: Cambridge University Press.

Petrini, O. 1991. Fungal endophytes of tree leaves. In *Microbial ecology of leaves*, ed. J. H. Andrews and S. S. Hirano, pp. 179–197. New York: Springer-Verlag.

Petrini, O., and G. C. Carroll. 1981. Endophytic fungi in the foliage of some Cupressaceae in Oregon. *Sydowia* 34:135–148.

Petrini, O., and M. Dreyfuss. 1981. Endophytische Pilze vom epiphytischen Araceae, Bromeliaceae und Orchidaceae. *Sydowia* 34:135–148.

Petrini, O., and E. Müller. 1979. Pilzliche Endophyten am Beispiel von *Juniperus communis* L. Sydowia 32:224–251.

Petrini, O., J. Stone, and F. E. Carroll. 1982. Endophytic fungi in evergreen shrubs in western Oregon—a preliminary study. *Canadian Journal of Botany* 60:789–796.

Pinto, L. S. R. C., J. L. Azevedo, J. O. Pereira, M. L. C. Vieira, and C. A. Labate. 2000. Symptomless infection of banana and maize by endophytic fungi impairs photosynthetic efficiency. *New Phytologist* 147:609–615.

Posada, F. J., F. E. Vega, and S. A. Rehner. 2003. *Beauveria* as a possible coffee endophyte. Annual meeting of the Society for Invertebrate Pathology, Burlington, Vermont, July 26–31, p. 47.

Rao, M. R., M. P. Singh, and R. Day. 2000. Insect pest problems in tropical agroforestry systems: Contributory factors and strategies for management. *Agroforest Systems* 50:243–277.

Rawlins, J. E. 1984. Mycophagy in Lepidoptera. In *Fungus-insect relationships: Perspectives in ecology and evolution*, ed. Q. Wheeler and M. Blackwell, pp. 382–423. New York: Columbia University Press.

Redman, R. S., D. D. Dunigan, and R. J. Rodriguez. 2001. Fungal symbiosis from mutualism to parasitism: who controls the outcome, host or invader? *New Phytologist* 151:705–716.

Rodrigues, K. F. 1994. The foliar fungal endophytes of the Amazonian palm *Euterpe oleracea. Mycologia* 86:376–385.

Rollinger, J. L., and J. H. Langenheim. 1993. Geographic survey of fungal endophyte community composition in leaves of coastal redwood. *Mycologia* 85:149–156.

Rygiewicz, P. T., and C. T. Andersen. 1994. Mycorrhizae alter quality and quantity of carbon allocated below ground. *Nature* 369:58–60.

Sagers, C.L., and P. D. Coley. 1995. Benefits and costs of defense in a neotropical shrub. *Ecology* 76:1835–1843.

Saikkonen, K., S. H. Faeth, M. Helander, and T. J. Sullivan. 1998. Fungal endophytes: A continuum of interactions with host plants. *Annual Review of Ecology and Systematics* 29:319–343.

Sampson, K. 1933. The systematic infection of grasses by *Epichloë typhina* (Pers.) Tul. *Transactions of the British Mycological Society* 18:30–47.

Schardl, C. L. 1996. *Epichloë* species: fungal symbionts of grasses. *Annual Review of Phytopathology* 34:109–130.

Schulz, B., U. Wanke, S. Draeger, and H. J. Aust. 1993. Endophytes from herbaceous plants and shrubs—effectiveness of surface sterilization methods. *Mycological Research* 97:1447–1450.

Schulz, B., C. Boyle, S. Draeger, A. K. Rommert, and K. Krohn. 2002. Endophytic fungi: a source of novel biologically active secondary metabolites. *Mycological Research* 106:996–1004.

Shah, P. A. and J. K. Pell. 2003. Entomopathogenic fungi as biological control agents. *Applied Microbiology and Biotechnology* 61:413–423.

Sherwood-Pike M., J. K. Stone, and G. C. Carroll. 1986. *Rhabdocline parkeri*, a ubiquitous foliar endophyte of Douglas fir. *Canadian Journal of Botany* 64:1849–1855.

Sieber, T. N. 1989. Endophytic fungi in twigs of healthy and diseased Norway spruce and white fir. *Mycological Research* 92:322–326.

Sieber, T. N., and C. E. Dorworth. 1994. An ecological study about assemblages of endophytic fungi in *Acer macrophyllum* in British Columbia: In search of candidate mycoherbicides. *Canadian Journal of Botany* 72:1397–1402.

Siegel, M. R., G. C. M. Latch, L. P. Bush, N. F. Fannin, D. D., Rowan, B. A. Tapper, C. W. Bacon, and M. C. Johnson. 1990. Fungal endophyte-infected grasses: Alkaloid accumulation and aphid response. *Journal of Chemical Ecology* 16:3301–3315.

Siemens, D. H., S. H. Garner, T. Mitchell-Olds, and R. M. Callaway. 2002. Cost of defense in the context of plant competition: *Brassica rapa* may grow and defend. *Ecology* 83:505–517.

Southcott, K. A., and J. A. Johnson. 1997. Isolation of endophytes from two species of palm, from Bermuda. *Canadian Journal of Microbiology* 43:789–792.

Stanley, S. J. 1992. Observations on the seasonal occurrence of marine endophytic and parasitic fungi. *Canadian Journal of Botany* 70:2089–2096.

Stanosz, G. R., J. T. Blodgett, D. R. Smith, and E. L. Kruger. 2001. Water stress and *Sphaeropsis sapinea* as a latent pathogen of red pine seedlings. *New Phytologist* 149: 531–548.

Stanosz, G. R., D. R. Smith, M. A. Guthmiller, and J. C. Stanosz. 1997. Persistence of *Sphaeropsis sapinea* on or in asymptomatic shoots of red and jack pines. *Mycologia* 89:525–530.

Stone, J. K. 1985. Foliar endophytes of *Pseudotsuga menziesii* (Mirb.) Franco. Cytology and physiology of the host-endophyte relationship. PhD dissertation, University of Oregon.

Stone, J. K. 1987. Initiation and development of latent infections by *Rhabdocline parkeri* on Douglas fir. *Canadian Journal of Botany* 65:2614–2621.

Stone, J. K. 1988. Fine structure of latent infections by *Rhabdocline parkeri* on Douglas fir, with observations on uninfected epidermal cells. *Canadian Journal of Botany* 66:45–54.

Stone, J. K., C. W. Bacon, and J. F. White, Jr. 2000. An overview of endophytic microbes: endophytism defined. In *Microbial endophytes*, ed. C.W. Bacon and J. F. White, pp. 3–29. New York: Marcel Dekker.

Suryanarayanan, T. S., T. S. Murali, and G. Venkatesan. 2002. Occurrence and distribution of fungal endophytes in tropical forests across a rainfall gradient. *Canadian Journal of Botany* 80:818–826.

Todorova, S. I., C. Cloutier, J. C. Cote, and D. Coderre. 2002. Pathogenicity of six isolates of *Beauveria bassiana* (Balsamo) Vuillemin (Deuteromycotina, Hyphomycetes) to *Perillus bioculatus* (F) (Hem., Pentatomidae). *Journal of Applied Entomology* 126:182–185.

Wagner, B. L., and L. C. Lewis. 2000. Colonization of corn, *Zea mays*, by the entomo-pathogenic fungus *Beauveria bassiana*. *Applied and Environmental Microbiology* 66:3468–3473.

Wilson, G. W. T., D. C. Hartnett, M. D. Smith, and K. Kobbeman. 2001. Effects of mycor-rhizae on growth and demography of tallgrass prairie forbs. *American Journal of Botany* 88:1452–1457.

Wilkinson, H. H., M. R. Siegel, J. D. Blankenship, A. C. Mallory, L. P. Bush, and C. L. Schardl. 2000. Contribution of fungal loline alkaloids to protection from aphids in a grass-endophyte mutualism. *Molecular Plant-Microbe Interactions* 13:1027–1033.

Zare, R., and W. Gams. 2001. A revision of *Verticillium* section Prostrata. IV. The genera *Lecanicillium* and *Simplicillium gen. nov. Nova Hedwigia* 73:1–50.

5

The Fungal Roots
of Microsporidian Parasites

Naomi M. Fast
Patrick J. Keeling

Microsporidia are a group of eukaryotic intracellular parasites that commonly infect animals, but they have also been found as parasites of two members of the alveolate protists: gregarine apicomplexa and ciliates (Vivier 1975; Sprague et al. 1992; Sprague and Becnel 1999). There are just more than a dozen microsporidian species known to infect humans, and these cause a variety of illnesses, including diarrhea and hepatitis (Weber et al. 1999; Franzen and Muller 2001). Human infections tend to involve immunocompromised hosts such as organ transplant recipients or AIDS patients (Weber et al. 1994; Weber and Bryan 1994). Microsporidia are also known to infect a number of other mammals, but the most common microsporidian infections are those of fish and arthropods (Becnel and Andreadis 1999; Shaw and Kent 1999). Within the arthropods, insect-infecting microsporidia are particularly common, and some of the best characterized microsporidia are insect pathogens. Of the roughly 1200 species of microsporidia currently defined, a significant proportion of these parasitize insects, with more than half of the described genera infecting an insect (Becnel and Andreadis 1999). These have been found to infect virtually all major groups of insects, and the parasites also are phylogenetically diverse, coming from at least three of the four major subdivisions of microsporidia, as well as from the more poorly understood basal lineages (Keeling and McFadden 1998). Interestingly, several theories for the origin of microsporidia suggest that they evolved from insect-parasitizing fungi (e.g., Cavalier-Smith 1998; Keeling et al. 2000), so the roots of this large and diverse group may trace back to an ancient entomopathogen, as will be discussed in more detail below.

The only life stage of microsporidia that is viable outside of the host cell is the spore, and this stage is also the most recognizable form of the parasite, making its features useful for diagnostic purposes. Microsporidian spores are small, ranging in size from 1 μm to 40 μm, and are generally ovoid in shape, although other shapes (e.g., rods and spheres) can be found in some species (Vávra and Larsson 1999). In a few instances, particularly in species that inhabit aquatic environments, the spores may be highly ornamented with surface extensions. Not all species possess the same spore type throughout their life cycles, and spore morphology may vary depending on the host or the stage of infection in an individual host (Vávra and Larsson 1999).

Despite the variation in spore size, shape, and ornamentation, there are a number of features that are consistent (fig. 5.1) (Vávra and Larsson 1999). The spore coat consists of two layers: a thin, dense, outer proteinaceous layer (exospore) and a thicker inner layer composed of chitin and protein (endospore). The surrounding coat layers protect the plasma membrane and sporoplasm of the unicellular parasite from physical and environmental damage. The sporoplasm lies within the plasma membrane and is almost entirely relegated to organelles associated with infection. These include the posterior vacuole, polar filament (or polar tube), and polaroplast (fig. 5.1). The polaroplast is an association of membranes located at the apex of the spore and is composed of an anterior region of highly organized membranous structures (the lamellar polaroplast) and a posterior region of loosely organized vesicles (the vesicular polaroplast). The polar filament (tube) also is positioned at the anterior of the spore, where it is anchored at the apex. From the anchoring disc the filament extends toward the posterior of the spore, and for approximately the lower one-half of its length it is coiled around the contents of the sporoplasm (fig. 5.1). The orientation and coiling of the polar filament is a highly conserved morphological feature at certain taxonomic levels. The number of coils and their arrangement with respect to one another can therefore be used to distinguish one microsporidian species from another (Sprague et al. 1992; Vávra and Larsson 1999). At the base of the spore, the polar filament (tube) ends proximal to the posterior vacuole. It is unclear whether the polar filament (tube) is continuous with or enters the posterior vacuole, or if the two organelles are physically distinct (Vinckier et al. 1993; Keohane and Weiss 1998; Vávra and Larsson 1999). In addition to these three infection-related organelles—the polaroplast, polar filament (polar tube), and posterior vacuole—the spore also contains a single nucleus or a diplokaryon (a condition in which two nuclei are appressed, or "back to back"), a number of nondescript

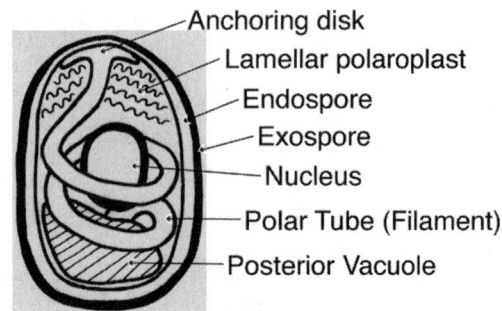

Figure 5.1. Schematic diagram of a microsporidian spore shows the surrounding coat layers (endospore and exospore). The sporoplasm within the plasma membrane contains the lamellar polaroplast, polar filament (or polar tube), posterior vacuole, and anchoring disk; these organelles are associated with infection.

vesicles, and ribosomes, that are sometimes closely packed into crystalline-like polyribosome structures (Vávra and Larsson 1999).

Microsporidian spores are triggered to germinate by a variety of signals that are not well understood (Undeen and Epsky 1990; Keohane and Weiss 1999). However, some clues about these triggers come from *in vitro* methods of germinating spores that involve changes in pH, osmolarity, and ion concentrations (Keohane and Weiss 1999). *In vitro* signals tend to be species specific, a situation that is likely also reflected *in vivo*. Regardless of the signal, the first sign of germination is a swelling of the polaroplast and posterior vacuole (Lom and Vávra 1963), resulting in an increased osmotic pressure within the spore (Kudo 1918; Oshima 1937; Undeen and Frixione 1990). It has been posited that the increased internal pressure arises as a result of the breakdown of the glucose disaccharide trehalose, because trehalase activity and glucose monomers have been detected at germination in several microsporidian species (Vander Meer and Gochnauer 1971; Undeen et al. 1987; Undeen and Frixione 1990; Undeen and Van der Meer 1994). However, these characteristics are not common to all microsporidian species, so it is perhaps more likely that trehalose breakdown is simply one of many steps associated with germination in those species where this activity is detected. An alternative model is one in which the calcium/calmodulin signaling pathway may play a role at the onset of germination (Keohane and Weiss 1998; Weidner et al. 1999). Regardless of the cause, microsporidian spores are known to possess aquaporins (water channels), which allow for the influx of water and concomitant swelling (Frixione et al. 1997). At some point osmotic pressure increases to such a degree that the anchoring disk attaching the polar filament to the apex of the spore breaks and the polar filament everts, becoming a tube. The dense, proteinaceous material that formed the core of the filament in the spore is located on the outer surface of the tube after eversion, forming the outer coating of the tube (Keohane and Weiss 1999). The everted tube can extend to distances as much as one hundred times the length of the spore; the eversion process can occur at velocities of 100 μm/s (Frixione et al. 1992). Continuing pressure within the spore then forces the sporoplasm through the fully everted polar tube so that if a potential host cell is within range of the germinating spore, the polar tube may strike and penetrate this host, and the infective sporoplasm is injected directly into the host cytosol. Following infection, the microsporidian spore contains the original membrane of the sporoplasm, and the parasite within the host possesses a new membrane likely derived from the polaroplast (Weidner et al. 1984; Undeen and Frixione 1991). This is a remarkable process, given that the sporoplasm moves from one end of the polar tube to the other in just 15–500 ms (Frixione et al. 1992).

Once a microsporidian has entered the host cell, the parasite generally undergoes a period of vegetative growth (merogony) followed by the formation of spores (sporogony). However, the variation in these processes between different microsporidian species is tremendous. Most microsporidian parasites multiply throughout merogony, but others also divide at sporogony. The infective apparatus develops after division, and enzyme labeling studies have indicated that the microsporidian infection machinery is derived from the Golgi and endoplasmic reticulum membrane systems (Vávra 1965; Takvorian and Cali 1994, 1996). The exospore and endospore layers develop last, and then the mature spores are released. In some cases, the spores are autoinfective and immediately germinate to infect different cells of the same host; in other cases the spores are passed into the environment to go on to infect new hosts. In many cases, the parasite induces significant changes in the organization of the host cell by surrounding itself with the host's mitochondria, nuclei, or endoplasmic reticulum (Hendrick et al. 1991; Vávra and Larsson 1999). In an extreme case of host hijacking by the microsporidian, the host cell is totally transformed into a large structure called a xenoma, a multinucleate cell formed by many rounds of nuclear division in the absence of cellular division. During formation of the xenoma, the parasites become ordered spatially from the core outward, depending on maturation stage (Weissenberg 1976; Canning et al. 1982; Larsson et al. 1996).

Identification of Microsporidia and Early Ideas about Their Origins

Microsporidia were first recognized by the symptoms of infection of the silkworm *Bombyx mori*, resulting in pébrine or pepper disease, which in the mid-nineteenth century all but destroyed the European silk industry. Nägeli (1857) named the causative agent *Nosema bombycis*; he included it among the schizomycete fungi, a group subsequently found to contain an assortment of microscopic organisms including bacteria and yeasts. In 1882, Balbiani removed *Nosema* from the schizomycetes based on further examination and created a new group, the Microsporidia (Balbiani 1882).

In the early 1900s, scientists began to grasp the breadth of microbial life and began to develop classification schemes to try to incorporate this diversity. Much of this work included distinguishing taxonomic groupings of single-celled eukaryotes, often referred to as protists or protozoa. The characteristic spore of microsporidia led to their inclusion in Sporozoa, a group that included other parasites with a spore life-stage. More specifically, microsporidia formed part of a subgroup within Sporozoa called Cnidosporidia, that also included actinomyxidia and myxosporidia (Kudo 1947). The microsporidia–actinomyxidia–myxosporidia affiliation was recognized for some time, although claims also were made that microsporidia should be considered a separate group based on evidence suggesting that their similarity to actinomyxidia and myxosporidia was superficial and on the identification of seemingly disparate features in microsporidia that were not shared by actinomyxidia and myxosporidia (Lom and Vávra 1962; Levine et al. 1980).

Ancient Eukaryotes: Archezoa and Early Molecular Phylogenies

In the early 1980s, Cavalier-Smith proposed a scheme for the origin of eukaryotes that renewed interest in the evolutionary position of microsporidia. His hypothesis focused on the lack of typical eukaryotic features in four protist groups, which he dubbed Archezoa (Cavalier-Smith 1983). The missing features included mitochondria, peroxisomes (or microbodies) and, in some cases, typical Golgi stacks, and

9+2 microtubule structures (i.e., flagella). In addition, the majority of Archezoa (including microsporidia) also lacked typical 80S eukaryotic ribosomes, and instead possessed 70S ribosomes, on par with those of prokaryotes (Ishihara and Hayashi 1968; Curgy et al. 1980). Cavalier-Smith proposed that these groups were the earliest diverging eukaryotic groups that had branched before the acquisition of mitochondria and the other structures. The Archezoa hypothesis claimed that the archezoan features represent primitive states, as these lineages diverged before the evolution of these characters. The four groups of amitochondrial protists placed in the Archezoa were Metamonada (e.g., *Giardia*), Parabasalia (e.g., *Trichomonas*), Archamoebae (e.g., *Entamoeba*), and Microsporidia (Cavalier-Smith 1983).

Shortly after the inception of the Archezoa, molecular sequences began to accumulate from these protists, and phylogenetic data were produced to address the issue. Vossbrinck et al. (1987) produced the first molecular sequence data from a microsporidian; their sequencing of the small subunit (SSU) rDNA from *Vairimorpha necatrix* revealed a highly unusual SSU gene that not only was much smaller than typical eukaryotic SSU rDNAs but also was extremely divergent in sequence. Phylogenetic analysis placed the *V. necatrix* sequence at the base of all eukaryotes, suggesting an ancient origin for microsporidia and providing support for the Archezoa hypothesis (Vossbrinck et al. 1987). Furthermore, additional analysis of the rRNAs of *V. necatrix* revealed that the 5.8S rRNA was fused to the large subunit (LSU) rRNA as it is in prokaryotic rRNAs, unlike those of any other eukaryote (Vossbrinck and Woese 1986). This arrangement in microsporidia was interpreted as a primitive eukaryotic feature—an evolutionary forerunner of the separated molecules in other eukaryotes. Based on the SSU phylogenetic results and seemingly prokaryotic nature of the fused LSU and 5.8S rRNAs, an early origin for the microsporidia seemed a reasonable interpretation. Moreover, when phylogenetic methods were applied to the other members of the proposed Archezoa, they also branched deeply in phylogenetic analyses, although this was debated in the case of the archamoebae.

As more molecules were developed for phylogenetic analysis, the hypothesized ancient position of microsporidia gained even more strength. Sequences coding for two components of the translation apparatus, elongation factor 1α (EF-1α) and elongation factor 2 (EF-2), were determined from the microsporidian *Glugea plecoglossi* (Kamaishi et al. 1996a,b). Phylogenetic analyses of these sequences clearly placed the microsporidian sequence at the base of all eukaryotes. Similarly, an analysis of the isoleucyl tRNA synthetase sequence from *Nosema locustae* also placed microsporidia at the base of eukaryotes (Brown and Doolittle 1995). However, despite the growing phylogenetic evidence supporting an ancient origin for microsporidia, doubts were raised based on a number of issues. There was concern about potential phylogenetic artifacts arising from the high level of divergence observed in most microsporidian sequences, manifesting itself as long branches in phylogenetic trees. Divergent sequences are often difficult to place in phylogenetic trees because they are drawn artifactually to other divergent sequences in the phylogeny, a phenomenon called long-branch attraction (Felsenstein 1978). As the branch leading to the outgroup in a phylogeny is often the longest branch on the tree, it is common for other long-branch sequences to be "drawn" to the base of the tree. Therefore, it became questionable whether microsporidia were truly ancient and branching at the base of eukaryotes or if their highly derived parasitic nature and divergent sequences were artifactually suggesting an ancient origin. Indeed, even the fused 5.8S-LSU rRNA was called into question, as it was noted that not only are microsporidian rRNAs divergent, but that they also have undergone numerous deletions—even in sequence regions that do not tend to vary in other organisms. Therefore, it was proposed that a deletion may have arisen in an rRNA operon processing site, resulting in a reversion to the fused 5.8S-LSU state (Cavalier-Smith 1993). These concerns notwithstanding, the bulk of the evidence up to the mid-1990s was strongly in favor of an early origin for microsporidia.

New Phylogenies Converge on a Fungal Alternative: Reassignment and Reinterpretation

Just more than 12 years after the initial Archezoa hypothesis was proposed, phylogenetic evidence began to accumulate indicating that the concerns regarding the evidence for an early origin of microsporidia were well justified. In 1996, two groups independently analyzed sequences for the cytoskeletal proteins α- and β-tubulin from a sampling of microsporidia that included *Encephalitozoon hellem*, *Nosema locustae*, and *Spraguea lophii*; the microsporidian sequences did not branch with other Archezoa, but instead diverged with the fungi (Edlind et al. 1996; Keeling and Doolittle 1996). The support for this grouping was quite strong—in α-tubulin phylogenies as high as 96%. However, microsporidian tubulin sequences are also divergent, as are fungal homologs, leading to the concern that the microsporidia–fungi relationship seen in tubulin trees could also result from long-branch attraction (Keeling and Doolittle 1996). Microsporidia and most fungi lack 9+2 microtubule structures at all stages of their life history, and this condition is generally correlated with a high degree of divergence in tubulin sequences. In the original analyses, only ascomycete and basidiomycete fungi were included, and both groups lack 9+2 structures. Therefore, tubulin sequences of flagellated chytrid fungi are more conserved and allowed for a direct test of long-branch attraction. Using the short-branch chytrid sequences as the only fungal representatives in the analysis, the long-branch microsporidian sequences still formed a sister group with the fungi (Keeling et al. 2000). This analysis served as strong evidence that the microsporidia–fungi relationship seen in tubulin phylogenies is not an artifact of long-branch attraction.

Based on the tubulin results, other aspects of microsporidian cellular and molecular biology were reexamined. Although EF-1α phylogenies placed microsporidia at the base of eukaryotes with strong support, a unique 12-amino acid insertion within the EF-1α (gene sequence is shared by microsporidian, fungal, and animal sequences (Kamaishi et al. 1996a). In addition, a characteristic two-amino acid insertion in animal and fungal homologs of the glycolytic enzyme enolase also is found in the microsporidian homolog. Unique insertions and deletions in sequences are often good indicators of relationships, as these shared characters can indicate a common ancestry because they are less likely to occur by convergence, although this is not always the case (Keeling and Palmer 2001). In fact, these insertions had been key in determining that animals and fungi are each other's nearest relatives (Baldauf and Palmer 1993; Baldauf et al. 2000). In addition to these specific sequence details, the presence of RNA components of the spliceosome also predicted a later origin for microsporidia. The spliceosome consists of both protein and RNA components and is responsible for the removal of introns from messenger RNA. Expression and structural prediction data proposed that functional U6 and U2 snRNAs are present in *Nosema locustae* (Fast et al. 1998). *Vairimorpha necatrix* also was found to possess a highly divergent U2 snRNA gene (DiMaria et al. 1996). Based on the assumption that introns arose after mitochondrial acquisition, Archezoa were thought to lack introns. This initially appeared to be true because no introns were found, although relatively few genes from microsporidia had been sequenced at the time. However, based on the presence of components of the spliceosome, it was reasoned that such machinery would not be maintained in the absence of introns and that introns were likely present at a low density. This prediction was later borne out when introns were found in genes from *Encephalitozoon cuniculi* (Biderre et al. 1998; Katinka et al. 2001).

In addition to the tubulin phylogenies, other molecular sequences began to be examined from microsporidia, and many corroborated a later origin for microsporidia. The LSU rDNA sequence from *Encephalitozoon cuniculi* was sequenced, and although its divergence resulted in a very long branch in the phylogenetic tree, it did not branch at the base of eukaryotes (Peyretaillade et al. 1998a). The significance of the specific branching position is questionable (the microsporidian branches with the alveolates), but the fact that the microsporidian sequence was not basal is noteworthy. The authors also examined the predicted structure of the *E. cuniculi* LSU rRNA and concluded that it is eukaryotic in nature, but is reduced in nature (Peyretaillade et al. 1998a). These results contradicted any suggestions that the LSU rRNA is part of a primitive, prokaryotelike fusion with the 5.8S rRNA and instead suggested that these characteristics are secondary derivations. Analysis of the sequence of the TATA box binding protein (TBP or TFIID) from *Nosema locustae* also suggested a late origin for the microsporidia (Fast et al. 1999). Moreover, TBP phylogenies possessed a consistent, albeit weak, topology where the microsporidian sequence was the sister to the fungal TBP homologs. At approximately the same time as the TBP analysis was undertaken, researchers began to focus on the amitochondrial nature of microsporidia and examined the sequences of mitochondrion-derived heat-shock protein (hsp)70 gene. The nature and implications of these discoveries are discussed later in this chapter.

Sequences of hsp70 genes from *Nosema locustae*, *Vairimorpha necatrix*, and *Encephalitozoon cuniculi* were determined independently at approximately the same time, and their phylogenetic affinities varied depending on the taxon sampling and method of analysis (Germot et al. 1997; Hirt et al. 1997; Peyretaillade et al. 1998b). All maximum likelihood (ML) analyses, however, revealed an affiliation between microsporidia and animals + fungi, although a specific relationship between the microsporidian hsp70 sequence and those of fungi was recovered only in ML analyses including *Vairimorpha* and *Nosema* homologs (Germot et al. 1997; Hirt et al. 1997; Peyretaillade et al. 1998b). Therefore, these analyses indicate that microsporidia are not among the earliest diverging eukaryotes and again reinforce the conclusion that there is a relationship between microsporidia and fungi.

Although support grew for a relationship between microsporidia and fungi, no molecular phylogeny supported this relationship so well as that exhibited by the tubulin genes. In 1999, Hirt and colleagues sequenced the largest subunit of RNA polymerase II (RPB1) from two microsporidia: *Nosema locustae* and *Vairimorpha necatrix*. Using sophisticated phylogenetic methods and testing the data with both fast-evolving and invariant sites removed, the microsporidian sequences branched firmly with the fungal representatives with high bootstrap support values ranging from 86% to 92%, a result that also was consistent with statistical tests of alternative tree topologies (Hirt et al. 1999).

In addition to providing strong support for the sisterhood of microsporidia and fungi based on RPB1, Hirt et al. (1999) reassessed the EF-1α and EF-2 data. Their aim was to see if the ancient origin for microsporidia still held when the molecular sequences were analyzed with updated phylogenetic methods and to investigate whether there are potential sources of phylogenetic artifact inherent in the microsporidian elongation factor sequences. In the case of EF-2, base compositional bias and long-branch attraction were determined to be responsible for the deep-branching position in the original analysis (Hirt et al. 1999). This conclusion was reached when archaebacterial outgroup sequences were removed (to deal with common amino acid biases) and the fast evolving sites also were removed (to avert some aspects of long branch attraction), and the sisterhood of microsporidia and fungi was recovered, albeit with weak bootstrap support. For EF-1α, pairwise comparisons of codon positions 1 and 3 between the microsporidian sequence and each other eukaryotic sequence revealed a high degree of substitution saturation (Hirt et al. 1999). Codon position 2 also showed a dramatic degree of substitution when compared with that of other eukaryotes (Hirt et al. 1999). Because any change in the second position of a codon codes for a different amino acid, the high rate of substitution at this position in the microsporidian EF-1α clearly indicates that the microsporidian EF-1α is evolving differently. Judging from these differences, it is unlikely that EF-1α phylogenies accurately resolved the position of microsporidia, the basal position occupied by the microsporidian sequence in EF-1α trees could in large part be due to substitution saturation. Clearly, the elongation factor data were problematic.

A similar situation was found in a reanalysis of rDNA sequences. A thorough analysis of among-site rate variation was undertaken for LSU rDNA sequences and, although previous analyses of microsporidian LSU sequences had predicted a later origin for this group, a specific relationship with fungi was not detected (Peyretaillade et al. 1998a; Van de Peer et al. 2000). However, when the substitution rate calibration method was used to compute evolutionary distances, the LSU sequences from *Encephalitozoon cuniculi* and *Nosema apis* did branch within the fungi (Van de Peer et al. 2000). Statistical tests of topologies grouping microsporidia with other eukaryotes were significantly worse than trees uniting them with fungi, with the exception of the topology in which microsporidia branched with the ciliates. The original phylogenetic analysis including a microsporidian LSU sequence also placed the microsporidian with the alveolates (a group that includes ciliates, apicomplexa, and dinoflagellates), perhaps because these taxa also have higher evolutionary rates for their rDNA sequences (Peyretaillade et al. 1998a). All in all, the reanalyses of these markers suggested that issues of evolutionary rates have caused serious problems in phylogenetic reconstruction using many microsporidian sequences. Although only α-tubulin, β-tubulin, and RPB1 phylogenies strongly supported a relationship between microsporidia and fungi, the congruence of multiple data sets strengthened the position. The demonstration that the basal position of the microsporidia was an artifact turned the tide of opinion. Microsporidia were no longer considered to be ancient eukaryotes, but instead highly derived fungi. For a summary of published phylogenetic results including microsporidian sequences, see figure 5.2.

Additional evidence continued to clarify the nature of the relationship between microsporidia and fungi. With the completion of the *Encephalitozoon cuniculi* genome, phylogenetic evidence continued to accumulate in support of the microsporidia—fungi relationship (Katinka et al. 2001). In particular, analysis of four protein-coding genes (seryl-tRNA synthetase, transcription initiation factor IIB, a GTP-binding protein, and the A subunit of the vacuolar ATPase) supported a relationship between microsporidia and fungi with bootstrap support values of 70% to 92% (Katinka et al. 2001). Yet another genome analysis assessed 103 protein

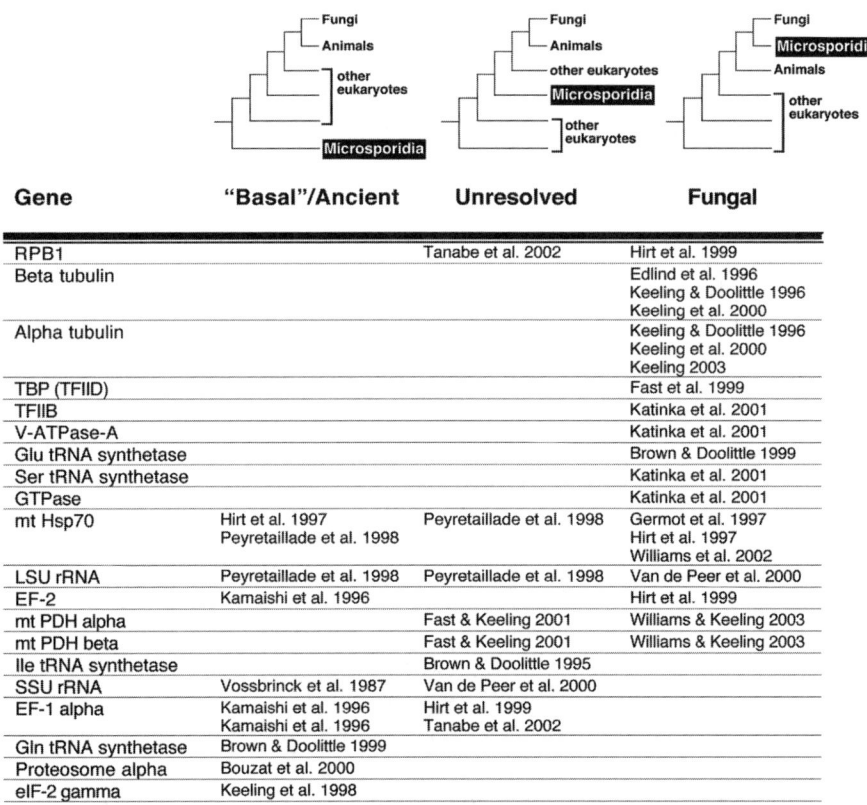

Gene	"Basal"/Ancient	Unresolved	Fungal
RPB1		Tanabe et al. 2002	Hirt et al. 1999
Beta tubulin			Edlind et al. 1996 Keeling & Doolittle 1996 Keeling et al. 2000
Alpha tubulin			Keeling & Doolittle 1996 Keeling et al. 2000 Keeling 2003
TBP (TFIID)			Fast et al. 1999
TFIIB			Katinka et al. 2001
V-ATPase-A			Katinka et al. 2001
Glu tRNA synthetase			Brown & Doolittle 1999
Ser tRNA synthetase			Katinka et al. 2001
GTPase			Katinka et al. 2001
mt Hsp70	Hirt et al. 1997 Peyretaillade et al. 1998	Peyretaillade et al. 1998	Germot et al. 1997 Hirt et al. 1997 Williams et al. 2002
LSU rRNA	Peyretaillade et al. 1998	Peyretaillade et al. 1998	Van de Peer et al. 2000
EF-2	Kamaishi et al. 1996		Hirt et al. 1999
mt PDH alpha		Fast & Keeling 2001	Williams & Keeling 2003
mt PDH beta		Fast & Keeling 2001	Williams & Keeling 2003
Ile tRNA synthetase		Brown & Doolittle 1995	
SSU rRNA	Vossbrinck et al. 1987	Van de Peer et al. 2000	
EF-1 alpha	Kamaishi et al. 1996 Kamaishi et al. 1996	Hirt et al. 1999 Tanabe et al. 2002	
Gln tRNA synthetase	Brown & Doolittle 1999		
Proteosome alpha	Bouzat et al. 2000		
eIF-2 gamma	Keeling et al. 1998		

Figure 5.2. Summary of alternative microsporidian origins based on published phylogenetic analyses.

sequences and found that, in concordance with previous studies, rate variation proved to be a problem in the majority of the phylogenies, but 19 proteins did support the microsporidia—fungi relationship when maximum likelihood optimality criteria were used (Thomarat 2000; Vivarès et al. 2002). Therefore, all available evidence, including evidence from a complete microsporidian genome, supports the relationship between microsporidia and fungi.

Fungal Sister or Bona Fide Fungus?

Although the mounting evidence convincingly indicated a relationship between microsporidia and fungi, in most cases phylogenies included only a single microsporidian sequence and a small smattering of ascomycete fungi. In addition, shared features (e.g., insertions in enolase and EF-1α sequences, presence of chitin and trehalose) unite microsporidia with animals and fungi, but not specifically with fungi (Van der Meer and Gochnauer 1971; Undeen et al. 1987; Kamaishi et al. 1996a; Katinka et al. 2001). With such a low level of sampling it is impossible to discern between two possibilities: (1) microsporidia are a sister-group to fungi, and (2) microsporidia branch from within the fungal radiation. At first glance this may appear to be a question of semantics, yet there is an immense difference between evolving from a protist ancestor of fungi and evolving from a true fungus. For comparison, if we pretended that there were a question about human evolution, the distinction would be akin to determining if humans are more closely related to animals or to choanoflagellates (the closest sister-group to animals)—no one would argue that there is a difference.

The best sampled molecules that have been used to address this question are the tubulins. The original analyses referred to in the previous section included few microsporidia and only representatives of the ascomycete and basidiomycete fungi (Edlind et al. 1996; Keeling and Doolittle 1996). Ascomycetes and basidiomycetes are derived, closely related fungal phyla. The early analyses did not include the other two phyla that are perceived to be basal, the zygomycetes and chytrids. Taxon sampling was broadened first for β-tubulin, by including sequences from several zygomycete and chytrid taxa and more microsporidians (Keeling et al. 2000). These phylogenetic analyses recovered the expected fungal relationships, where ascomycetes and basidiomycetes are sisters, and zygomycetes and chytrids are basal. In addition, the microsporidia branched within the fungal radiation, and not as sisters to it. Their specific position within the fungi was not well supported but showed a weak tendency to branch either as the sister to ascomycetes or within the zygomycetes, depending on taxa sampled in the analyses (Keeling et al. 2000). Additional microsporidian, zygomycete, and chytrid α- and β-tubulins were sampled in a subsequent analysis based on both individual and combined data sets (Keeling 2003). In the two individual trees, the basic relationships among the fungal phyla were resolved. Zygomycetes were paraphyletic in both individual tubulin trees, and the microsporidia formed a strong monophyletic group branching from within the zygomycetes.

Microsporidia showed some affiliation for entomophthoralean zygomycete sequences in both α- and β-tubulin trees. For α-tubulin, microsporidia grouped in a well-supported clade (82% bootstrap) including the Zoopagales (*Syncephalis*) and Entomophthorales (*Conidiobolus* and *Entomophaga*). For β-tubulin, the microsporidia branched with one member of the Entomophthorales (*Conidiobolus coronatus*), but the rest of the entomophthoralean representatives were in a clade with the mucoralean zygomycetes. Because the α- and β-tubulin data were largely overlapping, they were combined to increase the number of characters available for assessment. In the combined tree, the fungal topology was similar to that of previous studies, and the monophyly of fungi and microsporidia, and of ascomycetes, basidiomycetes, chytrids, and microsporidia individually were strongly supported. The combined tree also recovered polyphyletic zygomycetes with the microsporidia branching from within this fungal group (fig. 5.3). Mirroring the topology of the α-tubulin tree, the combined tree also recovered a microsporidia–Entomophthorales–Zoopagales clade that was moderately well supported by bootstrapping and statistical tests of alternative topologies.

The more recent analyses clearly supported a hypothesis of evolution of microsporidia from a bona fide fungus and suggested that microsporidia are derived fungi possibly related to the Entomophthorales or Zoopagales (Keeling 2003). Although it is premature to state such an affiliation with certainty, it is tempting to speculate about additional potential similarities between the groups. Both groups of zygomycetes (Entomophthorales and Zoopagales) include insect, other invertebrate animal, and fungal parasites. Use of invertebrate hosts (Tanabe et al. 2000) also occurs in the putatively basal lineages of microsporidia. In addition, a potential correspondence has been noted in the manner in which the entomophthoralean *Conidiobolus* disperses spores and in the eversion of the microsporidian polar tube (Keeling et al. 2000). However, caution should be taken when interpreting these seeming similarities. Speculation has been based on simple morphologies of polar tubes of microsporidia and of Actinomyxidia and Myxosporidia and the apical spore body of the harpellalean trichomycete zygomycetes and the polar tube of microsporidia (Kudo 1947; Lom and Vávra 1962; Cavalier-Smith 1998). Actinomyxidia and Myxosporidia have been ruled out as relatives of microsporidia, and there is no evidence at this time to support a specific relationship between the zygomycete group Harpellales and the microsporidia, although this should not be ruled out at this early stage.

Tubulin is not without problems as a phylogenetic marker, especially when it comes to the nonflagellated fungi; however, it is currently our best sampled molecule and therefore provides our best phylogenetic estimates to date. Bearing this in mind, there is still some skepticism regarding the exact nature of the relationship between microsporidia and fungi. Partial EF-1α and RPB1 sequences for zygomycetes and chytrids led several workers to conclude that microsporidia are not degenerate fungi and that a sister relationship between microsporidia and fungi still has not been resolved (Tanabe et al. 2002). As we have reviewed, EF-1α probably is not a useful phylogenetic marker for this question because of the fast evolutionary rate and site saturation. Tanabe et al. (2002) mentioned the 11–12 amino acid insertion in EF-1α that unites microsporidia and animals + fungi and also described a two amino acid deletion that is specific to all fungi but is absent in the two microsporidian sequences known. Using a parsimony argument, the authors claim that this indel argues against microsporidia evolving from within fungi. They

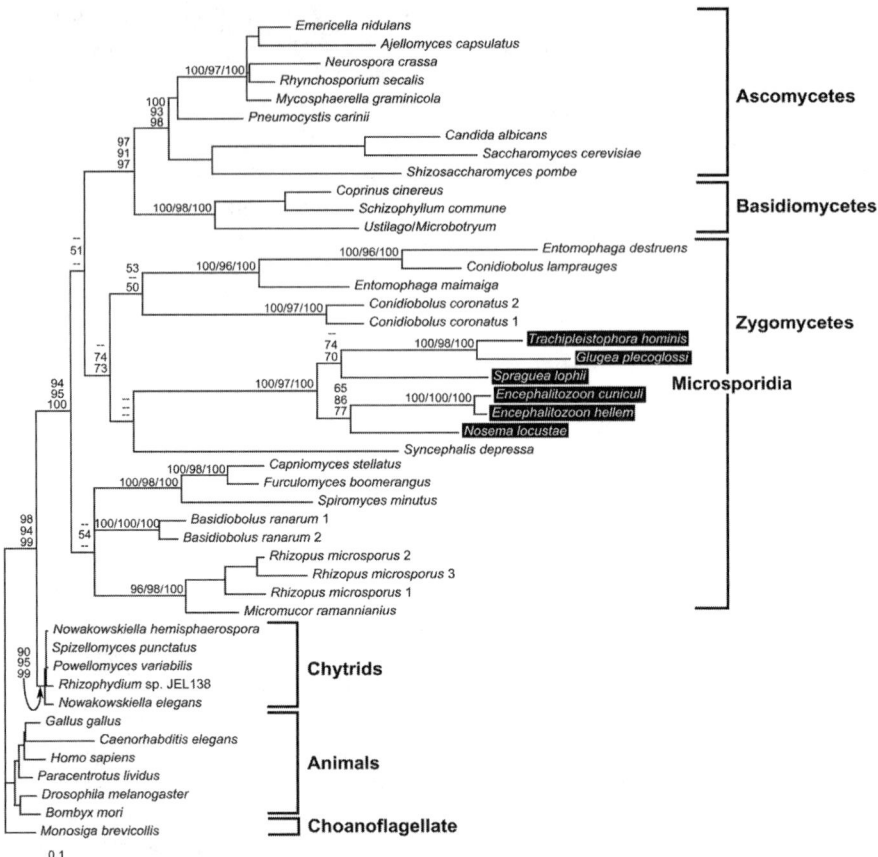

Figure 5.3. Phylogeny of combined α- and β-tubulin sequences from microsporidia, fungi, animals, and choanoflagellates. A protein maximum likelihood tree was inferred from the alignment used in Keeling (2003). Bootstrap proportions are shown for 100 resampled replicates analyzed with protein maximum likelihood, weighted neighbor-joining, and Fitch-Margoliash methods.

reasoned that if the deletion occurred in the ancestor of all fungi, it is not likely that the insertion was gained independently in microsporidia. Absence of the deletion would group microsporidia with the majority of other eukaryotes. However, the authors did point out variation in the indel status among some ciliate EF-1α sequences. In the RPB1 phylogenies discussed, the backbone of the tree was weakly supported, and fungal monophyly received weak support (Tanabe et al. 2002). As the authors mentioned, they restricted their analysis to the characters available from partial PCR products (301 characters) that may not provide the required resolution. The original RPB1 analysis that showed strong support for a microsporidia–fungi relationship used full-length sequences and 760 characters (Hirt et al. 1999). Moreover, the statistical tests of alternative topologies used in this study did not exclude a relationship between microsporidia and fungi for either molecule. Altogether, these phylogenetic results are equivocal. In contrast, the two amino acid fungal-specific indel that unites fungi to the exclusion of microsporidia is provocative, as it is at odds with the apparently strong results emerging from the tubulin analyses that may be biased by the highly divergent nature of the tubulin sequences. It will be interesting to see how the EF-1α indel holds up as sequences of more microsporidia and fungi become available and are analyzed thoroughly.

Cellular Organelles and Microsporidian Origins: Cryptic Mitochondria

When microsporidia were viewed as Archezoa, our understanding of their origin and of their amitochondriate nature were inextricably linked, and it was clear that their metabolism was different from other eukaryotes. Experiments with pure and germinated microsporidian spore material found no evidence for the mitochondrial pathways, tricarboxylic acid (TCA) cycle or oxidative phosphorylation (Dolgikh et al. 1997; Weidner et al. 1999). In addition, early ultrastructural studies could not identify any structure that appeared to be a typical mitochondrion (Vávra 1965). However, when analyses of α- and β-tubulin phylogenies suggested that microsporidia were related to fungi, their amitochondriate nature was questioned, since this evidence suggested strongly that the ancestors of microsporidia (i.e., the ancestor of fungi, or animals + fungi) must have had mitochondria.

A mitochondrial heritage for microsporidia was supported using an approach that had successfully detected such a history in *Entamoeba* and *Trichomonas*, two other lineages originally predicted to be amitochondriate (Clark and Roger 1995; Bui et al. 1996; Germot et al. 1996; Roger et al. 1996; Mai et al. 1999; Tovar et al. 1999). Most mitochondrial proteins are derived from the α-proteobacterial endosymbiont that gave rise to the organelle, but the majority of the genes encoding these proteins have been transferred to the host nuclear genome. The genes are transcribed and translated on cytosolic ribosomes, and their products are post-translationally targeted to the mitochondrion using an N-terminal extension called a transit peptide (Williams and Keeling 2003). One of the nucleus-encoded, mitochondrion-targeted proteins is Hsp70, mentioned earlier. Hsp70 is a chaperone that aids in stabilizing newly made proteins and assists in their movement across membranes. Different copies of Hsp70 function in different compartments of the cell; there are cytosolic, endoplasmic reticular, mitochondrial, and chloroplast copies. The mitochondrial and chloroplast copies are most closely related to α-proteobacterial and cyanobacterial sequences, respectively, due to the nature of the organellar proto-endosymbionts (Williams and Keeling 2003). Mitochondrial cpn60 and hsp70 genes were the target of a search for mitochondria transformed beyond recognition. Three research groups examined four microsporidian species independently at approximately the same time and discovered that they all possessed nuclear-encoded copies of mitochondrial hsp70 (Germot et al. 1997; Hirt et al. 1997; Peyretaillade et al. 1998b). Phylogenetic analysis grouped the sequences in a well-supported clade containing mitochondrial sequences of other eukaryotes (fig. 5.4); this clade was,

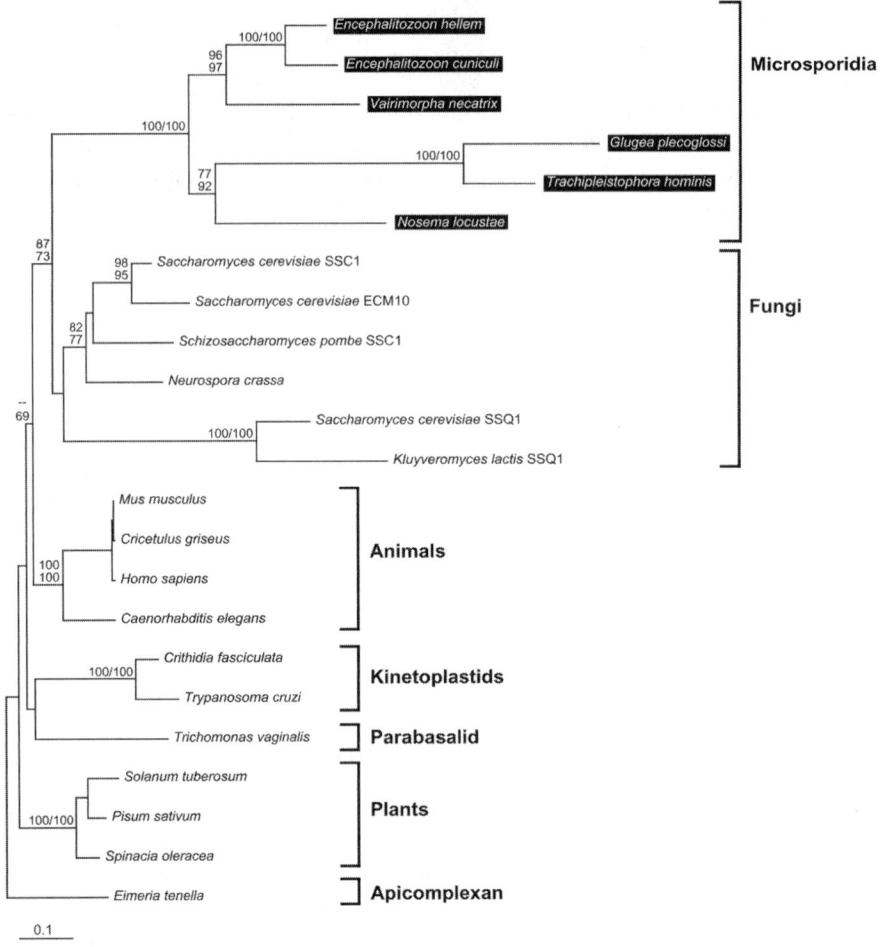

Figure 5.4. Phylogeny of eukaryotic mitochondrial Hsp70 sequences. A weighted neighbor-joining tree is shown. Bootstrap values greater than 65% are shown for weighted neighbor-joining and protein maximum likelihood replicates. Gamma-corrected distances were used as described in Fast and Keeling (2001). Phylogenies were inferred based on the alignment of Williams et al. (2002).

in turn, related to α-proteobacterial homologs. The presence of a gene of mitochondrial origin in microsporidia was interpreted to mean that microsporidia are secondarily amitochondriate, evidence in agreement with a derived placement of the group (Germot et al. 1997; Hirt et al. 1997; Peyretaillade et al. 1998b). All members of the Archezoa were found to be secondarily amitochondriate, and there is currently no evidence for any extant ancestrally amitochondriate eukaryotic lineage (reviewed in Williams and Keeling 2003).

The presence of relic mitochondrial genes in microsporidia has clarified their mitochondrial ancestry but has left unanswered questions about the fate of the organelle. As nucleus-encoded proteins that function in the mitochondrion and possess an N-terminal transit peptide that targets them to the organelle, the hsp70 sequences from microsporidia were inspected thoroughly for evidence of such a signal. Results were ambiguous, however, as the N-terminal extensions were extremely short (6–11 amino acids) when compared to mitochondrial targeting signals of other eukaryotes (Germot et al. 1997; Hirt et al. 1997; Peyretaillade et al. 1998b).

Gene sequences continued to accumulate suggesting the presence of a mitochondrion. Alpha and beta subunits of the pyruvate dehydrogenase complex E1 (PDH-E1) were isolated from the microsporidian *Nosema locustae* and identified in the completed genome of *Encephalitozoon cuniculi* (Fast and Keeling 2001; Katinka et al. 2001). Pyruvate dehydrogenase is a key enzyme complex in eukaryotic mitochondria and is responsible for oxidative decarboxylation of pyruvate upon entry into the mitochondrion, converting it to acetyl-CoA for use in the TCA cycle. There are three main subunits of the complex (E1, E2, E3). PDH-E1 decarboxylates pyruvate and converts it to the active aldehyde intermediate, 2-α-hydroxyethyl-thiamine pyrophosphate (HETPP), which is then converted into acetyl-CoA by the actions of PDH-E2 and PDH-E3. The presence of the PDH-E1 subunits was the first indication of retention of mitochondrial metabolism by microsporidia (Fast and Keeling 2001). However, *Encephalitozoon* lacks genes for the E2 and E3 subunits, indicating that PDH in microsporidia is operating in a manner unique among eukaryotes (Katinka et al. 2001). This is particularly true as other organisms that appear to be secondarily amitochondriate do not possess PDH, but instead decarboxylate pyruvate with a different enzyme, pyruvate:ferredoxin oxidoreductase (PFOR). Since both PFOR and PDH use HETPP as an intermediate, and biochemical and structural similarities have been noted between the enzymes, it is possible that microsporidian PDH is acting in a similar manner to PFOR. PFOR is an iron–sulfur protein that transfers electrons from pyruvate to ferredoxin. Therefore, it is possible that microsporidian PDH is creating HETPP to reduce a different iron–sulfur protein, which can go on to reduce ferredoxin. Other evidence for a pathway such as this comes from the *Encephalitozoon* genome, which contains genes for iron–sulfur center assembling proteins of α-proteobacterial origin, as well as ferredoxin and ferredoxin:NADH oxidoreductase (Katinka et al. 2001). Although these results clearly indicate the maintenance of mitochondrial-derived metabolism in microsporidia, they provide no direct evidence for the presence of the organelle itself. Indeed, no N-terminal targeting signals could be detected on the PDH subunits or the iron-sulfur assembling proteins (Fast and Keeling 2001).

Clear evidence for a mitochondrial organelle came with immunolocalization of Hsp70 in the microsporidian *Trachipleistophora hominis* (Williams et al. 2002). Specific antibodies raised against the putative mitochondrial Hsp70 were used in conjunction with immunofluorescence to identify several small, discretely stained bodies in the meront (dividing) life-stage of *Trachipleistophora*. Further examination with electron microscopy and immunogold localization experiments on thin sections identified Hsp70 within double membrane-bound vesicles. More accurate measurements of the structures were possible with electron microscopy and showed them to be approximately 90 nm long and 50 nm wide. At such a minute size, these mitochondrionlike organelles are the smallest among eukaryotic mitochondria

(Williams et al. 2002). Even the smallest known eukaryote, a picoplanktonic alga, possesses mitochondria that are substantially larger (Chretiennot-Dinet et al. 1995).

The abundance of mitochondria in *Trachipleistophora* meronts may indicate the important metabolic role they play within the cell (Williams et al. 2002). Clues about the role could be deduced from the complete genome of *Encephalitozoon cuniculi*. Mitotracker Red, which detects the membrane potential across the membranes of typical eukaryotic mitochondria, does not identify the microsporidian mitochondrial-like organelles, and there are no genes encoding F_0F_1 ATPases in the *Encephalitozoon* genome (Katinka et al. 2001; Williams et al. 2002). Genes for the electron transport chain and TCA cycle also are absent, in line with previous biochemical work (Dolgikh et al. 1997; Weidner et al. 1999; Katinka et al. 2001). The presence of PDH, the glycerol-3-phosphate shuttle, and ferredoxin strongly suggest the presence of pyruvate metabolism and electron shuttling. The presence of iron-sulfur cluster assembly proteins in the *Encephalitozoon* genome also is consistent with a potentially central role for the organelle, perhaps assembling iron-sulfur clusters for export to the cytosol. Although the determination of the exact metabolic function of the microsporidian mitochondrion awaits further research, it is clear that its function is different from typical mitochondria, and its role is substantially different from other amitochondriate protists such as *Giardia*, *Entamoeba*, and *Trichomonas* (see Williams and Keeling 2003 for a review of cryptic mitochondria in these groups).

Although there is now physical evidence for a mitochondrionlike organelle in microsporidia and a fairly clear picture of the metabolic activities taking place within the *E. cuniculi* organelle, the many facets of targeting to the organelle remain unknown. Given that all of the hsp70 sequences and most of the other *Encephalitozoon* sequences thought to encode proteins targeted to the organelle lack any appreciable N-terminal extensions (Germot et al. 1997; Hirt et al. 1997; Peyretaillade et al. 1998b; Katinka et al. 2001; Fast and Keeling 2001; Williams et al. 2002), and only a few of the proteins typically found in the translocation complex have been identified in the *Encephalitozoon* genome (Katinka et al. 2001), it would appear that organellar targeting occurs by an unusual variant of the standard targeting system. From their biochemistry to their ultrastructure, all aspects of microsporidian biology have been reduced, and perhaps organelle targeting signals in microsporidia are no different.

Future Prospects: Comparative Genomics of Microsporidia

The complete sequence of the genome of *Encephalitozoon cuniculi* (~2.9 Mbp) provided a wealth of information on the evolutionary origins of microsporidia by greatly increasing the number of phylogenetic markers that specify a fungal origin for microsporidia (Katinka et al. 2001). Early insights gained from the genome have not been limited to implications about the fungal nature of microsporidia; the content of the *E. cuniculi* genome has revealed a great deal about many aspects of microsporidian biology, and its structure is also of interest in its own right. Microsporidian genomes are very small by eukaryotic standards; the smallest is estimated to be 2.3 Mbp, making it smaller than many bacterial genomes (Biderre et al. 1994; Peyretaillade et al. 1998a). Microsporidian genomes, having evolved from larger fungal genomes, are a model for genome reduction and compaction. For example, the distance between adjacent genes is very short (the mean intergenic distance in *E. cuniculi* is only 129 bp), and even the genes themselves are on average significantly shorter than their homologs in *Saccharomyces* (Katinka et al. 2001).

The detailed analysis of the *E. cuniculi* genome is only just beginning, but it is already clear that this information will have a major impact on microsporidian research, and the future of microsporidian genomes holds greater promise. With such small genomes and high density of coding regions, the microsporidia are an ideal target for eukaryotic comparative genomics. Indeed, genome sequence surveys of *Nosema locustae* and *Spraguea lophii* have already been carried out (Hinkle et al. 1997; Fast and Keeling 2001), and the *N. locustae* genome is expected to be completed in the near future. These are important sources of information about the process and effects of nuclear genome reduction and compaction at its most extreme. While *E. cuniculi* and *N. locustae* are not very closely related microsporidia, they both have small genomes, so they are well suited for comparisons. On the other hand, a number of microsporidia have genomes that are many times larger than those of *E. cuniculi* or *N. locustae*. Exploring the structure and content of these more complex genomes offers perhaps the most exciting look at the pressures shaping genome reduction in these parasites and how they evolved from their fungal ancestors.

Acknowledgments   We thank B. A. P. Williams for the Hsp70 alignment. Microsporidian research in the Keeling laboratory is supported by a New Investigator Award in Pathogenic Mycology from the Burroughs-Wellcome Fund and a grant from the Canadian Institutes of Health Research (CIHR). N.M.F. is supported by postdoctoral fellowships from the CIHR and the Michael Smith Foundation for Health Research (MSFHR). P.J.K. is a scholar of the Canadian Institute for Advanced Research (CIAR) Evolutionary Biology Program, CIHR, and MSFHR.

Literature Cited

Balbiani, G. 1882. Sur les microsporidies ou sporospermies des articules. *Comptes Rendus de l'Association des Sciences* 95:1168–1171.

Baldauf, S. L., and J. D. Palmer. 1993. Animals and fungi are each other's closest relatives: Congruent evidence from multiple proteins. *Proceedings of the National Academy of Sciences of the USA* 90:11558–11562.

Baldauf, S. L., A. J. Roger, I. Wenk-Siefert, and W. F. Doolittle. 2000. A kingdom-level phylogeny of eukaryotes based on combined protein data. *Science* 290:972–977.

Becnel, J. J., and T. G. Andreadis. 1999. Microsporidia of insects. In *The microsporidia and microsporidiosis*, ed. M. Wittner and L. M. Weiss, pp. 447–501. Washington, DC: ASM Press.

Biderre, C., G. Méténier, and C. P. Vivarès. 1998. A small spliceosomal-type intron occurs in a ribosomal protein gene of the microsporidian *Encephalitozoon cuniculi*. *Molecular Biochemistry and Parasitology* 94:283–286.

Biderre, C., M. Pagès, G. Méténier, D. David, J. Bata, G. Prensier, and C. P. Vivarès. 1994.

On small genomes in eukaryotic organisms: Molecular karyotypes of two microsporidian species (Protozoa) parasites of vertebrates. *Comptes Rendus de l'Academie des Sciences III* 317:399–404.

Brown, J. R., and W. F. Doolittle. 1995. Root of the universal tree of life based on ancient aminoacyl-tRNA synthetase gene duplications. *Proceedings of the National Academy of Sciences of the USA* 92:2441–2445.

Bui, E. T., P. J. Bradley, and P. J. Johnson. 1996. A common evolutionary origin for mitochondria and hydrogenosomes. *Proceedings of the National Academy of Sciences of the USA* 93:9651–9656.

Canning, E. U., J. Lom, and J. P. Nicholas. 1982. Genus *Glugea* Thelohane, 1891 (Phylum Microspora): redescription of the type species *Glugea anomala* (Moniez, 1887) and recognition of its sporogonic development within sporophorous vesicles (pansporoblastic membranes). *Protistologica* 18:193–210.

Cavalier-Smith, T. 1983. A 6-kingdom classification and a unified phylogeny. In *Endocytobiology II: Intracellular space as oligogenetic*, ed. H. E. A. Schenk and W. S. Schwemmler, pp. 1027–1034. Berlin: Walter de Gruyter.

Cavalier-Smith, T. 1993. Kingdom protozoa and its 18 phyla. *Microbiologial Reviews* 57:953–994.

Cavalier-Smith, T. 1998. A revised six-kingdom system of life. *Biological Reviews* 73:203–266.

Chretiennot-Dinet, M. J., C. Courties, A. Vaquer, J. Neveux, H. Claustre, J. Lautier, and M. C. Machado. 1995. A new marine picoeukaryote: *Ostreococcus tauri* gen. et sp.nov. (Chlorophyta, Prasinophycae). *Phycologia* 34:285–292.

Clark, C. G., and A. J. Roger. 1995. Direct evidence for secondary loss of mitochondria in *Entamoeba histolytica*. *Proceedings of the National Academy of Sciences of the USA* 92:6518–6521.

Curgy, J. J., J. Vávra, and C. P. Vivarès. 1980. Presence of ribosomal RNAs with prokaryotic properties in Microsporidia, eukaryotic organisms. *Biologie Cellulaire* 38:49–51.

DiMaria, P., B. Palic, B. A. Debrunner-Vossbrinck, J. Lapp, and C. R. Vossbrinck. 1996. Characterization of the highly divergent U2 RNA homolog in the microsporidian *Vairimorpha necatrix*. *Nucleic Acids Research* 24:515–522.

Dolgikh, V. V., J. J. Sokolova, and I. V. Issi. 1997. Activities of enzymes of carbohydrate and energy metabolism of the spores of the microsporidian, *Nosema grylli*. *Journal of Eukaryotic Microbiology* 44:246–249.

Edlind, T. D., J. Li, G. S. Visvesvara, M. H. Vodkin, G. L. McLaughlin, and S. K. Katiyar. 1996. Phylogenetic analysis of beta-tubulin sequences from amitochondrial protozoa. *Molecular Phylogenetics and Evolution* 5:359–367.

Fast, N. M., and P. J. Keeling. 2001. Alpha and beta subunits of pyruvate dehydrogenase E1 from the microsporidian *Nosema locustae*: Mitochondrion-derived carbon metabolism in microsporidia. *Molecular Biochemistry and Parasitology* 117:201–209.

Fast, N. M., J. M. Logsdon, Jr., and W. F. Doolittle. 1999. Phylogenetic analysis of the TATA box binding protein (TBP) gene from *Nosema locustae*: Evidence for a microsporidia-fungi relationship and spliceosomal intron loss. *Molecular Biology and Evolution* 16:1415–1419.

Fast, N. M., A. J. Roger, C. A. Richardson, and W. F. Doolittle. 1998. U2 and U6 snRNA genes in the microsporidian *Nosema locustae*: Evidence for a functional spliceosome. *Nucleic Acids Research* 26:3202–3207.

Felsenstein, J. 1978. Cases in which parsimony and compatibility methods will be positively misleading. *Systematic Zooology* 27:401–410.

Franzen, C., and A. Muller. 2001. Microsporidiosis: Human diseases and diagnosis. *Microbes and Infection* 3:389–400.

Frixione, E., L. Ruiz, J. Cerbon, and A. H. Undeen. 1997. Germination of *Nosema algerae* (Microspora) spores: conditional inhibition by D2O, ethanol and Hg2+ suggests dependence of water influx upon membrane hydration and specific transmembrane pathways. *Journal of Eukaryotic Microbiology* 44:109–116.

Frixione, E., L. Ruiz, M. Santillan, L. V. de Vargas, J. M. Tejero, and A. H. Undeen. 1992. Dynamics of polar filament discharge and sporoplasm expulsion by microsporidian spores. *Cell Motility and the Cytoskeleton* 22:38–50.

Germot, A., H. Philippe, and H. Le Guyader. 1996. Presence of a mitochondrial-type 70-kDa heat shock protein in *Trichomonas vaginalis* suggests a very early mitochondrial endosymbiosis in eukaryotes. *Proceedings of the National Academy of Sciences of the USA* 93:14614–14617.

Germot, A., H. Philippe, and H. Le Guyader. 1997. Evidence for loss of mitochondria in Microsporidia from a mitochondrial- type HSP70 in *Nosema locustae*. *Molecular Biochemistry and Parasitology* 87:159–168.

Hendrick, R. P., J. M. Groff, and D. V. Baxa. 1991. Experimental infections with *Nucleospora salmonis* n. g., n. s.: An intranuclear microsporidium from chinook salmon (*Oncorhynchus tshawitscha*). *American Fisheries Society Newsletter* 19:5.

Hinkle, G., H. G. Morrison, and M. L. Sogin. 1997. Genes coding for reverse transcriptase, DNA-directed RNA polymerase, and chitin synthase from the microsporidian *Spraguea lophii*. *Biological Bulletin* 193:250–251.

Hirt, R. P., B. Healy, C. R. Vossbrinck, E. U. Canning, and T. M. Embley. 1997. A mitochondrial Hsp70 orthologue in *Vairimorpha necatrix*: Molecular evidence that microsporidia once contained mitochondria. *Current Biology* 7:995–958.

Hirt, R. P., J. M. Logsdon, Jr., B. Healy, M. W. Dorey, W. F. Doolittle, and T. M. Embley. 1999. Microsporidia are related to Fungi: Evidence from the largest subunit of RNA polymerase II and other proteins. *Proceedings of the National Academy of Sciences of the USA* 96:580–585.

Ishihara, R., and Y. Hayashi. 1968. Some properties of ribosomes from the sporoplasm of *Nosema bombycis*. *Journal of Invertebrate Pathology* 11:377–385.

Kamaishi, T., T. Hashimoto, Y. Nakamura, F. Nakamura, S. Murata, N. Okada, K.-I. Okamoto, M. Shimzu, and M. Hasegawa. 1996a. Protein phylogeny of translation elongation factor EF-1a suggests microsporidians are extremely ancient eukaryotes. *Journal of Molecular Evolution* 42:257–263.

Kamaishi, T., T. Hashimoto, Y. Nakamura, Y. Masuda, F. Nakamura, K. Okamoto, M. Shimizu, and M. Hasegawa. 1996b. Complete nucleotide sequences of the genes encoding translation elongation factors 1 alpha and 2 from a microsporidian parasite, *Glugea plecoglossi*: Implications for the deepest branching of eukaryotes. *Journal of Biochemistry* 120:1095–1103.

Katinka, M. D., S. Duprat, E. Cornillot, G. Méténier, F. Thomarat, G. Prenier, V. Barbe, E. Peyretaillade, P. Brottier, P. Wincker, F. Delbac, H. El Alaoui, P. Peyret, W. Saurin, M. Gouy, J. Weissenbach, and C. P. Vivarès. 2001. Genome sequence and gene compaction of the eukaryote parasite *Encephalitozoon cuniculi*. *Nature* 414:450–453.

Keeling, P. J. 2003. Congruent evidence from alpha-tubulin and beta-tubulin gene phylogenies for a zygomycete origin of microsporidia. *Fungal Genetics and Biology* 38:298–309.

Keeling, P. J., and W. F. Doolittle. 1996. Alpha-tubulin from early-diverging eukaryotic lineages and the evolution of the tubulin family. *Molecular Biology and Evolution* 13:1297–1305.

Keeling, P. J., M. A. Luker, and J. D. Palmer. 2000. Evidence from beta-tubulin phylogeny that microsporidia evolved from within the fungi. *Molecular Biology and Evolution* 17:23–31.

Keeling, P. J., and G. I. McFadden. 1998. The origins of microsporidia. *Trends in Microbiology* 6:19–23.

Keeling, P. J., and J. D. Palmer. 2001. Lateral transfer at the gene and subgenic levels in the evolution of eukaryotic enolase. *Proceedings of the National Academy of Sciences of the USA* 98:10745–10750.

Keohane, E. M., and L. M. Weiss. 1998. Characterization and function of the microsporidian polar tube: A review. *Folia Parasitologica* 45:117–127.

Keohane, E. M., and L. M. Weiss. 1999. The structure, function, and composition of the microsporidian polar tube. In *The microsporidia and microsporidiosis*, ed. M. Wittner and L. M. Weiss, pp. 196–224. Washington, DC: ASM Press.

Kudo, R. R. 1918. Experiments on the extrusion of polar filaments of cnidosporidian spores. *Journal of Parasitology* 4:141–147.

Kudo, R. R. 1947. *Protozoology*. Springfield, IL: Charles C. Thomas.

Larsson, J. I. R., D. Ebert, J. Vávra, and V. N. Voronin. 1996. Redescription of *Pleistophora intestinalis* Chatton, 1907, a microsporidian parasite of *Daphnia magna* and *Daphnia pulex*, with establishment of a new genus Glugoides (Microspora, Glugeigae). *European Journal of Protistology* 32:251–261.

Levine, N. D., J. O. Corliss, F. E. Cox, G. Deroux, J. Grain, B. M. Honigberg, G. F. Leedale, A. R. d Loeblich, J. Lom, D. Lynn, E. G. Merinfeld, F. C. Page, G. Poljansky, V. Sprague, J. Vavra, and F. G. Wallace. 1980. A newly revised classification of the protozoa. *Journal of Protozoology* 27:37–58.

Lom, J., and J. Vávra. 1962. A proposal to the classification within the subphylum Cnidospora. *Systematic Zoology* 11:172–175.

Lom, J., and J. Vávra. 1963. The mode of sporoplasm extrusion in microsporidian spores. *Acta Protozoology* 1:81–92.

Mai, Z., S. Ghosh, M. Frisardi, B. Rosenthal, R. Rogers, and J. Samuelson. 1999. Hsp60 is targeted to a cryptic mitochondrion-derived organelle ("crypton") in the microaerophilic protozoan parasite *Entamoeba histolytica*. *Molecular and Cellular Biology* 19:2198–2205.

Nägeli, K. 1857. Über die neue Krankheit der Seidenraupe und verwandte Organismen. *Botanica Zeitung* 15:760–761.

Oshima, K. 1937. On the function of the polar filament in *Nosema bombycis*. *Parasitology* 29:220–224.

Peyretaillade, E., C. Biderre, P. Peyret, F. Duffieux, G. Méténier, M. Gouy, B. Michot, and C. P. Vivarès. 1998a. Microsporidian *Encephalitozoon cuniculi*, a unicellular eukaryote with an unusual chromosomal dispersion of ribosomal genes and a LSU rRNA reduced to the universal core. *Nucleic Acids Research* 26:3513–3520.

Peyretaillade, E., V. Broussolle, P. Peyret, G. Méténier, M. Gouy, and C. P. Vivarès. 1998b. Microsporidia, amitochondrial protists, possess a 70-kDa heat shock protein gene of mitochondrial evolutionary origin. *Molecular Biology and Evolution* 15:683–689.

Roger, A. J., C. G. Clark, and W. F. Doolittle. 1996. A possible mitochondrial gene in the early-branching amitochondriate protist Trichomonas vaginalis. *Proceedings of the National Academy of Sciences of the USA* 93:14618–14622.

Shaw, R. W., and M. L. Kent. 1999. Fish microsporidia. In *The Microsporidia and microsporidiosis*, ed. M. Wittner and L. M. Weiss, pp. 418–446. Washingon, DC: ASM Press.

Sprague, V., and J. J. Becnel. 1999. Checklist of available generic names for Microsporidia with type species and type hosts. In *The Microsporidia and microsporidiosis*, ed. M. Wittner and L. M. Weiss, pp. 517–530. Washingon, DC: ASM Press.

Sprague, V., J. J. Becnel, and E. I. Hazard. 1992. Taxonomy of phylum Microspora. *Critical Reviews in Microbiology* 18:285–395.

Takvorian, P. M., and A. Cali. 1994. Enzyme histochemical identification of the Golgi apparatus in the microsporidian, *Glugea stephani*. *Journal of Eukaryotic Microbiology* 41:63S–64S.

Takvorian, P. M., and A. Cali. 1996. Polar tube formation and nucleoside diphosphatase activity in the microsporidian, *Glugea stephani*. *Journal of Eukaryotic Microbiology* 43:102S–103S.

Tanabe, Y., K. O'Donnell, M. Saikawa, and J. Sugiyama. 2000. Molecular phylogeny of parasitic Zygomycota (Dimargaritales, Zoopagales) based on nuclear small subunit ribosomal DNA sequences. *Molecular Phylogenetics and Evolution* 16:253–262.

Tanabe, Y., M. M. Watanabe, and J. Sugiyama. 2002. Are microsporidia really related to fungi? A reappraisal based on additional gene sequences from basal fungi. *Mycological Research* 106:1380–1391.

Thomarat, F. 2000. Phylogenetic analysis of the complete genome of the microsporidian *Encephalitozoon cuniculi*. PhD thesis, Universite Lyon, France.

Tovar, J., A. Fischer, and C. G. Clark. 1999. The mitosome, a novel organelle related to mitochondria in the amitochondrial parasite *Entamoeba histolytica*. *Molecular Microbiology* 32:1013–1021.

Undeen, A. H., L. M. Elgazzar, R. K. Vander Meer, and S. Narang. 1987. Trehalose levels and trehalase activity in germinated and ungerminated spores of *Nosema algerae* (Microspora: Nosematidae). *Journal of Invertebrate Pathology* 50:230–237.

Undeen, A. H., and N. D. Epsky. 1990. *In vitro* and *in vivo* germination of *Nosema locustae* (Microsporidia: Nosematidae) spores. *Journal of Invertebrate Pathology* 56:371–379.

Undeen, A. H., and E. Frixione. 1990. The role of osmotic pressure in the germination of *Nosema algerae* spores. *Journal of Protozoology* 37:561–567.

Undeen, A. H., and E. Frixione. 1991. Structural alteration of the plasma membrane in spores of the microsporidium *Nosema algerae* on germination. *Journal of Protozoology* 38:511–518.

Undeen, A. H., and R. K. Vander Meer. 1994. Conversion of intrasporal trehalose into reducing sugars during germination of *Nosema algerae* (Protista: Microspora) spores: A quantitative study. *Journal of Eukaryote Microbiology* 41:129–132.

Van de Peer, Y., A. Ben Ali, and A. Meyer. 2000. Microsporidia: Accumulating molecular evidence that a group of amitochondriate and suspectedly primitive eukaryotes are just curious fungi. *Gene* 246:1–8.

Van der Meer, J. W., and T. A. Gochnauer. 1971. Trehalose activity associated with spores of *Nosema apis*. *Journal of Invertebrate Pathology* 17:38–41.

Vávra, J. 1965. Étude au microscope électronique de la morphologie et du développement de quelques microsporidies. *Comptes Rendus de l'Academie des Sciences* 261:3467–3470.

Vávra, J., and J. I. R. Larsson. 1999. Structure of the microsporidia. In *The Microsporidia and microsporidiosis*, ed. M. Wittner and L. M. Weiss, pp. 7–84. Washington, DC: ASM Press.

Vinckier, D., E. Porchet, E. Vivier, J. Vávra, and G. Torpier. 1993. A freeze-fracture study of the Microsporidia (Protozoa: Microspora). II. The extrusion apparatus: Polar filament polaroplast, posterior vacuole. *European Journal of Biochemistry* 29:370–380.

Vivarès, C. P., M. Gouy, F. Thomarat, and G. Metenier. 2002. Functional and evolutionary analysis of a eukaryotic parasitic genome. *Current Opinions in Microbiology* 5:499–505.

Vivier, E. 1975. The Microsporidia of the protozoa. *Protistology* 11:345–361.

Vossbrinck, C. R., J. V. Maddox, S. Friedman, B. A. Debrunner-Vossbrinck, and C. R. Woese. 1987. Ribosomal RNA sequence suggests Microsporidia are extremely ancient eukaryotes. *Nature* 326:411–414.

Vossbrinck, C. R., and C. R. Woese. 1986. Eukaryotic ribosomes that lack a 5.8S RNA. *Nature* 320:287–288.

Weber, R., and R. T. Bryan. 1994. Microsporidial infections in immunodeficient and immunocompetent patients. *Clinical and Infectious Diseases* 19:517–521.

Weber, R., R. T. Bryan, D. A. Schwartz, and R. L. Owen. 1994. Human microsporidial infections. *Clinical Microbiology Reviews* 7:426–461.

Weber, R., D. A. Schwartz, and P. Deplazes. 1999. Laboratory diagnosis of microsporidiosis. In *The Microsporidia and microsporidiosis*, ed. M. Wittner and L. M. Weiss, pp. 315–362. Washington, DC: ASM Press.

Weidner, E., W. Byrd, A Scarbourough, J. Pleshinger, and D. Sibley. 1984. Microsporidian spore discharge and the transfer of polaroplast organelle into plasma membrane. *Journal of Protozoology* 31:195–198.

Weidner, E., A. M. Findley, V. Dolgikh, and J. Sokolova. 1999. Microsporidian biochemistry and physiology. In *The Microsporidia and microsporidiosis*, ed. M. Wittner and L. M. Weiss, pp. 172–195. Washington, DC: ASM Press.

Weissenberg, R. 1976. Microsporidian interactions with host cells. In *Comparative pathobiology, Vol. 1. Biology of the Microsporidia*, ed. L. A. Bulla and T. C. Cheng, pp. 203–238. New York: Plenum Press.

Williams, B. A., R. P. Hirt, J. M. Lucocq, and T. M. Embley. 2002. A mitochondrial remnant in the microsporidian *Trachipleistophora hominis*. *Nature* 418:865–869.

Williams, B. A., and P. J. Keeling. 2003. Cryptic organelles in parasitic protists and fungi. *Advances in Parasitology* 54:9–67.

6

Fungal Biotrophic Parasites of Insects and Other Arthropods

Alex Weir

Meredith Blackwell

Necrotrophic fungal parasites proliferate on the dead cells and tissues of the hosts they kill, a trait that suggests great potential for biological control. By comparison, biotrophic parasites require living hosts, and the most successful among them do not kill their hosts outright (Benjamin et al. 2004). For this reason there has been less interest in these rather obscure fungal parasites of insects and arthropods, including certain mites and millipedes. Biotrophic fungi seldom cause disease symptoms, and, moreover, they are so small that mycologists and entomologists rarely notice them. There has, however, been practical interest in some of these fungi because with increasing use of coccinellids in biological control, the beetles sometimes are discovered to be infected with biotrophic parasites. For this reason their distribution may be prohibited (Blackwell and Weir, unpublished obs.).

Several morphological traits are common to fungal biotrophic parasites of arthropods. These include an overall reduction in the size of the fungus body (*thallus*), the occurrence of the thallus on the surface of the host with haustoria penetrating the host cuticle, the loss of structures or even entire life-cycle states (e.g., sexual or, less commonly, asexual states), and convergent evolution of certain morphological traits that enhance arthropod associations (e.g., sticky spores). Dramatic changes in thallus appearance from rapid divergence have made it difficult or impossible to place some biotrophic ectoparasites among their nearest relatives on the basis of morphology. Roland Thaxter, the first mycologist to study minute fungal ectoparasites extensively, seldom was concerned with questions of their relationships to other fungi. Thaxter (1908) did suggest on one occasion, however, that one group (Laboulbeniomycetes, see below), might be related to hypocrealean ascomycetes, but more often he did not speculate on the relationships of these or the many other fungi he described from associations with terrestrial arthropods (Thaxter

1914, 1920). Instead, he suggested that the discovery of additional forms eventually would fill in morphological gaps, and questions of relationships would be answered. Thaxter could not envision the polymerase chain reaction (PCR), DNA sequencing, or phylogenetic analysis, the tools that would provide molecular characters from small fungi that, as in the case of many biotrophic parasites, do not grow readily in culture. Although details remain to be elucidated, significant progress has been made toward the goal of determining the relationships of many arthropod ectoparasitic fungi. As will be seen, the missing pieces to fill Thaxter's gaps were in some cases entire life-cycle states or the connection of known states as part of a single life cycle. One might assume that all recent progress in phylogenetics has been due to molecular studies, and while this is in large part true, life-history studies have made significant contributions, especially to inform taxon sampling for molecular studies.

Examples of progress in classification of ectoparasitic fungi include classification of the Laboulbeniomycetes as an independent class of the phylum Ascomycota; new placement of some enigmatic ectoparasites in Laboulbeniomycetes; connection of *Amphoromorpha* as a phoretic state in the life cycle of *Basidiobolus*; and the general recognition of the extreme degree of morphological divergence among certain closely related taxa. A more difficult question often remains: not what is the closest relative of an insect associate, but what are its more distant relatives. Phylogeny can be used to hypothesize character state changes that occurred throughout the evolution of a fungus, but only when we have a clearer idea of ancestral states. Are there missing fungi that would fill not only morphological, but also molecular gaps? Are they extinct, undiscovered, or unrecognized? Have we used all the morphological evidence obtainable for a fungus in the search for its ancestral states? Are we looking at DNA regions that have the level of resolution to answer the questions posed? Do such regions exist? The deeper branches of the trees often remain unresolved, and molecular data may be as disappointing as trying to discover homologous morphological traits.

Only one group of insect ectoparasites has been evolutionarily successful in terms of divergence into many clades. For this reason much of this chapter centers on the ascomycete class Laboulbeniomycetes, but it also includes some more poorly known fungi (table 6.1). This chapter emphasizes the influence that modern phylogenetic studies have had in advancing our knowledge of these fungi, but life histories, morphology, and host relations are included in the discussion.

Laboulbeniomycetes

Overview

The Laboulbeniomycetes (figs. 6.1–6.11) is a well-defined group of phylum Ascomycota consisting of more than 2000 species that have obligate biotrophic associations with arthropods or that are associated with arthropods for dispersal (Alexopoulos et al. 1996). These fungi have been classified in a variety of taxonomic groups, and they periodically were considered as related to floridean red algae

Table 6.1. Ectoparasitic fungi reported from insects, taxonomic status, and geographical ranges.

Reference and Ectoparasite	Taxonomic Status	Geographical Range[a]
Thaxter (1914)		
Hormiscium myrmecophilum	Unknown relationship	Amazon (T) and Portugal[b]
Muiogone chromopteri	Unknown relationship	Cameroon (T)
Muiaria gracilis	Unknown relationship	Cameroon (T)
Muiaria lonchaeana	Unknown relationship	Cameroon (T)
Muiaria armata	Unknown relationship	Sarawak (T)
Muiaria repens	Unknown relationship	Cameroon (T)
Chantransiopsis decumbens	Unknown relationship	Java (T)
Chantransiopsis stipatus	Unknown relationship	Java (T)
Chantransiopsis xantholini	Unknown relationship	Cambridge, Massachusetts (T)
Amphoromorpha entomophila	*Basidiobolus* spp.	Widespread?
Spegazzini (1918)		
Thaxteriola infuscate	*Pyxidiophora* anamorph	Cambridge, Massachusetts (T?)
Thaxteriola subhyalina	*Pyxidiophora* anamorph	Argentina (T?)
Entomocosma laboulbenioides	*Pyxidiophora* anamorph	Argentina, USA, Massachusetts
Amphoropsis minuta	*Pyxidiophora* anamorph	(T?)
Amphoropsis subminuta	*Pyxidiophora* anamorph	(T?)
Amphoropsis media	*Pyxidiophora* anamorph	(T?)
Myriopodophila argentina	*Pyxidiophora* anamorph	(T?)
Chantransiopsis bonaerensis	*Tetrameronycha bonaerensis*	Africa, West Indies[c]
Chantransiopsis platensis	Unknown	Argentina (T), Ecuador[b]
Thaxter (1920)		
Cantharosphaeria chilensis	Trichosphaeriaceae?	Chile
Termitaria snyderi	*Kathistes* clade	NA, SA, Eu, As[c]
Termitaria coronata	*Kathistes* clade?	NA, SA, Eu, Af, Au
Muiogone medusae	Unknown relationship	Cameroon (T?)
Muiaria curvata	Unknown relationship	Panama (T?)
Muiaria fasciculata	Unknown relationship	Cameroon (T?)
Coreomycetopsis oedipus	Laboulbeniomycetes?	NA
Thaxteriola nigromarginata	*Pyxidiophora* anamorph	Java (T?)
Endosporella diopsidis	*Pyxidiophora* anamorph	Cameroon (T?)
Laboulbeniopsis termitarius	Laboulbeniomycetes	NA, SA, Eu, Af, As
Amphoromorpha blattina	*Basidiobolus* spp.	Widespread?
Enterobryus compressus	*Eccrinales* (= *Passalomyces*)	SA
Aposporella elegans	Unknown relationship	Cameroon (T)
Aposporella gracilis	Unknown relationship	Cameroon (T)
Colla (1929)		
Mattirolella silvestrii	*Kathistes* clade?	SA
Heim (1951)		
Antennopsis gallica	Unknown relationship	NA, Eu
Buchli (1960)		
Antennopsis grassei	Unknown relationship	Af
Antennopsis gayi	Unknown relationship	NA, Af, As
Khan and Kimbrough (1974)		
Mattirolella crustosa	*Kathistes* clade?	NA
Balazy and Wisniewski (1974)		
Aegeritella superficialinus	Unknown relationship	Eu

(*continued*)

Table 6.1. (continued)

Reference and Ectoparasite	Taxonomic Status	Geographical Range[a]
Majewski and Wisniewski (1978)		
Acariniola spp.	*Pyxidiophora* spp.	NA (T?)
Blackwell and Kimbrough (1978)		
Hormiscioideus filamentosus	Unknown relationship	Brazil
Blackwell et al. (1980)		
Termitariopsis cavernosa	*Termitaria* clade?	Panama
Kimbrough and Lenz (1982)		
Termitaria macrospora	*Kathistes* clade?	Au (T)
Termitaria rhombicarpa	*Kathistes* clade?	Au (T)
Termitaria longiphialidis	*Kathistes* clade?	Au (T)

Many fungi remain known from the type collection only (T). In some cases the exact species previously reported is not known, so a specific proposed relationship or collection record cannot be verified, and this information is noted (?).

[a]Af, Africa; As, Asia; Au, Australia; Eu, Europe; NA, North America; SA, South America.
[b]Santamaria (1995).
[c]Rossi and Blackwell (1990).

by several generations of leading mycologists. The floridean link was appealing because of superficial similarities in several traits for each group such as sexual reproduction with involvement of trichogyne and spermatia and pit connections and plugged simple septa in ascomycetes. Most mycologists today, however, accept Laboulbeniomycetes as fungi, although some continue to place them in several phyla other than Ascomycota, most recently among Zygomycota (Cavalier-Smith 1998, 2001).

Thaxter (1908) recognized the Laboulbeniomycetes as unusual ectoparasitic ascomycetes that lacked mycelium, and his early classification placed the species in three major groups, suborder Laboulbeniineae with two families (Peyritschiellaceae, Laboulbeniomycetaceae) and the suborder Ceratomycetaceae, although he omitted the family name within this group. This classification persisted for many years until a major revision based on morphology and development led to a revised classification (Tavares 1985). Another innovation was the phylogenetic placement based on life history and DNA sequence comparisons that linked the Laboulbeniomycetes firmly to filamentous ascomycetes (Blackwell and Malloch 1989a; Blackwell 1994; Weir and Blackwell 2001b). The Laboulbeniomycetes may have found a more stable, if not fully resolved, place as a separate class of ascomycetes (fig. 6.1). Currently the class is divided into two orders, Laboulbeniales and Pyxidiophorales. Tavares (1985) recognized four families within the Laboulbeniales: Laboulbeniomycetaceae, Ceratomycetaceae, Euceratomycetaceae, and Herpomycetaceae. A fifth family, Pyxidiophoraceae (Pyxidiophorales), later was placed in the group (see Eriksson et al. 2003; http://www.umu.se/myconet/Myconet.html). It is expected that DNA will continue to provide characters that can be used to examine the relationships within the group. The entire ectoparasitic thallus of Laboulbeniales is derived from enlargement and subsequent cell division of the two-celled ascospore (fig. 6.2).

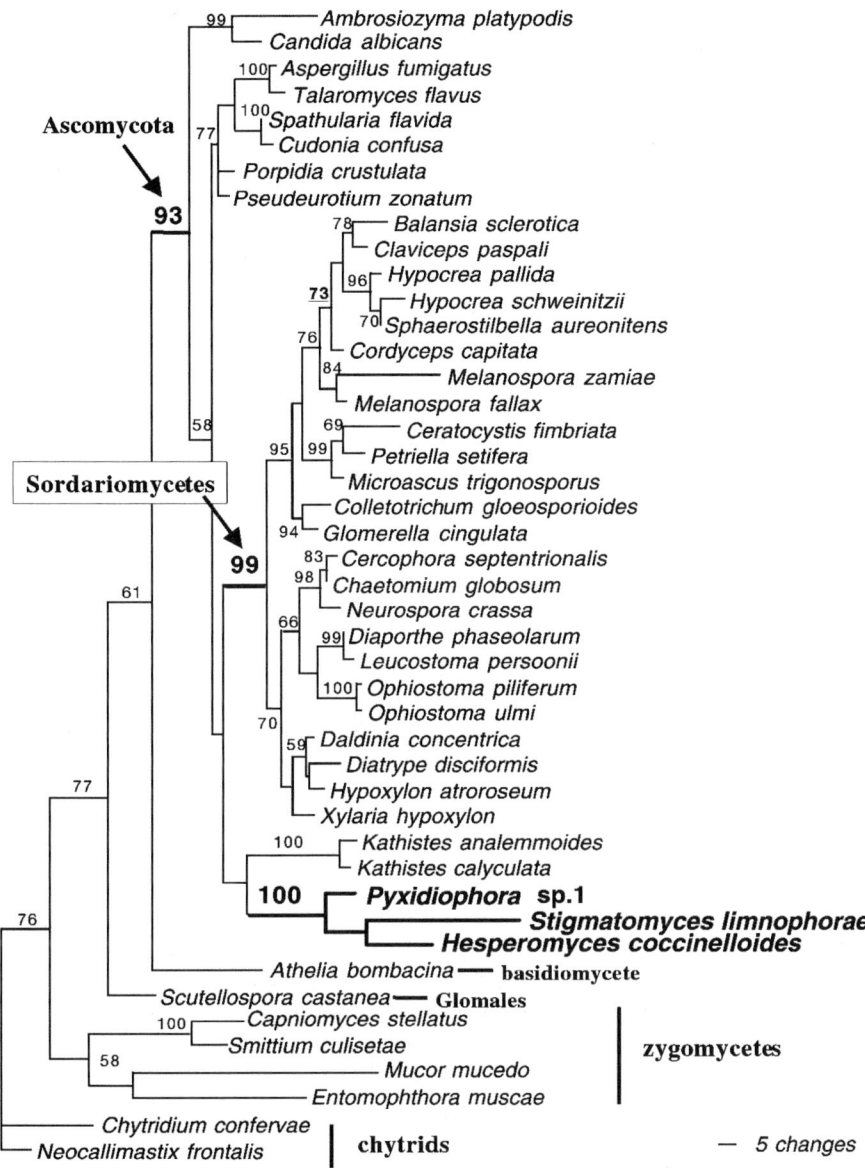

Figure 6.1. Phylogenetic analysis using parsimony criteria with chytrids as designated outgroup taxa placed Laboulbeniomycetes as a distinct class of Ascomycota (see taxa in boldface type). Note the placement of *Kathistes* spp., sister taxa to *Termitaria* (not included, but see Blackwell et al. 2003) in a clade with Laboulbeniomycetes. The relationship between the Laboulbeniomycetes clade and *Kathistes* is not supported, but both groups are strongly supported as being excluded from the Sordariomycetes (pyrenomycetes or perithecial ascomycetes). Reference and outgroup taxa include chytrids, zygomycetes, Glomales, and basidiomycete. (After Weir and Blackwell 2001b).

123

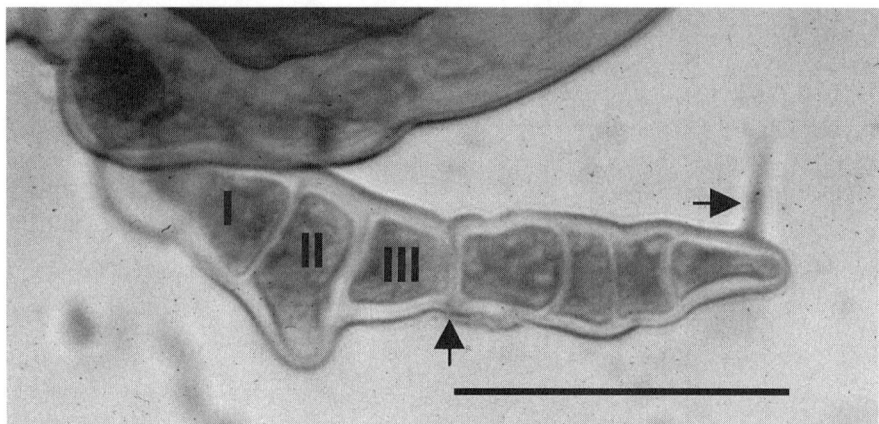

Figure 6.2. Brightfield light micrograph of a young thallus of *Arthrorynchus nycteribae* at a 10-celled stage. The original septum that divided the two-celled ascospore remains visible (vertical arrow) during early stages of thallus development. Cells of the receptacle (I, II, and III) are shown, and the initiation of perithecium development is marked by the outgrowth of cell II. The left-pointing arrow marks the remnant of the attenuated ascospore tip (somewhat out of focus) that can be recognized for a long time during development of the primary appendage. Growth and development of the perithecium will overcome the other parts of the developing thallus, and the primary appendage will be offset. At maturity the primary appendage will be the functional antheridial appendage. The enlarged basal cell I is the point of attachment and haustorium penetration of the host integument. Scale bar = 35 μm. (Photomicrograph by M. Blackwell.)

The thallus of Laboulbeniales (figs. 6.3–6. 7) is unusual for a fungus because it has determinate growth. The male of dioecious forms may consist of as few as three cells, and some of the more complex thalli may consist of several thousand cells. Plumelike structures and triggers associated with the thalli are elaborate in some taxa and are thought to assist in ascospore release (fig. 6.8). When a potential host contacts a mature thallus, sticky spores are released and adhere to it. Below the surface of the arthropod cuticle (fig. 6.5), absorption through a peglike or rootlike haustorium provides nutritional resources for the thallus; however, the fungi only rarely have been reported to cause damage to their hosts (Tavares 1985). Dispersal to initiate the next round of the life cycle is by autoinfection, the spreading of spores by a host to a new area of its body; by direct infection from one insect to another; or, apparently in some cases, by indirect infection from the insect habitat.

The concept of the Laboulbeniomycetes as derived from within a clade of mycelial ascomycetes including Pyxidiophorales came when members of *Pyxidiophora* were compared to the recognized Laboulbeniomycetes (Blackwell and Malloch 1989a). Species of *Pyxidiophora* possess mycelia and phoretic asexual states (anamorphs) that use arthropods for dispersal. The species are saprobic or, more often, mycoparasitic in a wide variety of substrates that include dung, decaying plant and algal matter, and bark beetle habitats (Malloch and Blackwell 1993). More

Figure 6.3. Thalli of *Laboulbenia idiostoma* on the antenna of a chrysomelid host. About 15 thalli are mixed among the antennal setae. The inflated structures are the perithecia, and antheridial appendages (arrow) can be distinguished at the base of the perithecia. The individual cells of the thallus are not well defined in scanning electron microscopy because of an apparent coating over the thalli. They do, however, show the overall shape of the major structures beautifully. Scale bar = 100 μm. (Scanning electron micrograph by A. Weir.)

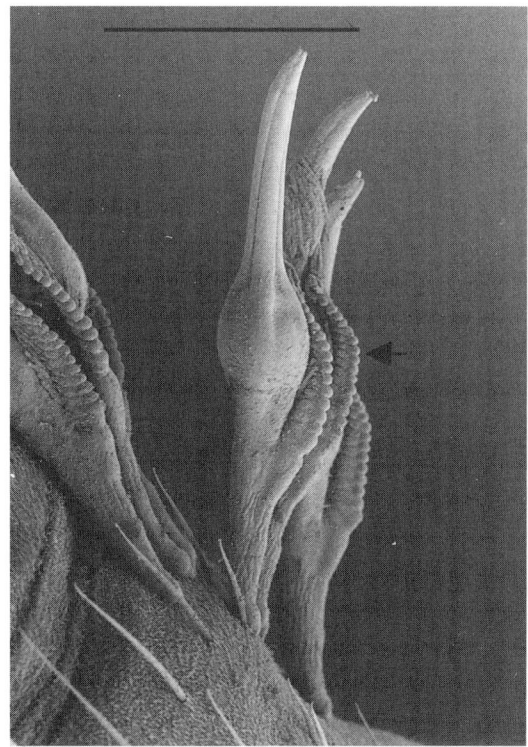

Figure 6.4. Several thalli of *Stigmatomyces crassicollis* on the sternites of a dipteran host. Perithecia and antheridial (arrow) appendages can be distinguished easily; the attachment cell, the site of haustorium penetration, can be seen against the insect integument. Note that the main axis, including the primary antheridial appendage, has been offset by enlargement of the perithecium. Scale bar = 100 μm. (Scanning electron micrograph by A. Weir.)

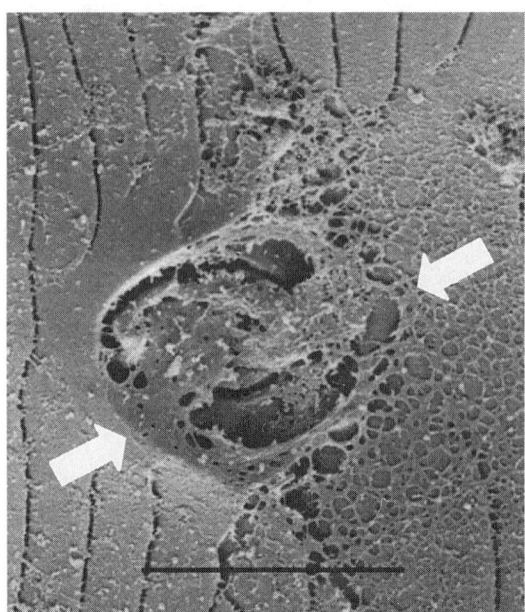

Figure 6.5. The upper surface of the host integument after removal of a thallus of *Rhanchomyces philonthinus*. Note the remnants of the mucilaginous matrix (lower right-pointing arrow) and the uplifted outer cuticular layers of the host (upper left-pointing arrow). Scale bar = 100 μm. (Scanning electron micrograph by A. Weir.)

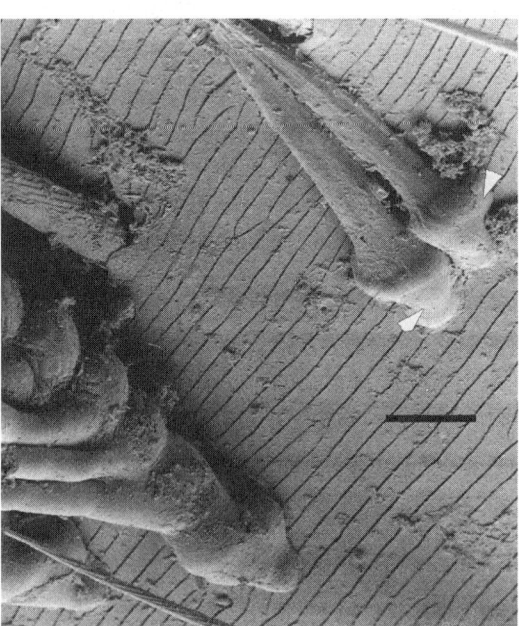

Figure 6.6. A pair of germinating ascospores of *Rhachomyces philonthimus* with swollen attachment cells (arrows). Mature thalli are above. Scale bar = 100 μm. (Scanning electron micrograph by A. Weir.)

Figure 6.7. Cluster of antheridia of *Laboulbenia idiostoma* (arrow). The perithecia of several thalli are shown as well. Scale bar = 10 μm. (Scanning electron micrograph by A. Weir.)

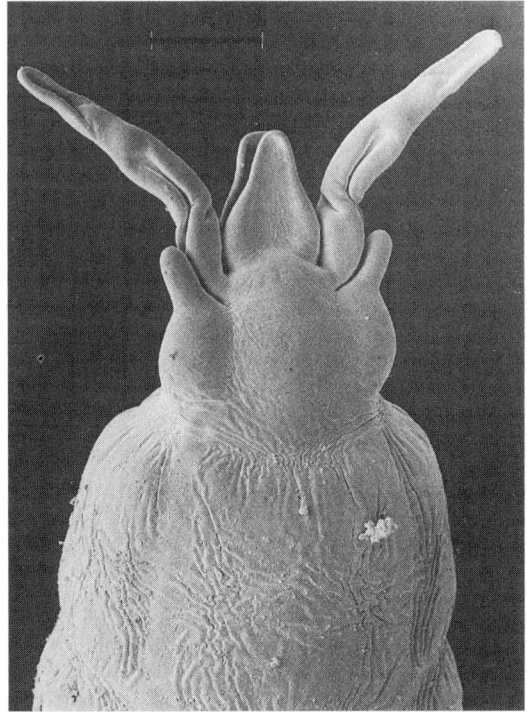

Figure 6.8. Perithecial apex of *Hesperomyces virescens* showing porpoiselike "lips" of the wall cells and two pairs of trigger-like apical appendages. Scale bar = 10 μm. (Scanning electron micrograph by A. Weir.)

127

detailed evidence for the taxonomic placement is discussed later in this chapter. Pyxidiophoras may at first glance appear to have life cycles less tied to arthropods than do other Laboulbeniomycetes, but they require arthropods for rapid, targeted dispersal to fresh habitats and may actually parasitize the arthropod vector. Among the Laboulbeniomycetes, only a few species of *Pyxidiophora* have been induced to complete their life cycles in culture, most commonly in association with a fungal host (Malloch and Blackwell 1993).

Species of *Pyxidiophora* (figs. 6.9, 6.10) usually appear in the very early stages of ecological succession in their various habitats, and they help change the character of the habitat. For this reason these fungi require a means of rapid, targeted dispersal involving a highly mobile insect that is attracted to a suitable fresh medium for the next round of the life cycle. Mites also are essential because they are in more intimate contact with the mature *Pyxidiophora* ascospores than are the insects. Phoretic mites prey on invertebrate animals in the right place at the right time: near *Pyxidiophora* perithecia at exactly the time of ascospore maturity. The ascospores adhere to the mites (fig. 6.9), and the mites shuttle the fungal spores to the insect (Blackwell et al. 1986a; Blackwell and Malloch 1989b). Attached ascospores may undergo cell divisions and conidium formation during travel to or soon after arrival at the new substrate and arrive ready to flourish in the new habitat. A complex phoretic anamorph developed from an ascospore is unknown in other ascomycete life cycles. *Pyxidiophora* species also produce four known mycelial anamorphs in their life cycles before perithecium production, but it is the addition of an ascospore-derived specialized anamorph in the *Pyxidiophora*–Laboulbeniales ancestor that may have had a profound effect on the large number of taxa that may have diversified with their insect hosts.

Parasitic life cycles with two obligate hosts are unusual but not unknown among fungi (Blackwell and Malloch 1989b; Malloch and Blackwell 1993). In addition to

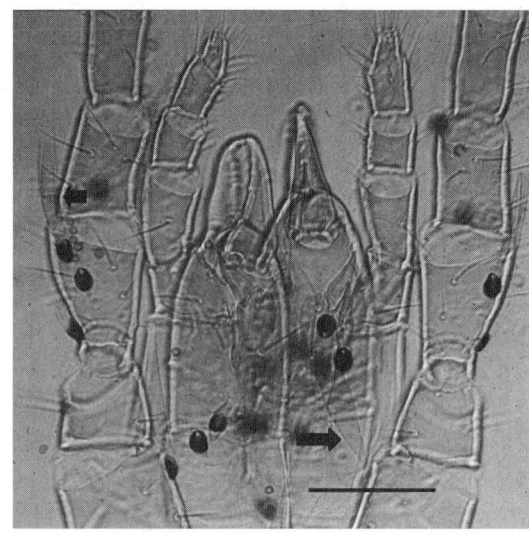

Figure 6.9. *Pyxidiophora kimbroughii* ascospores (two at arrows) on mite disperser. The arrow at the left indicates the septum that divides the hyaline spore into two cells. Darkened attachment cells are easily seen with the aid of a dissecting microscope at 100×. Scale bar = 50 μm. (Photomicrograph by M. Blackwell.)

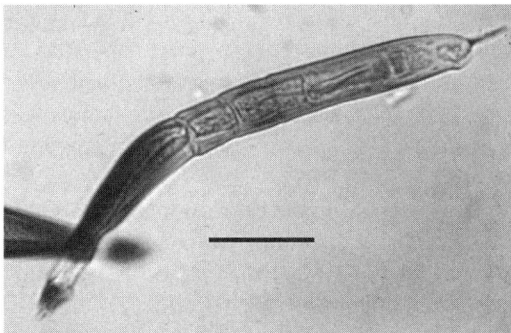

Figure 6.10. *Endosporella diapsis* from a slide mounted by Roland Thaxter; it is probably the anamorph of a species of *Pyxidiophora*. Scale bar = 25 μm. (Photomicrograph by M. Blackwell.)

Pyxidiophora species, which usually are mycoparasites as well as parasites of the mites that disperse them, other examples include a few chytrids, such as species of *Coelomomyces* that parasitize mosquitoes and copepods, and the rust fungi that attack completely different plant hosts such as ferns and conifers (Alexopoulos et al. 1996). Two-host life cycles are referred to as "heteroecious." Dependence on two hosts is known among nonfungal organisms, and Malloch (1995) used the term "heteroxenous" from the zoological literature, when he discussed the advantages of a multiple-host life style for many groups he gave as examples.

Occurrence and Distribution

Host Specificity

Members of the Laboulbeniomycetes are known from a wide range of ecosystems, although few species have been described from the more arid regions of the world, reflecting the paucity of potential hosts or dispersers in such habitats and the low number of observations in these habitats. As a direct consequence of their obligate relationship with their hosts, certain Laboulbeniomycetes are believed to exhibit a high degree of host specificity. Examination of published host–Laboulbeniales lists (Frank 1982; Huldén 1983; Weir 1996) primarily from temperate regions shows generally high levels of host specificity displayed by Laboulbeniomycetes when compared to most other groups of parasites that exploit arthropods. A relatively small proportion of the fungal parasites have been reported from more than one host, but host association data are too fragmentary for any firm conclusions to be reached. A related question for which more information is needed concerns delimitation of the broad-ranging species. Taxa currently are delimited using morphological characters that may not resolve at the level needed to detect variation over broad distributions.

In Poland, the most intensively studied geographical region for Laboulbeniomycetes (Majewski 1994), the insect host range for many parasites appears to be restricted taxonomically and often encompasses only species that belong to the same genus or group of closely related genera. At odds with these findings is the realization that some species, for example *Laboulbenia vulgaris*, have broad geographic distributions, ranging from tropical to cool temperate regions (Hammond 1995). Such broad ranges are almost unknown for the arthropod host species and suggest the occurrence of morphologically similar but genetically and physiologically distinct cryptic species of Laboulbeniomycetes. Further detailed genetics studies will be required to clarify some of the issues regarding species concepts and degree of specificity raised here.

Experimental evidence from the classic cross-inoculation studies of Richards and Smith (1954, 1955a,b, 1956) indicated that host specificity was exhibited by some species of *Herpomyces* on cockroaches they infected in a laboratory study. Laboratory-based experiments also supported a degree of host specificity of *Laboulbenia slackensis* on carabid beetles (De Kesel 1995, 1996). Factors important to an understanding of host relations include whether additional potential hosts are absent in the environment. For example, in nature *Laboulbenia slackensis* is restricted to a single host, *Pogonus chalceus*, the sole carabid inhabitant of coastal salt marshes in Belgium, whereas in the laboratory it infected a broader range of additional carabid hosts. Experimental infection of noncarabid hosts was unsuccessful, indicating host genetic influence. De Kesel (1995, 1996) suggested that environmental factors also influenced successful host infection, although it appears that the natural host distribution has resulted from reinforced ecological isolation.

Position Specificity

Laboulbeniomycete fungi show position specificity—the restricted distribution of thalli on specific parts of the host body. This phenomenon was first observed by Peyritsch (1875), who noted that *Stigmatomyces baeri* usually developed to maturity on the dorsum of the female fly host and on the ventral surface of the male. Position phenomena of this type are frequently encountered and readily explained by assuming that ascospore transmission from one host individual to another occurs during copulation. Transmission during mating does not, however, explain many other examples of position specificity found among laboulbeniomycete parasites because both sexes of the host often are equally heavily infested at the same position on the integument. In the most striking example, 16 species of the genus *Chitonomyces* have been described at different positions on the African whirligig beetle *Orectogyrus specularis* (Coleoptera, Gyrinidae) (Thaxter 1926). Whether these are all valid species or whether they represent different growth forms of the same species remains to be determined.

Sex-of-Host Specificity

Benjamin and Shanor (1952) described a third type of specificity phenomenon, sex-of-host specificity. In their work on the parasites of the carabid *Bembidion picipes*, six fungal species were distinguished; two species were found only on male beetles, and a third species apparently was restricted to females. Modern taxonomic work suggests that the hosts are part of a complex of several species, but the basic finding of skewed infection ratios appears to be correct and is being addressed using DNA sequences (M. Hughes, per. comm.).

In a striking example, *Triainomyces hollowayanus* displays all three types of specificity. It infects only one of four potential millipede host species in New Zealand, and only female millipedes are infected (Rossi and Weir 1998). Perhaps more remarkable, *T. hollawayanus* thalli develop only at the basal region of the second pair of legs of the females. The basis for the strict sex-of-host and position specificity is unknown, but several possibilities have been suggested: transmission during an exclusively female behavior that has yet to be observed or a failure for the fungus to germinate and develop on potential male hosts (Rossi and Weir 1998).

The nature of both position and sex-of-host specificity phenomena has been questioned by Scheloske (1976), who pointed out that some Laboulbeniales species previously were regarded as distinct on the basis of their different growth forms. The morphological differences, however, were due to microhabitat differences, including those of the insect bodies or modification by insect activities such as grooming. This effect has been recognized in several Laboulbeniales, and one of the best-documented examples is *Herpomyces stylopage* with thalli that develop on the antennae of roaches. On thin-walled setae of the antennae, the fungal parasites have thalli with a shield of cells that develop from the receptacle, while on the thicker-walled parts of the antennae a shield is not produced (Tavares 1985).

Hosts

Coleoptera. About 80% of the described species of Laboulbeniales parasitize beetles. Infection is, however, taxonomically clumped (Weir and Hammond 1997) within members of the 12 currently recognized beetle superfamilies that are hosts for these fungi. The available host data indicate that predacious adephagan (e.g., Caraboidea, Dytiscoidea) and staphylinoid beetles in moist habitats predominate as hosts. Other beetles belonging to several major lineages, most notably the Curculionoidea and other primarily phytophagous or plant-associated species, are seldom hosts for Laboulbeniales in temperate regions (Weir and Hammond 1997). Some differences in such a host distribution, however, have been identified for Sulawesi, a moist tropical site (Weir and Hammond 1997). It may be that chrysomeloids (principally Alticinae and Eumolpinae) and other plant-associated taxa are proportionally better represented as hosts in tropical than in temperate regions.

Diptera. Only a small proportion (10%) of currently recognized Laboulbeniomycetes parasitize adult flies. Most of the species are members of the genus *Stigmatomyces* that occur worldwide on a wide range of Brachycera (horseflies, robberflies) and other flies associated with aquatic or moist habitats such as Ephydridae (shore or brine flies). Other fungal genera parasitic on brachycerid flies include *Laboulbenia* and *Dimeromyces*, both with wide host ranges among arthropods, and *Ilytheomyces* and *Rhizomyces* of much more restricted occurrence, with all 15 species of *Ilytheomyces* being found only on a single genus of Ephydridae and species of *Rhizomyces* occurring only on African Diopsidae. The dipterous Pupipara (bat flies and keds) also are parasitized by Laboulbeniomycetes, including species of *Arthrorhynchus* on Old World bat flies (Nycteribiidae) and *Gloeandromyces* on New World bat flies (Streblidae).

Hemiptera (Heteroptera). Most Hemiptera workers follow an arrangement recognizing seven suborders: Enicocephalomorpha, Dipsocoromorpha, Gerromorpha, Leptopodomorpha, Nepomorpha, Cimicomorpha, and Pentatomomorpha (Henry and Froeschner 1988). Laboulbeniomycetes are known to occur on Gerromorpha, Nepomorpha, Cimicomorpha, and Pentatomomorpha. More than 70 species in the fungal genera *Coreomyces, Cupulomyces, Laboulbenia, Majewskia, Monandromyces, Polyandromyces, Prolixandromyces, Tavaresiella*, and *Triceromyces* are associated with these insects, the majority parasitizing Gerromorpha. This suborder includes mainly aquatic or semiaquatic families such as Hebridae and Hydrometridae.

Other Insecta. Other groups of insects, including Blattodea (e.g., cockroaches), Dermaptera (e.g., earwigs), Hymenoptera (e.g., ants), Isoptera (termites), Mallophaga (bird lice), Orthoptera (e.g., crickets), and Thysanoptera (e.g., thrips), are known to support Laboulbeniales infections. Some of the fungal genera (e.g., *Laboulbenia, Rickia*) found on these hosts occur on a wide range of arthropods; others are more restricted (e.g., *Filariomyces, Herpomyces*). All 25 described species of *Herpomyces* are restricted to cockroaches. *Herpomyces* is morphologically and developmentally very different from other Laboulbeniales. The species are also exceptional in that they are the only Laboulbeniales known to parasitize larval as well as adult insects (Richards and Smith 1955a).

Myriapoda. The occurrence of Laboulbeniales on millipedes was not confirmed until early this century, although structures possibly attributable to immature specimens of these fungi were observed and illustrated by Verhoeff (1897). There have been few subsequent reports, and representatives from only four genera are known on these hosts. Three of the genera, *Diplopodomyces* (on Callipodidae), *Triainomyces* (on Sphaerotheridae), and *Troglomyces* (on Julidae) are monotypic; the fourth, *Rickia*, is widespread on a wide range of hosts. Four species of *Rickia* parasitize millipedes of the families Julidae and Harpagophoridae. In general, infection foci are found on the appendages of millipedes, particularly the legs.

Acarina. Thalli representing 4 genera and 54 species of Laboulbeniales are associated with mesostigmatid mites (Tavares 1985). Most of the known acarine hosts are phoretic on other arthropods, particularly beetles, which may have provided a mechanism for dispersal and speciation. The four genera represented occur on a wide range of arthropod hosts, and frequently the parasite on the mite is also found on the integument of the mite's phoretic host. Examples of beetles or termites and phoretic mites are well known (Blackwell and Rossi 1986). In one extreme case a species of *Laboulbenia* is parastic on several mites and staphylinid and histerid beetles, all of which occur in the nests of *Eciton* ants. As we mentioned above, phoretic states of *Pyxidiophora* (Pyxidiophorales) also are developed from the ascospore, and these often can be found attached to phoretic mites both on insects in nature and in insect collections.

Fungi. No members of the Laboulbeniales are known to use fungal hosts. As mentioned earlier, species of *Pyxidiophora* (Pyxidiophorales) are fungal parasites in rapidly decomposing substrates. Although some taxa grow and develop ascospores in pure culture, most appear to rely on fungi for normal growth and sporulation. Some Pyxidiophoras are fusion biotrophs; fusion biotrophs form well-developed contact cells from short lateral branches that fuse with the host hyphae (Malloch and Blackwell 1993, ms. in prep.). In a few cases peglike haustoria have been observed to penetrate the integument of mites at the attachment point of the *Pyxidiophora* ascospore, indicating that *Pyxidiophora* has two entirely different hosts (fungus and arthropod) in distinct parts of its life cycle. Unlike the arthropod biotrophic association, the parasitism of fungi, at least in culture, may result in death of the host (Malloch and Blackwell, unpublished data).

Structure and Thallus Organization

Mycelial Laboulbeniomycetes (Pyxidiophorales)

Pyxidiophora is atypical among the Laboulbeniomycetes because the two-celled ascospores germinate directly by germ tubes or yeastlike cells with secondary development of a mycelium (Blackwell and Malloch 1989b). Mycelial development in *Pyxidiophora* is different from the condition in the Laboulbeniales in which the entire thallus develops from a two-celled ascospore. In some Pyxidiophoras the mycelium is reduced until a potential fungal host becomes associated with the mycoparasitic mycelium. *Chalara*-like, *Gabarnaudia*-like, *Graphium*-like, and others are among the wide variety of conidial forms reported from the species of *Pyxidiophora* that have been studied in culture or on natural substrates (Blackwell et al. 1986b; Malloch and Blackwell ms. in prep.). The most common of the named *Pyxidiophora* phoretic anamorphs, *Thaxteriola* (fig. 6.9) and *Acariniola*, were named before their connection with *Pyxidiophora* ascospores was recognized. The *Pyxidiophora* perithecium develops from a mycelium and the wall is unusual because it is composed of a single layer of cells. This type of perithecium is developmentally different from that of other Laboulbeniomycetes that develop from an ascospore and secondarily have perithecial walls that are two cells thick. The perithecium of *Pyxidiophora* and Laboulbeniales probably arose independently of the perithecium of the other pyrenomycetes (Sordariomycetes, Eriksson et al. 2003).

Laboulbeniomycetes Other Than *Pyxidiophora*: Laboulbeniales

The structure and organization of the thallus in the Laboulbeniales is different from that of all other fungi because ascospore germ tube formation has been suppressed, and the entire thallus develops by division and enlargement from the two cells of the ascospore. Although at first glance thallus development appears simple, patterns of cell structure and development have been interpreted differently (Thaxter 1896, 1908; Sugiyama 1973). Tavares (1985) developed the terminology used to name the cells of the thallus and their derivatives, and her usage is followed here. This section is intended as an introduction to the specialized literature on morphology of the Laboulbeniales.

Receptacle. The precise structure, organization, and relative position of the cells of the receptacle provide the main taxonomic characters used in the delimitation of genera. The receptacle develops from the cell of the ascospore that is attached to the insect. In the majority of known genera the receptacle consists of three cells denoted by Roman numerals: I (basal cell), II (suprabasal cell), and III (usually referred to as the uppermost cell of the receptacle) (fig. 6.2). Exceptions include genera in which there is no division into separate cells II and III (e.g., *Rhizopodomyces*), genera in which cell II subdivides either vertically, into a variable number of usually small cells (e.g., *Rickia*), or horizontally, forming a superposed series of usually flattened cells (e.g., *Blasticomyces*). Cell III also may become divided in a few genera, most notably in the large genus *Laboulbenia*, in which two additional receptacle cells are formed, denoted as cells IV and V.

Additionally, some genera develop secondary axes that lie outside the primary axis of the original ascospore. These structures are known as secondary receptacles. Atypical divisions of the primary and secondary receptacles also can occur and can include development of sterile, antheridial, and perithecial branches from any cell of the receptacle except cell I, the absence of certain cells, or altered relationships between cells.

Appendages. The primary appendage is formed by the division of the uppermost (unattached, shorter) cell of the ascospore (figs. 6.2–6.4, 6.6) and is usually a direct continuation of the primary receptacle axis, although it usually becomes offset as perithecium development progresses. An appendage may consist of as few as one or two cells (e.g., *Filariomyces*, *Rhizopodomyces*) or be much more extensive, as in *Corethromyces* and *Laboulbenia* (fig. 6.7). Antheridia borne on the appendage may be clustered (e.g., *Triceromyces*) or regularly distributed along the axis. Antheridia also may originate from small corner cells or develop within the cells. Sterile or antheridial branches arising from the lowermost ascospore segment are denoted as secondary appendages, and these may arise above or below the perithecia. Sterile branchlets also may develop directly from the external wall cells of the perithecium (e.g., *Helodiomyces*) and are referred to as perithecial appendages (Tavares 1985).

Antheridia. Antheridial characters remain important in current concepts of classification. Three distinct types of spermatial development have been distinguished: (1) exogenous with spermatia borne on intercalary cells or terminally from cells of the appendage, (2) simple endogenous in which spermatia are formed within a flask-shaped cell (fig. 6.7), and (3) compound endogenous with antheridial cells united into a compound structure so that their spermatia are discharged into a common chamber before exiting through a single opening. Exogenous spermatial production is unusual and is found mainly in those species associated with aquatic hosts; the simple endogenous type of antheridium is by far the one most commonly encountered in the Laboulbeniales. Compound endogenous antheridia are formed occasionally and are known only in representatives of the subfamilies Monoicomycetoideae and Peyritschielloideae. Although we assume that antheridia are functional in the Laboulbeniales, there is not yet any evidence to support or refute this contention.

Perithecium. The Laboulbeniales perithecium (figs. 6.3, 6.4, 6.7) is almost always formed as an outgrowth from the receptacle, usually from the suprabasal cell (II) (fig. 6.2). The typical perithecium consists of stalk cells (the primary stalk cell [VI] and the secondary stalk cell [VII]), and an internal procarp, or its derivative cells, surrounded and surmounted by one outer and one inner layer of wall cells (Tavares 1985). Perithecia develop either by upgrowth of wall cells around the carpogonium as in the Laboulbeniineae or by the intrusion of the carpogonium upward between the rows of wall cells as in the monogeneric Herpomycetineae. Within the Laboulbeniineae the basic difference in the perithecia is in the relationship of the stalk cells (VI and VII) to the primary axis of the thallus. Usually, three basal cells are formed above the stalk cells; cell m, the basal cell formed from cell VI (or its derivative VI' in the Ceratomyceteae), and cells n and n' both of which originate from cell VII. Division of cells n and n' results in the formation of three vertical rows of perithecial wall cells, with cell m dividing to form the fourth row (see Rossi and Weir 1998). The number and arrangement of perithecium wall cells are important taxonomic characters.

Phylogenetic Relationships of Laboulbeniomycetes

The Laboulbeniomycetes have had a confusing taxonomic history, beginning with their discovery and misinterpretation as abnormal cuticular outgrowths of insects. They also have been considered to be acanthocephalans, basidiomycetes, zygomycetes, ascomycetes, and close relatives of floridean red algae (for a complete history, see Weir and Blackwell 2001b). Although most recent classifications placed Laboulbeniomycetes in Ascomycota (Barr 1983), there has been reluctance by some to accept the group as ascomycetes, and in 1998 they were classified apart from the Ascomycota in the phylum Archemycota, class Zoomycetes (Cavalier-Smith 1998), largely on the basis of their attachment to arthropods. This placement ignored evidence from a study depicting ascosporogenesis (Hill 1977) and a phylogenetic analysis using partial small subunit (SSU) rDNA to place *Pyxidiophora* and *Rickia* sp. in an independent clade within ascomycetes using basidiomycetes as outgroup taxa (Blackwell 1994). A similar classification scheme excluding Laboulbeniomycetes from Ascomycota was repeated by Cavalier-Smith (2001).

   The Cavalier-Smith (1998) hypothesis placing Entomophthorales, Zoopagales, Harpellales, Asellariales, Laboulbeniales, and Pyxidiophorales in Zoomycetes was tested by phylogenetic analysis of molecular characters using parsimony criteria. The data analyzed consisted of 1.1 kb of the SSU rRNA gene (Weir and Blackwell 2001b) for taxa that included species of Laboulbeniales, *Pyxidiophora*, trichomycete and entomophthoralean zygomycetes, and chytrids as outgroup taxa. The Laboulbeniomycetes, including *Pyxidiophora*, received high bootstrap support as an independent clade of Ascomycota (fig. 6.1). Ten different hypotheses were tested using maximum likelihood analysis. The two best trees were those in which Laboulbeniomycetes were placed in "ascomycetes" with the ascomycete group unspecified and in which the Laboulbeniomycetes were grouped within Sordariomycetes (Eriksson et al. 2003). Four other trees placing Laboulbeniomycetes in various ascomycete groups were not significantly worse. Significantly worse trees were those that specified Laboulbeniomycetes as Dothideomycetes (loculoascomycete ascomycetes), as Zygomycota, as a group of Laboulbeniomycetes + Trichomycetes, and as a group of

Laboulbeniomycetes + Trichomycetes + Entomophthorales. Laboulbeniomycetes were never supported as a member of any group other than Ascomycota.

Relationships within Laboulbeniomycetes

Species of *Pyxidiophora* first were recognized as the nearest relatives of non-mycelial Laboulbeniales based on morphological and ecological evidence alone (Blackwell and Malloch 1989a), evidence that was acceptable for grouping *Pyxidiophora* with the Laboulbeniales (Eriksson and Hawksworth 1993). Morphological similarities in considering *Pyxidiophora* as a probable relative of Laboulbeniales include sequential ascus development, ascospore development, and unusual mature ascospore morphology; all appear identical at the light microscopic level. The two-celled ascospores of both groups are unusual in that they are long (up to 500 µm) and broader and longer at the end that exits the perithecium neck. This ascospore cell, destined to become the receptacle, is the site of development of a specialized, darkened attachment region by which the ascospore adheres and penetrates the host integument with a haustorium. Some species of *Pyxidiophora* have ascospores that divide to produce phoretic anamorphs of more that 50 cells that closely resemble the young thalli of Laboulbeniales (Blackwell and Malloch 1989b).

Since the original suggestion that *Pyxidiophora* and Laboulbeniales were related (Blackwell and Malloch 1989b), all molecular evidence strongly supports this placement (Blackwell 1994; Weir and Blackwell 2001b). In addition to *Thaxteriola* (fig. 6.9) and *Acariniola* dispersal anamorphs derived from ascospores that have been linked to *Pyxidiophora* in cultural studies (Blackwell and Malloch 1989b, 1990), *Endosporella* (fig. 6.10), *Amphoropsis*, *Myriopodophila*, and *Entomocosma* (table 6.1; Spegazzini 1918; Thaxter 1914, 1920) have a similar construction of linearly superposed cells and probably are also anamorphs of *Pyxidiophora* (Blackwell et al. 1986a; Blackwell 1994).

One relationship of special interest is that of *Laboulbeniopsis termitarius*, which has a minute thallus of three to four linearly superposed cells (Kimbrough and Gouger 1970). This species occurs on a wide variety of termites and was considered to be a member of the "Laboulbeniales Imperfecti" (Gäumann and Dodge 1928). Molecular evidence placed *L. termitarius*, which was reported to produce ascospores (Blackwell and Kimbrough 1976b) as the most reduced member of the Laboulbeniomycetes and as a sister taxon to *Pyxidiophora* (Henk et al. 2003). In contrast, *Coreomycetopsis oedipus*, a termite ectoparasite with a thallus of few superposed cells and an attachment region identical to *L. termitarius*, probably produces conidia. A close relationship between the two termite fungi has been suggested (Blackwell and Kimbrough 1976a, b; Blackwell 1994), and this implies inclusion of *C. oedipus* in Laboulbeniomycetes.

Within the remainder of the Laboulbeniomycetes, new phylogenetic evidence based on analysis of SSU rDNA sequences for 25 taxa, including 2 species of *Pyxidiophora* and *Laboulbeniopsis oedipus* (Henk et al. 2003), shows several evolutionary trends within the class. The Laboulbeniomycetes continue to be supported as monophyletic; two *Pyxidiophora* species were grouped together. Some groupings previously suggested to be monophyletic on the basis of morphology were not supported, and the preliminary data predict that new infragroup arrangements will be forthcoming.

Other Biotrophic Ectoparasites

In addition to the Laboulbeniomycetes and its recently linked taxon additions discussed above, there are other ectoparasites of arthropods described largely by Thaxter (1914, 1920) and Spegazzini (1918). These fungi can be divided into three groups on the basis of morphology: (1) several to 13 linearly superposed cells, most of which are believed to be anamorphs of Pyxidiophora (table 6.1), (2) darkly pigmented crusts (Termitaria and relatives), and (3) Thaxter's largely filamentous, pigmented "idiocentricities." Several other fungi mentioned in the literature as fungal associates of insects are included in table 6.1, but these are not believed to be parasitic (see below).

Termitaria

Species of *Termitaria*, *Mattirolella*, and *Termitariopsis* have similar morphologies (table 6.1). These ectoparasites cause disfiguring lesions on the integument of termites or, in the case of *Termitariopsis*, of ants. Species of *Termitaria* have a worldwide distribution occurring on diverse groups of termites (Kimbrough and Lenz 1982; Blackwell and Rossi 1986; Hojo et al. 2001). The two known species of *Mattirolella* also infect termites; each of the two species is known only from the type collection in Central or South America (Kimbrough and Thorne 1982). *Termitariopsis* is known only from the type collection on ants from Panama (Blackwell et al. 1980). The thallus of these fungi consists of a basal cell layer from which haustorial cells penetrate the insect cuticle. A darkly pigmented sporodochium occurs on the surface of the insect integument. DNA from portions of the nuclear-encoded SSU rDNA and the β-tubulin gene of *T. snyderi*, the best known of the *Termitaria* species, was analyzed with other insect-associated fungi with a surprising result. The closest relatives of *T. snyderi* were the ascomycetes, *Kathistes analemmoides* and *K. calyculata* (Blackwell et al. 2003). These ascomycetes do not share morphological traits with species of *Termitaria*. *Kathistes* species are primarily sexual, although they produce *sporidiomata*, inconspicuous structures of unknown function that are associated with perithecia in the substrate, but which are not similar to *T. snyderi* structures (Malloch and Blackwell 1990). Species of *Termitaria* are some of the best known of the small biotrophic fungal parasites on insects, perhaps because of the contrast of the dark sporodochia on the light-colored bodies of the termite hosts. We have a clearer picture of the relationships of species of *Termitaria*, but it is not completely satisfying because of the long branch separating this two-member clade from other ascomycetes and because of our inability to infer character state changes.

Other Ectoparasites: "Idiocentricities"

Other species described from insects, largely by Thaxter (1914, 1920) and Spegazzini (1918), were referred to by Thaxter (1914) as "idiocentricities" (table 6.1). These fungi have small filamentous thalli, usually with a differentiated holdfast cell and small haustoria. Although the thalli are similar with their darkly pigmented bases and multiple filaments, it is not known if these fungi compose a monophyletic group. Several of the fungi in this group specialize on termites, including three species of *Antennopsis* (fig. 6.11), the best known genus collected multiple times (Buchli 1966; Blackwell and Rossi 1986). Genera such as *Hormiscium*, *Muiogone* (fig. 6.12), *Muiaria* (fig. 6.13), and *Chantransiopsis* are poorly known (Thaxter 1914, 1920). Often only alcohol-preserved arthropods were available to describe these fungi; examination and culture in artificial media of fresh material should advance our knowledge.

Non-parasites: *Amphoromorpha, Enterobryus,* and *Cantharosphaeria*

Some of the arthropod fungi listed in table 6.1 are not parasites. One commonly observed example is the genus *Amphoromorpha*. For many years the biology of the saclike thalli described as *Amphoromorpha* was unknown. We now know that the thalli of *Amphoromorpha* are capilliconidia of *Basidiobolus ranarum* or other fungi that produce such structures (fig. 6.14). Fungi producing capilliconidia include certain Zygomycota in several genera of Entomophthorales and Zoopagales

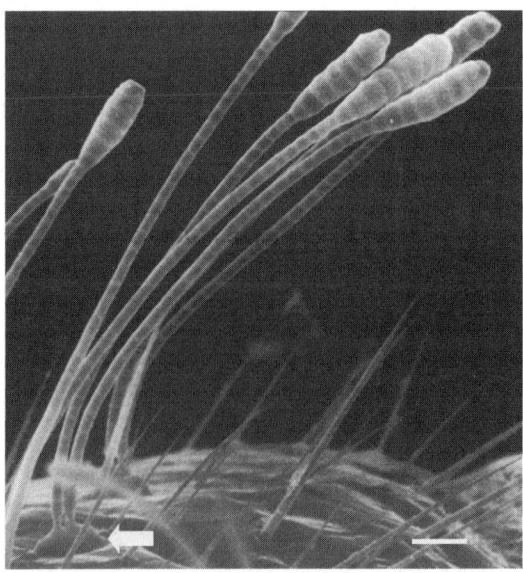

Figure 6.11. Species of the genus *Antennopsis* are restricted to termite hosts. Arrow indicates the three- to four-celled attachment structure. Scale bar = 20 μm. (Scanning electron micrograph by M. Blackwell.)

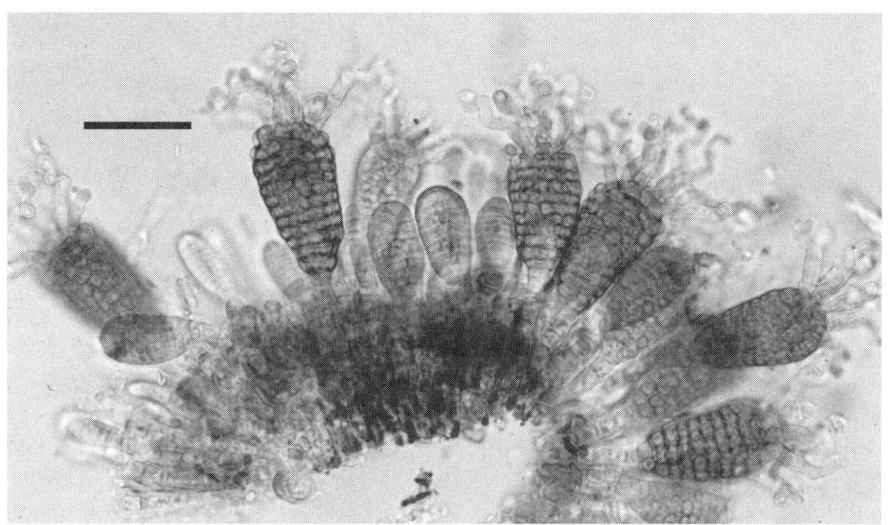

Figure 6.12. Multicellular conidia of *Muiogone medusae* at different stages of maturity. Curled appendages sometimes bear secondary conidia. Photograph of specimen mounted and drawn by Thaxter (1920). Scale bar = 30 μm. (Photomicrograph by M. Blackwell.)

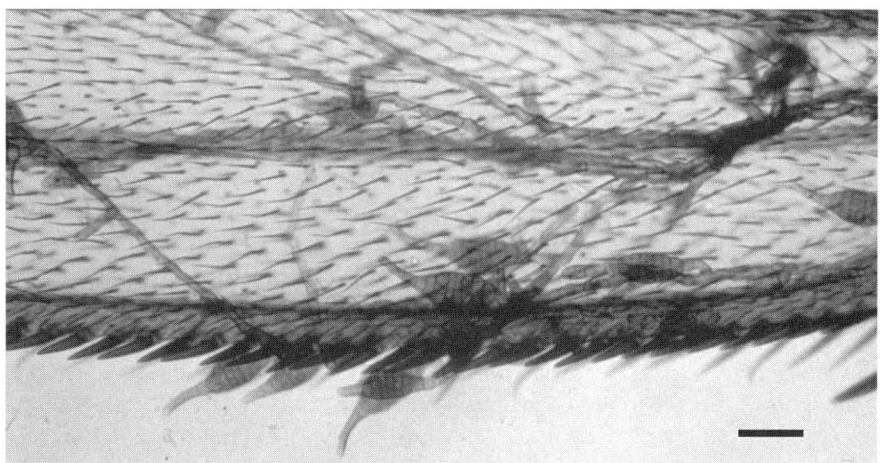

Figure 6.13. *Muiaria repens* on wing of fly host, mounted and illustrated by Thaxter (1914). Scale bar = 15 μm. (Photomicrograph by M. Blackwell.)

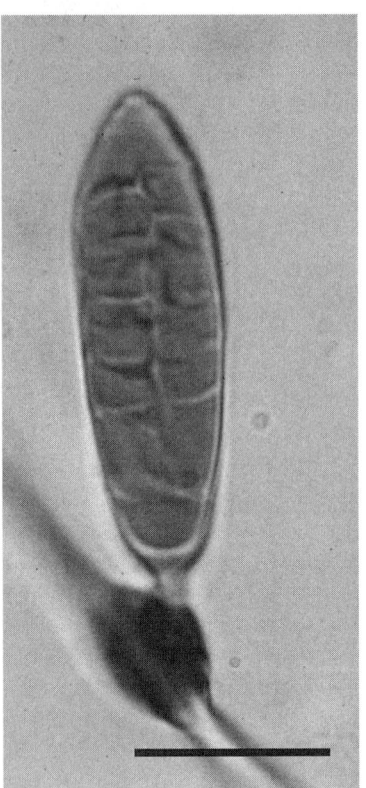

Figure 6.14. "Synthetic" *Amphoromorpha* from experiment in which mites were placed in a culture of *Basidiobolus ranarum*. Capilliconidia adhered to the mites, and over several days the adhesive darkened. Scale bar = 20 μm. (Photomicrograph by M. Blackwell and D. Malloch.)

(Blackwell and Malloch 1989c, 1991). Capilliconidia are asexual spores filled with cytoplasm that travels through a long slender conidiophore as if by capillary action. At maturity the conidiophore is evacuated of cytoplasm, and the spore has traits that enhance dispersal such as an adhesive droplet at the distal end and a region of dehiscence from the conidiophore. When an arthropod or another object, including a growing hypha, touches the spore, the spore adheres to it. The adhesive material darkens after attachment to an insect. Capilliconidia are phoretic spores, but because they are not thought to produce haustoria, they are not considered to be parasites. Larger capilliconidia often cleave internally (fig. 6.14), but germination of the cleaved segments has not been observed. It is not clear how many species of *Amphoromorpha* exist because of the few informative characters and the great size variation in capilliconidia of *Basidiobolus*.

Cantharosphaeria chilensis grows on the surface of insects, but its nutritional mode (parasitic or saprobic) is unclear. Müller and von Arx (1962) suggested that this species should be placed in *Eriosphaeria*, but they did not make the transfer. *Cantharosphaeria chilensis* is, however, listed as a synonym of *Eriosphaeria* (Hawksworth et al. 1995; Kirk et al. 2001), but transfer of a fungus that grows on debris on an insect to a group of wood-inhabiting perithecial ascomycetes needs scrutiny.

Another species described by Thaxter (1920) is a trichomycete insect-gut fungus (Zygomycetes, Eccrinales). Thaxter's *Enterobryus compressus* was reclassified in a new genus, *Passalomyces compressus* (Lichtwardt et al. 1999). Although Thaxter (1920) suspected that the fungus was a parasite, there is no evidence to suggest that this view is correct.

Summary and Future Directions

Great progress has been made in the last 15 years toward the classification and phylogeny of terrestrial arthropod biotrophic ectoparasites. The progress came despite dramatic evolutionary divergence in morphologies and life histories, sometimes with the loss of entire life cycle states. Furthermore, convergent traits have led some to incorrect conclusions about the relationships of the fungi. Our current understanding has come not only from the application of molecular methods, but also from relatively low tech but highly informative life-history studies. The following points summarize the state of our knowledge of the biology, taxonomy, and phylogeny of the obscure arthropod-associated fungi discussed in this chapter.

- A variety of fungi are specialized as biotrophic ectoparasites of terrestrial arthropods.
- Fungal biotrophs are morphologically simple and reduced in size, usually requiring a minimum of 100× magnification and back lighting for recognition on the host. For this reason they are poorly known by both mycologists and entomologists. Sizes range from < 1mm for Laboulbeniomycetes and *Termitaria* at the larger end of the spectrum to the smallest size of 30–50 μm for the dispersal anamorphs of *Basidiobolus* and *Pyxidiophora*.
- Laboulbeniomycetes is the most diverse group in terms of morphology and species numbers, with about 2000 described species. The Laboulbeniomycetes, now firmly linked to filamentous ascomycetes, is one of the few groups that have lost the ability to reproduce asexually. Loss of sexual reproduction is the more common loss of a life-history state among other arthropod parasitic and fungi in general.
- The addition of *Pyxidiophora*, *Laboulbeniopsis*, and the diverse phoretic anamorphs of *Pyxidiophora* increases the known diversity of Laboulbeniomycetes and clarifies the relationships of several biotrophic arthropod parasites.
- Molecular evidence indicates that certain fungal biotrophs do not share morphological features with their closest non–arthropod-associated relatives. Examples are the presence of mycelium in *Pyxidiophora* and its absence in Laboulbeniales. *Termitaria* is asexual and *Kathistes* is sexual.
- Some fungi previously considered to be parasites are dispersal states in the life cycles of well-known fungi and are not parasitic (e.g., *Amphoromorpha*, *Stylopage*).

New phylogenetic understanding and improved techniques have renewed interest in these fungi and their interactions with their hosts. As we have mentioned, many are small and may be difficult to locate and observe on a host. The small size also has caused difficulty in DNA extraction and amplification using PCR (Weir and Blackwell 2001a), and until now DNA sequences have come from pooled samples of several thalli. So far only rRNA gene regions with high copy number have been successfully amplified, but the development and application of additional molecular methods will provide the means to address questions concerning population structure and host specificity, especially the investigation of possible parallel differentiation of fungal and insect populations. Phylogenetic studies of the speciose Laboulbeniales will be needed to test the current classification hypotheses. Moreover, intricate morphological character-state modifications of the thallus and innovations in life histories (e.g., loss of asexual state, dramatic host shifts) will be tracked in phylogenetic studies.

Literature Cited

Alexopoulos, C. J., C. W. Mims, and M. Blackwell. 1996. *Introductory mycology*. New York: John Wiley & Sons.

Balazy, S., and J. Wisniewski. 1974. *Aegeritella superficialis* gen. et sp. nov.—epifityczny grzyb na mrówkach z rodzaju *Formica* L. *Prace Komisji Nauk Roln. i Komisji Nauk Leên. PTPN, Poznan* 38:3–15.

Barr, M. E. 1983. The ascomycete connection. *Mycologia* 75:1–13.

Benjamin, R. K., and L. Shanor. 1952. Sex of host specificity and position specificity of *Laboulbenia* on *Bembidion picipes*. *American Journal of Botany* 39:125–131.

Benjamin, R. K., M. Blackwell, I. Chapela, R. A. Humber, K. G. Jones, K. A. Klepzig, R. W. Lichtwardt, D. Malloch, H. Noda, R. A. Roeper, J. W. Spatafora, and A. Weir. 2004. The search for diversity of insects and other arthropod associated fungi. In *Biodiversity of fungi: Standard methods for inventory and monitoring*, ed. G. M. Mueller, G. F. Bills, and M. Foster, pp. 395–433. New York: Academic Press.

Blackwell, M. 1994. Minute mycological mysteries: The influence of arthropods on the lives of fungi. *Mycologia* 86:1–17.

Blackwell, M., J. R. Bridges, J. C. Moser, and T. J. Perry. 1986a. Hyperphoretic dispersal of a *Pyxidiophora* anamorph. *Science* 232:993–995.

Blackwell, M., D. Henk, and K. G. Jones. 2003. Extreme morphological divergence: phylogenetic position of a termite ectoparasite. *Mycologia* 95:987–992.

Blackwell, M., and J. W. Kimbrough. 1976a. A developmental study of the termite-associated fungus *Coreomycetopsis oedipus*. *Mycologia* 68:551–558.

Blackwell, M., and J. W. Kimbrough. 1976b. Ultrastructure of the termite-associated fungus *Laboulbeniopsis termitarius*. *Mycologia* 68:541–550.

Blackwell, M., and J. W. Kimbrough. 1978. *Hormiscioideus filamentosus* gen. et sp. nov., a termite-infecting fungus from Brazil. *Mycologia* 70:1273–1280.

Blackwell, M., and D. Malloch. 1989a. *Pyxidiophora*: A link between the Laboulbeniales and hyphal ascomycetes. *Memoirs of the New York Botanical Garden* 49:23–32.

Blackwell, M., and D. Malloch. 1989b. *Pyxidiophora*: Life histories and arthropod associations of two species. *Canadian Journal of Botany* 67:2552–2562.

Blackwell, M., and D. Malloch. 1989c. Similarity of *Amphoromorpha* and secondary capilliconidia of *Basidiobolus*. *Mycologia* 81:735–741.

Blackwell, M., and D. Malloch. 1990. Discovery of a *Pyxidiophora* with *Acariniola*-type ascospores. *Mycological Research* 94:415–417.

Blackwell, M., and D. Malloch. 1991. Life history and arthropod dispersal of a coprophilous *Stylopage*. *Mycologia* 83:360–366.

Blackwell, M., T. J. Perry, J. R. Bridges, and J. C. Moser. 1986b. A new species of *Pyxidiophora* and its *Thaxteriola* anamorph. *Mycologia* 78:607–614.

Blackwell, M., and W. Rossi. 1986. Biogeography of fungal ectoparasites of termites. *Mycotaxon* 25:581–601.

Blackwell, M. R. Samson, and J. W. Kimbrough. 1980. *Termitariopsis cavernosa*, gen. et sp. nov., a sporodochial fungus ectoparasitic on ants. *Mycotaxon* 12:97–104.

Buchli, H. 1960. Une nouvelle espèce de campignon parasité du genre *Antennopsis* IEM sur les termites de Madagascar. *Compte Rendus Hebdomadaires Seances de l'Academie des Sciences* 250:3365–3367.

Buchli, H. 1966. Notes sur les parasites fongiques des Isoptères. *Revue d'Ecologieet de Biologie du Sol* 3:589–610.

Cavalier-Smith, T. 1998. A revised six-kingdom system of life. *Biological Reviews* 73:203–266.

Cavalier-Smith, T. 2001. What are fungi? In *Mycota*, vol. VII, part A, ed. D. J. McLaughlin, E. G. McLaughlin, and P. A. Lempke, pp. 3–37. New York: Springer Verlag.

Colla, S. 1929. Su alcuni funghi parassiti delle termiti. *Bollettino di Zoologia (Portici)* 22:39–48.

De Kesel, A. 1995. Population dynamics of *Laboulbenia clivinalis* Thaxter (Ascomycetes, Laboulbeniales) and sex-related thallus distribution on its host *Clivina fossor* (Linnaeus, 1758) (Coleoptera, Carabidae). *Bulletin et annales de la Société royale d'entomologie de Belgique* 131:335–348.

De Kesel, A. 1996. Host specificity and habitat preference of *Laboulbenia slackensis*. *Mycologia* 88:565–573.

Eriksson, O. E., H.-O. Baral, R. S. Currah, K. Hansen, C. P. Kurtzman, G. Rambold, and T. Laessøe eds. 2003. Outline of Ascomycota—2003. *Myconet* 9:1–89

Eriksson, O. E., and D. L. Hawksworth. 1993. Outline of the Ascomycetes-1993. *Systema Ascomycetum* 12:51–257.

Frank, J. H. 1982. The parasites of the Staphylinidae (Coleoptera). A contribution towards an encyclopedia of the Staphylinidae. *Bulletin of the Florida Agricultural Experiment Station* 824:1–118.

Gäumann, E. A., and C. W. Dodge. 1928. *Comparative morphology of the fungi*. New York: McGraw-Hill.

Hammond, P. M. 1995. Described and estimated species numbers: An objective assessment of current knowledge. In *Microbial diversity and ecosystem function,* ed. D. Alsopp, D. L. Hawksworth, and R. R. Colwell, pp. 29–71. Wallingford, UK: CABI Bioscience.

Hawksworth, D. L., P. M. Kirk, B. C. Sutton, and D. N . Pegler. 1995. A*insworth and Bisby's dictionary of the fungi*, 8th ed. Egham, UK: International Mycological Institute.

Heim, R. 1951. Mémoire sur l'*Antennopsis*, ectoparasite du termite de Saintonge. *Bulletin de la Société mycologique de France* 67:336–364.

Henk, D. A., A. Weir, and M. Blackwell. 2003. *Laboulbeniopsis*, an ectoparasite of termites newly recognized as a member of the Laboulbeniomycetes. *Mycologia* 95:561–564.

Henry, T. J., and R. C. Froeschner, eds. 1988. *Catalog of the Heteroptera or true bugs of Canada and the continental United States.* New York: Brill Academic Publishers.

Hill, T. W. 1977. Ascocarp ultrastructure of *Herpomyces* sp. (Laboulbeniales) and its phylogenetic implications. *Canadian Journal of Botany* 55:2015–2032.

Hojo M., T. Miura, K. Maekawa, R. Iwata, and A. Yamane. 2001. *Termitaria* species (Termitariales, Deuteromycetes) found on Japanese termites (Isoptera). *Sociobiology* 38:327–342.

Huldén, L. 1983. Laboulbeniales (Ascomycetes) of Finland and adjacent parts of the U.S.S.R. *Karstenia* 23:31–136.

Khan, S. R., and J. W. Kimbrough. 1974. Taxonomic position of *Termitaria* and *Mattirorella* (entomogenous deuteromycetes). *American Journal of Botany* 61:395–399.

Kimbrough, J. W., and R. Gouger. 1970. Structure and development of the fungus *Laboulbeniopsis termitarius*. *Journal of Invertebrate Pathology* 16:205–213.

Kimbrough, J. W., and M. Lenz. 1982. New species of *Termitaria* (Termitariales, Deuteromycetes) on Australian termites (Isoptera). *Botanical Gazette* (Crawfordsville) 143:262–272.

Kimbrough, J. W., and B. L. Thorne. 1982. Structure and development of *Mattirolella crustosa* (Termitariales, Deuteromycetes) on Panamanian termites. *Mycologia* 74:201–209.

Kirk, P. M., P. F. Cannon, and J. C. David. 2001. *Ainsworth and Bisby's dictionary of the fungi*, 9th ed. Wallingford, UK: CABI Bioscience.

Lichtwardt, R. W., M. M. White, M. J. Cafaro, and J. K. Misra. 1999. Fungi associated with passalid beetles and their mites. *Mycologia* 91:694–702.

Majewski, T. 1994. Laboulbeniales of Poland. *Polish Botanical Studies* 7:1–466.

Majewski, T., and J. Wisniewski. 1978. Records of parasitic fungi of Thaxteriolae group on sub-cortical mites. *Mycotaxon* 7:508–510.

Malloch, D. 1995. Fungi with heteroxenous life histories. *Canadian Journal of Botany* (supplement) 73:S1334–S1342.

Malloch, D., and M. Blackwell. 1990. *Kathistes*, a new genus of pleomorphic ascomycetes. *Canadian Journal of Botany* 68:1712–1721.

Malloch, D., and M. Blackwell. 1993. Life histories of three undescribed species of *Pyxidiophora* occurring on beached marine algae [abstract]. Annual meeting of the Mycological Society of America, Athens, Georgia, June 1993.

Müller, E., and J. A. von Arx. 1962. Die Gattungen der didymosporen Pyrenomyceten. *Beiträge zur Kryptogammenflora der Schweiz* 11:1–922.

Peyritsch, J. 1875. Über Vorkommen und Biologie von Laboulbeniacéen. *Sitzungsberichte der Kaiserlichen AkademieWissenschaften. Mathematische-Naturwissenschaftliche Classe*. Abt. 1 72:377–385.

Richards, A. G., and M. N. Smith. 1954. Infection of cockroaches with *Herpomyces* (Laboulbeniales). III. Experimental studies on host specificity. *Botanical Gazette* [Crawfordsville] 116:195–198.

Richards, A. G., and M. N. Smith. 1955a. Infection of cockroaches with *Herpomyces* (Laboulbeniales). I. Life history studies. *Biological Bulletin of the Marine Biological Laboratory* 108:206–218.

Richards, A. G., and M. N. Smith. 1955b. Infection of cockroaches with *Herpomyces* (Laboulbeniales). IV. Development of *H. stylopage* Spegazzini. *Biological Bulletin of the Marine Biological Laboratory* 109:306–315.

Richards, A. G., and M. N. Smith. 1956. Infection of cockroaches with *Herpomyces* (Laboulbeniales). II. Histology and histopathology. *Annals of the Entomological Society of America* 49:85–93.

Rossi, W., and M. Blackwell. 1990. Fungi associated with African earwigs and their relationship to South American forms. *Mycologia* 82:138–140.

Rossi, W., and A. Weir. 1998. *Triainomyces*, a new genus of Laboulbeniales on the pill millipede, *Procyliosoma tuberculatum* from New Zealand. *Mycologia* 90:282–289.

Santamaria, S. 1995. Sobre alguns fongs rars recollectats en insectes vius. *Revista Sociedad Catalana de Micologia* 18:137–150.

Scheloske, H.-W. 1976. *Eusynaptomyces benjaminii* spec. Nova (Ascomycetes, Laboulbeniales) und seine Anpassungen an das Fortpflanzungsverhalten seines Wirtes *Enochrus testaceus* (Coleoptera, Hydrophilidae). *Plant Systematics and Evolution* 126:267–285.

Spegazzini, C. 1918. Observaciones microbiologicas. *Anales de la Sociedad Científica Argentina* 85:311–323.

Sugiyama, K. 1973. Species and genera of the Laboulbeniales (Ascomycetes) in Japan. *Ginkgoana* 2:1–97.

Tavares, I. I. 1985. Laboulbeniales (Fungi, Ascomycetes). *Mycologia Memoir* 9:1–627.

Thaxter, R. 1896. Contribution towards a monograph of the Laboulbeniaceae. Part I. *Memoirs of the American Academy of Arts and Science* 12:187–429.

Thaxter, R. 1908. Contribution toward a monograph of the Laboulbeniaceae.Part II. *Memoirs of the American Academy of Arts* 13:217–469.

Thaxter, R. 1914. On certain fungus-parasites of living insects. *Botanical Gazette* [Crawfordsville] 58:235–253.

Thaxter, R. 1920. Second note on certain fungus-parasites of living insects. *Botanical Gazette* [Crawfordsville] 69:1–27.

Thaxter, R. 1926. Contribution towards a monograph of the Laboulbeniaceae. Part IV. *Memoirs of the American Academy of Arts* 15:427–580.

Verhoeff, K. W. 1897. Beiträge zur vergleichenden Morphologie, Gattungs- und Artsystematik der Diplopoden, mit besonderer Berücksichtigung derjenigen Siebenbürgens. *Zoologischer Anzeiger* 20:97–125.

Weir, A. 1996. A preliminary host-parasite list of British Laboulbeniales (Fungi; Ascomycotina). *The Entomologist* 115:50–58.

Weir, A., and M. Blackwell. 2001a. Extraction and PCR amplification of DNA from minute ectoparasitic fungi. *Mycologia* 93:802–806.

Weir, A., and M. Blackwell. 2001b. Molecular data support the Laboulbeniales as a separate class of Ascomycota, Laboulbeniomycetes. *Mycological Research* 105:715–722.

Weir, A., and P. Hammond. 1997. Laboulbeniales on beetles: Host utilization patterns and species richness of the parasites. *Biodiversity and Conservation* 6:701–719.

Part II

Fungi Mutualistic with Insects

7

Reciprocal Illumination
A Comparison of Agriculture in Humans and in Fungus-growing Ants

Ted R. Schultz
Ulrich G. Mueller
Cameron R. Currie
Stephen A. Rehner

I have seen great surprise expressed in horticultural works at the wonderful skill of gardeners, in having produced such splendid results from such poor materials; but the art has been simple, and, as far as the final result is concerned, has been followed almost unconsciously.

Charles Darwin
On the Origin of Species

The fungus-growing ants of the tribe Attini (subfamily Myrmicinae) rely on the cultivation of fungi for food. The cultivated fungi are the sole source of nutrition for the larvae and the principal source of nutrition for the adults. All of the approximately 210 described attine ant species occur exclusively in the New World. Because the Attini are monophyletic and because no other ants are known to cultivate fungi, fungiculture is thought to have arisen a single time in ants. Attine ant fungiculture is perhaps the most unusual example of the more general phenomenon of ant agriculture, which has originated many times. Diverse ant species and clades cultivate mutualistic plants by removing weeds, eliminating pests, planting seeds, and providing soil and manure; other ant species herd, protect, and even breed mutualistic aphids and other homopterans (Hölldobler and Wilson 1990; Schultz and McGlynn 2000). No doubt many general ecological patterns and principles could be elucidated by comparing the full range of ant and human agriculture. This chapter provides the first step in such an exercise by focusing on the much more limited comparison of the agricultural systems of fungus-growing ants and humans.

Most fungus-growing ant species, including the most "primitive," belong to the eight genera collectively known as the lower Attini (fig. 7.1, table 7.1). Lower attines are mostly inconspicuous, cryptic species with relatively small colonies of a dozen to a thousand individuals and small to moderate-sized fungus gardens (Price et al.

149

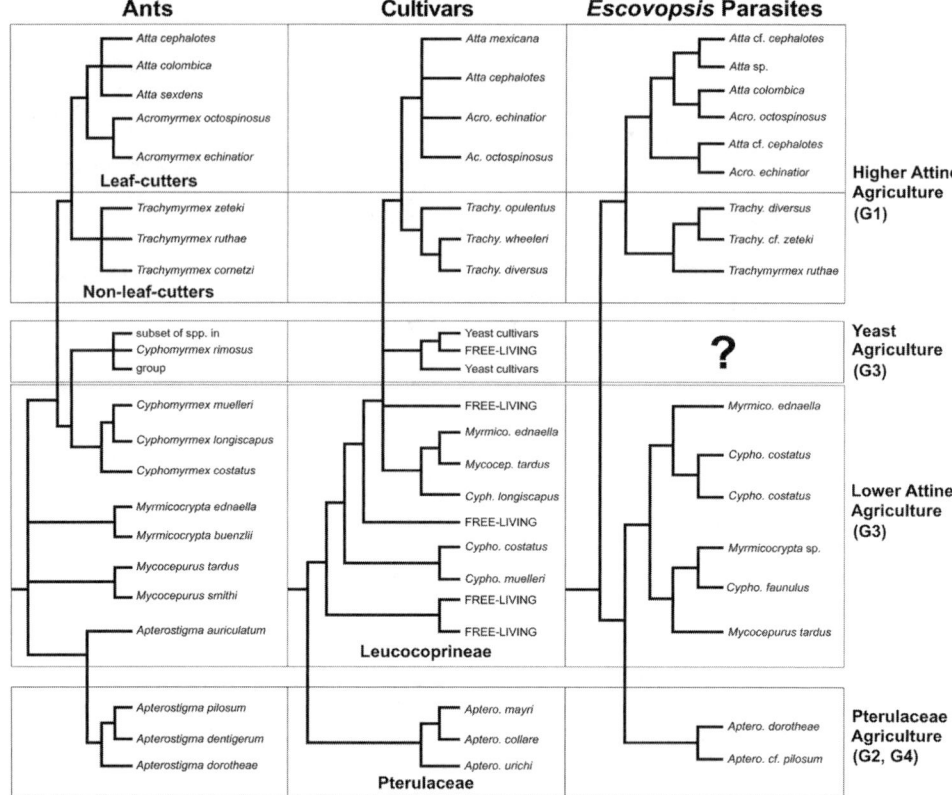

Figure 7.1. The four principal attine agricultural systems, juxtaposing the phylogenies of the attine ants, their domesticated fungi, and their *Escovopsis* garden parasites. The congruence of the topologies at more ancient phylogenetic levels indicates that these organisms have coevolved. G1, G2, G3, and G4 are names used for the respective domesticate groups in some previous publications.

2003). Attine agriculture reaches its most obvious culmination in the leaf-cutting ants, consisting of 43 species in the genera *Acromyrmex* and *Atta*. Leaf-cutting ants are the dominant herbivores of New World tropical forests and savannahs, and the greatest bane of Neotropical agriculture (Cherrett 1986; Fowler et al. 1986a). Colonies of some *Atta* species may contain up to 7 million individuals and persist for 20 years (Fowler et al. 1986b; Price et al. 2003).

Because leaf-cutting ants impact the environment significantly and because their nests and foraging columns are highly conspicuous, humans have paid special attention to them since prehistoric times. They play a major role, for example, in the ancient Mayan creation myth, the *Popul Vuh* (Wheeler 1907; Tedlock 1985). Although most early observers assumed that the ants directly consumed their cut leaves (Buckley 1860), the nineteenth-century English naturalist Henry Walter Bates disagreed, concluding instead that "the leaves are used to thatch the mounds to keep

Table 7.1. The fungus-growing ant genera and subgroups, with approximate numbers of described ant species and their known associated fungal domesticate groups.

Ant Group	Ant Genus	No. of Species	Fungal Domesticate[a]
Lower Attini	*Myrmicocrypta*	24	Lower attine Leucocoprineae
	Mycocepurus	5	(G3)
	Apterostigma (basal spp.)	13	
	Mycetarotes	4	
	Mycetophylax	6	
	Mycetosoritis	5	
	Cyphomyrmex (basal spp.)	22	
	Apterostigma (derived spp.)	34	Pterulaceae (G2, G4)
	Cyphomyrmex (derived spp.)	38	Yeast Leucocoprineae (G3)
Uncertain placement	*Mycetoagroicus*	3	Unknown
Higher Attini	*Trachymyrmex*	40	Higher attine Leucocoprineae
	Sericomyrmex	19	(G1)
	Acromyrmex	27	
	Pseudoatta (social parasites)	1	
	Atta	16	

[a]G1, G2, G3, and G4 are names used for the respective domesticate groups in some previous publications.

out the deluging rains and protect their broods within" (Bates 1863, p. 12). In a striking example of synchronous scientific discovery, Thomas Belt (1874) and Fritz Müller (1874) independently discovered the true purpose of leaf cutting. In the words of Belt: "I believe the real use [the ants] make of [the leaves] is as a manure, on which grows a minute species of fungus, on which they feed;—that they are, in reality, mushroom growers and eaters" (Belt 1874, p. 79).

In this chapter, we ask whether and to what extent analogous ecological forces have shaped the symbioses between humans and their domesticated plants and animals on the one hand and the symbioses between attine ants and their fungi on the other. We also ask whether knowledge about human agricultural evolution can inform and structure attine biological research and, conversely, whether the study of attine biology can inform human agricultural practice. Hundreds of extant and many more extinct species of attine ants have, after all, successfully practiced a stable and sustainable agricultural strategy for approximately 50 million years (Mueller et al. 2001), whereas various populations of the single human species have practiced agriculture for a maximum of 10,000 years (Smith 1998a).

The Attine Agricultural Symbiosis

The fungi cultivated by the majority of attine species are parasol mushrooms in the monophyletic tribe Leucocoprineae (Agaricaceae) (fig. 7.1, table 7.1; Heim 1957; Hervey et al. 1977; Chapela et al. 1994; Mueller et al. 1998). The Leucocoprineae contains the genera *Leucocoprinus* and *Leucoagaricus*, as well as a few species currently assigned to *Lepiota* (Mueller et al. 1998; Johnson 1999). The salient features of the *Leucocoprinus* life cycle are summarized in figure 7.2. The earliest diverging clades within the attine ant genus *Apterostigma* also cultivate fungi from this group (Villesen et al. in press), but all species in one derived *Apterostigma* clade cultivate species of the distantly related coral fungi (Pterulaceae) (Munkacsi and McLaughlin 2001; Villesen et al. in press). Based on cultural characters (Chapela et al. 1994) and on phylogenetic analyses of ribosomal RNA genes (Chapela et al. 1994; Mueller et al. 1998; Rehner et al. unpublished data), the attine fungi are currently divided into four major groups (table 7.1, fig. 7.1): (1) typical lower attine fungi, thought to be least diverged from the ancestral attine domesticate (figs. 7.3, 7.4; Mueller et al. 1998); (2) yeastlike fungi, a derived, monophyletic subgroup within the lower attine fungi that grow as yeast morphs when associated with ants rather than as typical attine mycelial morphs (figs. 7.5, 7.6; Wheeler 1901, 1907;

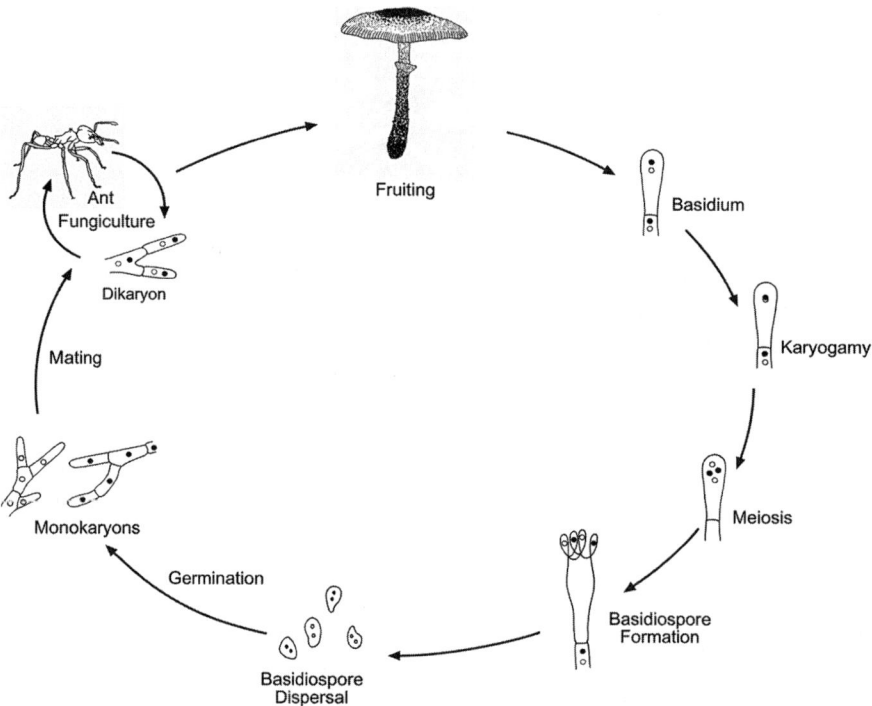

Figure 7.2. The life cycle of *Leucocoprinus*, a typical hymenomycete (true mushroom-forming fungus). The principal cellular and morphological structures produced in the hymenomycete life cycle include: (1) dikaryotic mycelium, containing two physically separate, haploid nuclei, sometimes considered functionally diploid; (2) fruiting bodies (mushrooms) produced from dikaryotic mycelia; (3) basidia, specialized sex cells formed on the gill surfaces and the site of long-delayed nuclear fusion (karyogamy) and meiosis; and (4) basidiospores—haploid meiospores, usually air-dispersed, that germinate to form a monokaryotic and haploid mycelium. Mycelia fuse (plasmogamy or mating) with sexually compatible monokaryons to form a dikaryotic mycelium.

Figure 7.3. A free-living (feral) lower-attine mushroom (Leucocoprineae) growing in the leaf litter in Panama. (Photograph by U. G. Mueller.)

Figure 7.4. Lower attine agriculture: the fungus garden of *Cyphomyrmex faunulus* constructed on the underside of a rotten log in São Gabriel, Amazonas, Brazil. (Photograph by T. R. Schultz.)

153

Figure 7.5. A free-living (feral) attine yeast agriculture mushroom (Leucocoprineae) growing in the leaf litter in Panama. (Photograph by U. G. Mueller.)

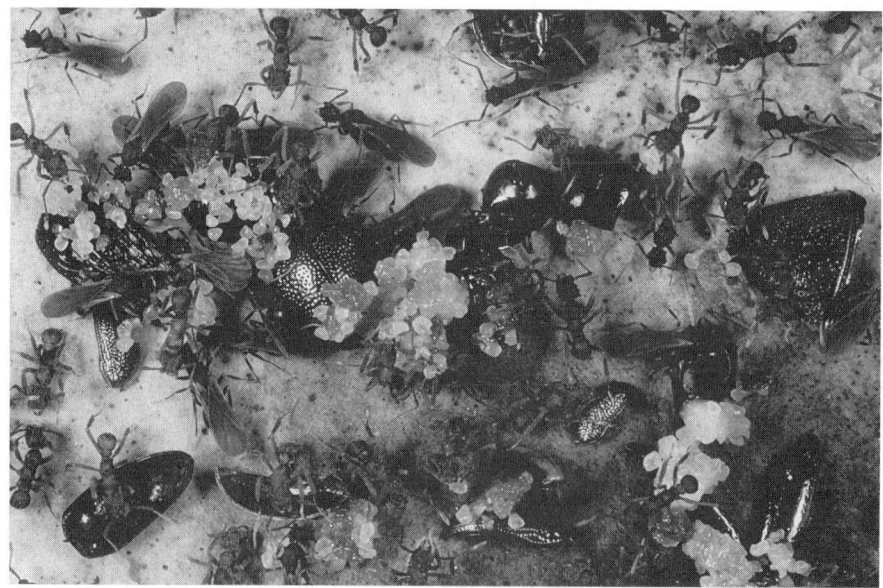

Figure 7.6. Attine yeast agriculture: the yeast fungus garden of *Cyphomyrmex salvini* taken from a cavity in a rotten log at La Selva, Costa Rica. (Photograph by T. R. Schultz.)

154

Weber 1972, 1982; Mueller et al. 1998); (3) the highly derived, monophyletic higher-attine fungi cultivated by the higher Attini, including the leaf-cutting ants, originating from a lower-attine–like leucocoprineaceous ancestor (figs. 7.7–7.9; Rehner et al. unpublished data); and (4) the attine pterulaceous fungi, divided into two monophyletic sister groups: the nonveiled pterulaceous fungi, cultivated by an apparently paraphyletic group of ants in the genus *Apterostigma*, and the veiled pterulaceous fungi, cultivated by a monophyletic group of ants in the genus *Apterostigma* that weave the aerial hyphae into a characteristic tentlike veil that surrounds the garden (figs. 7.10, 7.11; Villesen et al. in press).

Upon leaving the maternal nest, an attine daughter queen carries within her infrabuccal pocket a pellet of natal-nest cultivar, which serves as the nucleus for her new garden (von Ihering 1898; Huber 1905a,b). This behavior leads to the clonal propagation of garden fungi from parent to daughter nests, at least over short evolutionary time periods. The pattern of strict ant–fungus co-cladogenesis expected from this garden-founding behavior is disrupted over longer evolutionary time periods, however, because lower attine colonies occasionally replace their domesticates with free-living fungi and because both lower and higher attine ants replace their domesticates with fungi obtained from other attine ant colonies (Mueller et al. 1998; Bot et al. 2001; Green et al. 2002; Rehner et al., unpublished data).

The majority of attine leucocoprineaceous and pterulaceous gardens are infected by a highly specialized "crop disease" caused by species of the ascomycete fungal genus *Escovopsis* that so far are known only from attine fungus gardens (Currie et al. 1999a, 2003a; Currie 2001a,b). Low-level, chronic *Escovopsis* infections diminish garden and colony growth rates. At times, *Escovopsis* can also overrun and decimate

Figure 7.7. A higher attine mushroom (Leucocoprineae), growing from the surface of a nest of the leaf-cutting ant *Acromyrmex disciger.* (Reprinted from Möller 1893.)

Figure 7.8. Higher-attine agriculture: the fungus garden of *Trachymyrmex septentrionalis*, collected from a subterranean nest in Long Island, New York, USA. Clusters of gongylidia are visible on the garden surface. (Photograph by T. R. Schultz.)

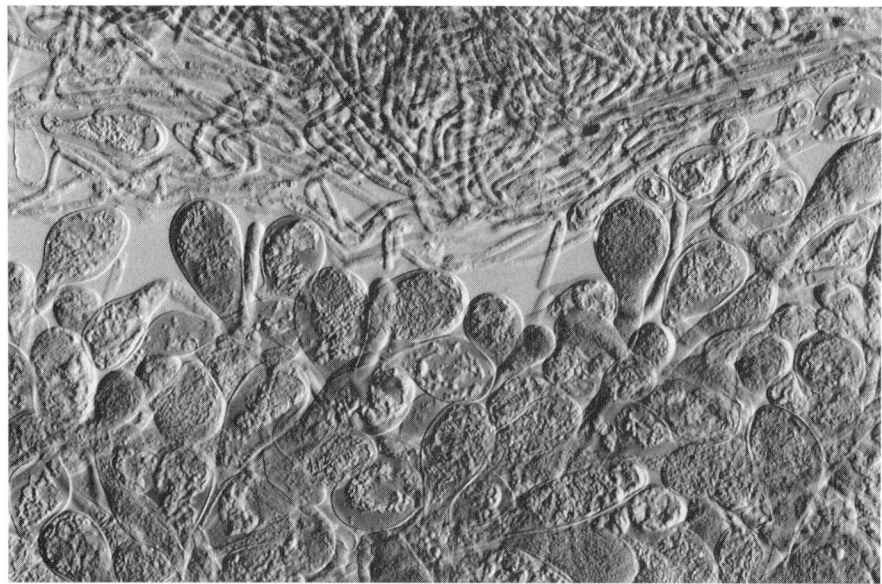

Figure 7.9. Lower half: gongylidia, the nutritious swollen hyphal tips produced by higher attine domesticated fungi and harvested by the ants for food; from the fungus garden of *Atta cephalotes*. Upper half: typical mycelium from the same garden. (Photograph by U. G. Mueller.)

Figure 7.10. Basidiocarps of *Pterula typhuloides*, a free-living species in the Pterulaceae (coral fungi), the family closely related to the domesticates of the pterulaceous-cultivating *Apterostigma* attine ant species, growing on decaying leaves of an unidentified dicot tree on Bordeaux Mountain, St. John, U.S. Virgin Islands. (Photograph by T. J. Baroni.)

gardens, usually resulting in the deaths of both ant and fungal cultivar symbionts (Currie 2001a,b). Ants are able to detect *Escovopsis* spores and hyphae and to remove them by "weeding" (Currie and Stuart 2001); in the leaf-cutting ants, specialized "garbage caste" workers handle garden refuse and have minimal contact with other castes, presumably to prevent the spread of *Escovopsis* within ant colonies (Hart and Ratnieks 2002). In the myrmecological equivalent of biological pest control, attine ants culture actinomycete bacteria of the family Pseudonocardiaceae in specialized locations on their exoskeletons (Currie et al. 1999b, 2003b). The actinomycetes

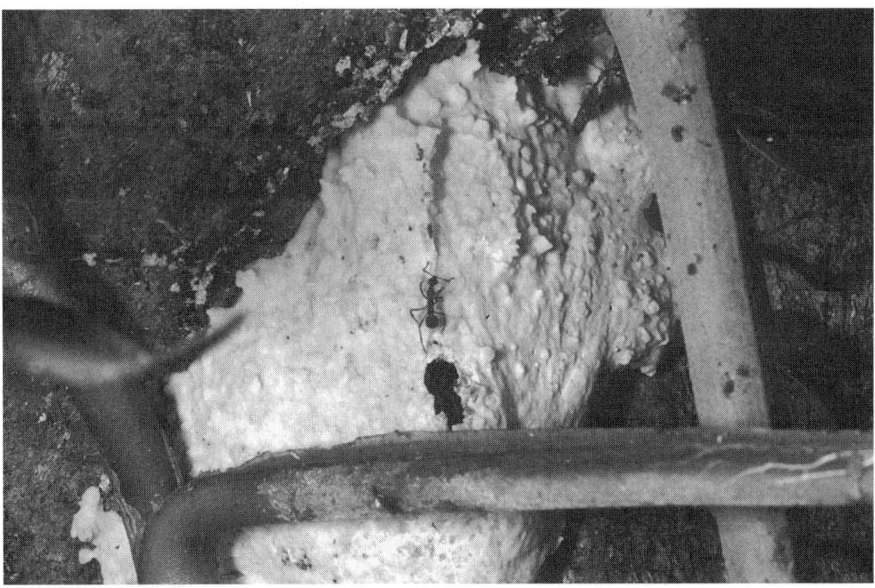

Figure 7.11. Attine pterulaceous agriculture: the fungus garden of *Apterostigma collare*, constructed in a tree in La Selva, Costa Rica, and surrounded by a mycelial veil constructed by the ants. The veil is characteristic of one of the two subgroups of pterulaceous attine fungal domesticates. (Photograph by T. R. Schultz.)

produce an antibiotic with specific action against *Escovopsis* (Currie et al. 1999b, 2003b). Because the study of attine crop diseases is an entirely new area of research, it is likely that additional fungus-garden pathogens await discovery.

The relationship between attine ants and their associated fungi has been variously regarded as either parasitism, in which the ants benefit at the expense of the fungi, or as mutualism, in which both partners benefit. In either case, the benefit to the ants is clear enough: the colony cannot survive without a fungus garden. The benefit to the fungus is less obvious, however, leading to the widespread assumption, also made about human domesticates, that attine garden fungi are essentially enslaved by the ants (Mueller 2002). This enslavement scenario implies a significant conflict of interest in which the garden fungus is continuously trying to escape from the symbiosis, especially through fruiting (i.e., forming mushrooms), and in which the ants actively suppress the formation of fruiting bodies through constant policing and pruning of mycelium (Autuori 1940; Muchovej et al. 1991; Fisher et al. 1994a,b; Mueller 2002). Evidence in support of this scenario includes the observation that fruiting bodies are absent from well-populated nests but that fruiting may occur in nests in which ant populations are diminished or absent (Mueller 2002). The mutualistic scenario, in contrast, argues that the association significantly increases the fitness of the attine fungus, relative to its fitness in the free-living state, in at least three ways: (1) by increasing its representation in the next generation, (2) by increasing its geographic distribution due to dispersal by foundress queens, and (3) by providing protection from parasites and pathogens (e.g., *Escovopsis*) due to various activities of the ants. Even under a mutualistic scenario, however, some subset of the separate evolutionary interests of ants and fungi are likely to be in conflict (see Mueller 2002). For example, because fungi are dispersed only by queens, a biased sex ratio favoring females serves the interests of the fungi, but not necessarily the interests of the ants. Conflicts of interest are related to issues of control and enslavement, which are discussed below.

Agricultural Evolution

Human agriculture arose independently at least nine times during the past 10,000 years (table 7.2). The resulting agricultural systems differ in many ways, most notably in the particular species of domesticated plants and animals. In spite of these differences, some researchers have proposed features shared by all systems. Some have also proposed general conditions that may have propelled some societies to make the transition from the ancestral strategy of hunting-gathering, in which humans obtain all of their nutrition from wild plant and animal sources, to the derived strategy of agriculture, in which humans obtain a significant proportion of their nutrition from domesticated plants and animals (e.g., Sauer 1952; Flannery 1973; Rindos 1984; MacNeish 1991; Harlan 1992; Diamond 1997; Smith 1998a).

Generalizations about human agricultural evolution are complicated by the observation that hunting-gathering and agriculture are two extremes in a complex continuum of food acquisition strategies (Smith 1998a,b, 2001a,b; Pringle 1998). In fact, many stable human societies have employed (and, in some cases, continue

Table 7.2. Nine independent origins of human agriculture, with dates of origin and primary domesticates.

Region	Date	Plant Domesticates	Animal Domesticates
Near East (Fertile Crescent)	10,000 BP	Wheat, barley, lentils	Sheep, goat, pig, cattle
New Guinea	10,000 BP	Banana, taro	Pig
Mesoamerica	9000 BP	Maize, beans, squash	Turkey
Southern China (Yangtze River)	8500 BP	Rice	Pig, water buffalo, chicken, silkworm
Northern China (Yellow River)	8000 BP	Millet	Pig, water buffalo, chicken, silkworm
South Central Andes	7000 BP	Quinua, potato, beans	Llama, guinea pig
Lowland Neotropics	7000 BP	Yams, manioc, arrowroot, beans, peanuts, peach palm	None
Eastern United States	5000 BP	Sunflower, goosefoot, squash	None
Sub-Saharan Africa	5000 BP	Millet, sorghum, African rice	Cattle, guinea fowl

From Diamond (1997), Piperno and Pearsall (1998), Smith (1998a), Denham et al. (2003), and Neumann (2003).

159

to employ) successful strategies that include various combinations of hunting and gathering, management of local environments, and management of domesticated plants and/or animals. In some cases these domesticates were regularly imported from the wild and thus remained unmodified relative to wild populations (e.g., goats in the ancient Near East; Zeder and Hesse 2000); in others, the domesticates were modified due to human-mediated (artificial) selection (e.g., squash in Mesoamerica; Smith 1997).

Unlike most human agricultural systems (table 7.2), which incorporate multiple, distantly related domesticates, ant colonies are dependent on a single crop. A given attine ant species is remarkably faithful to a particular subclade of closely related fungi within the four groups indicated in table 7.1. Although different colonies of an attine ant species may cultivate different variants (whether these are species or subspecific strains is unknown) within its associated fungal subclade, as far as is known, an ant colony cultivates a single clonal monoculture at any given time. Thus, whereas human agriculturalists rely on multiple domesticates, and no single human domesticate provides a complete diet (table 7.2), attine ant colonies obtain their nutrition from a single fungal clone; the adult diet may be supplemented, however, with leaf sap, nectar, fruit juices, and possibly other food sources (see below) encountered by the foraging adults (Littledyke and Cherrett 1976; Quinlan and Cherrett 1979; Bass and Cherrett 1995; Oliveira et al. 1995; Murakami and Higashi 1997; Leal and Oliveira 2000).

The two most frequently cited advantages of human agriculture are, first, the energy savings compared to hunting and gathering and, second, the relative reliability and predictability of the agricultural food resource (Hayden 1995; Diamond 1997; Piperno and Pearsall 1998). These advantages do not necessarily lead to a state of complete reliance on agriculture, as is demonstrated by the many historical and extant stable-state human societies that practice mixed food-acquisition strategies. Some studies suggest that in resource-rich environments, mixed strategies and perhaps even pure hunting-gathering may require less effort than a strategy of complete reliance on agriculture (Boserup 1965; Lee 1968; Lee and DeVore 1968; Sahlins 1968; Pimentel and Hall 1989; Harlan 1992). However, the issue of the relative labor costs versus nutritional returns of hunting-gathering, mixed strategies, and agriculture is unresolved and remains the subject of continuing research and debate (Piperno in press).

As human populations have increased and wild resources have become limited, agriculture has, as a matter of historical record, replaced hunting-gathering in most human societies (Smith 2001a,b). In contrast, foraging has remained a necessary component of the attine ant food-acquisition strategy because attine fungal domesticates are saprophytic biomass consumers (unlike the plant domesticates of humans, but more like human animal domesticates), and ants must forage to obtain that biomass for their fungi. Various studies (e.g., Turner 1974; Quinlan and Cherrett 1979; Bass and Cherrett 1995) suggest that attine ants expend as much foraging energy and import more biomass than do equivalently sized colonies of predatory/omnivorous hunter-gatherer ants, and that the net yield in ant biomass per unit foraging effort may be similar to that of the hunter-gatherers, at least for the lower Attini. If so, then, of the two cited advantages accruing to human agriculturalists, the second, reliability, must be more important for attine agriculture. Growing fungi allows attine ants to occupy a niche unoccupied by other ants, thereby releasing them from direct competition for protein sources such as live prey and dead arthropods. Unlike humans, who have always relied on animals and plants as sources of food, initially consuming and later cultivating fungal intermediates allowed the ancestors of the Attini to access a food source previously inaccessible to ants, that of living and dead plant tissue. In collecting insect frass and small plant parts for garden substrate, lower attine ants in fact compete for food with fungal and bacterial detritivores rather than with predators and scavengers; they must locate and use these resources before they have been colonized and degraded by microbes. In collecting large volumes of living plant material for garden substrate, leaf-cutting higher attine ants have shifted from competing with detritivores for dead vegetable material to competing with vertebrate and invertebrate herbivores for living plant tissue.

Associating with attine ants also represents a major shift in food acquisition strategy for the leucocoprineaceous fungi: instead of relying on passive dispersal, they use an agent that actively locates, concentrates, and prepares suitable substrates before they are degraded by competitor microbes. In the case of the higher attine fungi, the symbiosis has provided access to an entirely new, previously unavailable resource: living vegetation. Living leaves and other plant parts are normally protected from fungal invasion by a variety of defenses, including waxy coats and other physical barriers; higher attine ants remove those barriers by extensively preparing substrates for consumption by their fungal domesticates (Cherrett et al. 1989). Similar strategies of using animal vectors for directed dispersal to suitable resources have evolved many times among fungi (e.g., pollinator-dispersed rusts; Webber and Gibbs 1989), and it is not implausible that many fungal groups, including the Leucocoprineae, use ants as vectors for dispersal to competitor-free resources.

Agricultural Evolution: Origins

Like humans, the ancestral food-acquisition strategy of ants is that of omnivorous hunter-gatherer. In contrast to human agriculture, attine ant agriculture had a single origin sometime around 45–65 million years ago, and all extant attine ants are descended from a single agricultural pioneer (Schultz and Meier 1995; Meier and Schultz 1996; Wetterer et al. 1998; Mueller et al. 2001). Although an intermediate state probably once existed, in which proto-Attini were facultative associates of symbiotic fungi, no such extant species are known. Detailed ecological investigation may eventually identify a nonattine guild of facultative ant fungivores, as hypothesized by Mueller et al. (2001). Alternatively, ant lineages loosely associated with fungi may have gone extinct during attine evolution because they were outcompeted by early attine ant lineages with tighter coevolutionary associations. A similar competitive sweep has occurred in humans, so that today most human mid-range, low-level food-production economies have been supplanted by agriculture (Smith 2001a,b).

Agriculture was absent during the first 90,000 years of human (*H. sapiens*) history. Yet it has arisen multiple times during the past 10,000 years. This dramatic change in human food-acquisition behavior is sometimes attributed to the global shift toward increasingly benign and stable climatic conditions following the Pleistocene glaciation, which tended to favor agriculture, as well as changes in regional climatic conditions in the major agricultural centers of origin (Piperno and Pearsall, 1998; Smith 1998a). Climate change of a different and more drastic sort also has been suggested as a condition predicating ant agriculture. Citing Janzen (1995), Mueller et al. (2001) speculated that the "nuclear winter" following the Cretaceous-Tertiary extinction event of 65 million years ago may have favored detritivores and precipitated the curious, possibly simultaneous origins of attine fungiculture, restricted to the New World, and termite fungiculture, restricted to the Old World (Mueller et al. 2001). In the words of Janzen (1995, p. 785):

> What animals are most likely to survive a serious nuclear winter? Those whose food in some form does not directly depend on immediate photosynthesis. That is to say, those that eat dormant seeds and insects, those that eat decaying organic matter (especially nongreen plant parts), and those that eat these eaters. And especially those that are very good at finding small particulate bits of these resources, scattered and dwindling until sunlight again can penetrate the clouds in amounts sufficient for serious vegetation growth. That is to say, seed- and detritus-eating invertebrates and the invertebrates and small vertebrates that eat them and each other.

One obvious requirement for the origin of agriculture is that humans must have lived for extended periods of time in the vicinity of, and regularly encountered while foraging, suitable domesticates, such as plants and animals that were potentially useful to humans and that possessed traits that, in aggregate, preadapted them for domestication (Diamond 1997; Smith 1998a). This requirement would appear to have been met by the ancestral attine ant as well, which, like the majority of tropical rainforest ants, foraged in the leaf litter, where it frequently encountered both the vegetative mycelium and fruiting bodies (mushrooms) of leucocoprineaeceous fungi.

In their original forms, proto-domesticates may or may not have been useful to humans. In either case, according to the "camp follower" scenario of agricultural origins (see below), they were likely to have been preadapted to thrive in human-disturbed habitats (Flannery 1973; Bye 1981; Rindos 1984; Harlan 1992; Diamond 1997; Smith 1998a). In the case of immediately useful plants, human hunter-gatherers could have incidentally broadcast the seeds (or other propagules) into disturbed areas around their villages, thrown them away with uneaten refuse into garbage piles, or consumed and subsequently deposited them as human waste. Plants that thrived in such microenvironments would become "camp followers," growing in relatively greater abundances in the vicinity of humans. Alternatively, nonuseful, camp-following weeds preadapted for disturbed habitats could have invaded human settlements on their own, without the aid of humans. In either case, the camp-follower scenario has two requirements: (1) Proto-domesticates must have thrived in one or more microenvironments associated with human-disturbed habitats, and (2) they must have been immediately or potentially useful to humans. Camp-following plants that were not immediately useful may initially have been ignored, tolerated, or even removed by humans. Due to their continuing proximity to humans over time, however, any even minimally useful variants were (consciously or unconsciously) favored, marking the beginning of domestication and human-mediated selection (de Tapia 1992, citing Bye 1981). Even if, as suggested by some researchers (D. Piperno, pers. comm.), camp-following proto-domesticates were not the same species that were subsequently domesticated by humans, it is possible that humans acquired the knowledge and skills that eventually led to agriculture through such early interactions.

Examples of camp-following mutualisms occur in primates, including humans. *Cebus* monkeys feed on the fruit and distribute the seeds of *Gustavia* trees and, by also feeding on the buds, influence the fruiting pattern of the trees to their advantage (Oppenheimer and Lang 1969). Stands of baobab trees (*Adansonia* spp.) in Africa are closely associated with occupied or deserted human villages. Humans eat the fruit and use the leaves as potherbs, the bark as a source of fiber, and the large hollow boles as reservoirs for the storage of water (Harlan 1992). Fruit trees occur in corridors that line the paths used by Congo Pygmies, presumably due to casual discarding of fruit pits and defecation while traveling (Laden 1992, cited in Hayden 1995).

The camp-follower hypothesis shares similarities with a number of hypotheses proposed for the origin of the attine ant–fungus mutualism (reviewed by Mueller et al. 2001). Ant-disturbed microenvironments in which the leucocoprineaceous proto-domesticate most likely thrived include both the nest refuse pile and leaf litter or other substrates adjacent to the nest. The infrabuccal pellet hypothesis proposed by Bailey (1920) and further developed by Mueller et al. (2001) suggests that the proto-domesticate was useful to the ants early in the association, and this hypothesis thus provides a mechanism for active transport of the proto-domesticate by the ants into the vicinity of the nest. The infrabuccal pellet hypothesis is based on the observation that all ants accumulate food particles and detritus in their infrabuccal pockets, a specialized pouch in the bottom of the mouth that filters out and accumulates solid particles picked up during grooming or strained out from liquid food. These infrabuccal pellets have been shown to contain a significant proportion of fungal spores and hyphae (Bailey 1920; Letourneau 1998). Infrabuccal pellets are expelled by individual ants at a rate of about one per day in colony refuse piles inside or outside the nest or, by foragers, at random locations in the vicinity of the nest (Quinlan and Cherrett 1978a; Febvay and Kermarrec 1981; Mueller et al. 2001; Little et al. 2004).

Because spores and hyphae in the pellets are viable, infrabuccal pellets provide a plausible mechanism for the vegetative dispersal of fungi (Wheeler and Bailey 1920), and it is possible that some fungi have capitalized on this mechanism by making their hyphae attractive to ant foragers (Mueller et al. 2001). Although their function remains to be demonstrated, the mycelium of at least some lower-attine fungi possess hyphal swellings (Möller 1893; Urich 1895; Weber 1972; Mueller 2002) that may be homologous with the gongylidia of higher attine fungi (fig. 7.9). The latter are preferentially harvested and eaten by higher attine ants (Quinlan and Cherrett 1978b, 1979; Angeli-Papa and Eymé 1985). If leucocoprineaceous proto-gongylidia function as ant attractants, they are evolutionarily convergent, vegetative analogs of elaiosomes, nutritious seed appendages that serve as an inducement for ant dispersal of the seeds of a variety of unrelated herbaceous plants (Serenander 1906; Handel et al. 1981; Beattie 1985; Handel and Beattie 1990a,b).

Aside from the refuse pile, a leucocoprineaceous proto-domesticate could have thrived in the substrate associated with the nest of the ancestral attine ant. It is plausible that the ancestral attine ant nested between leaves in the leaf litter, the typical substrate of leucocoprineaceous fungi. Nests of various species in the lower attine genera *Apterostigma*, *Cyphomyrmex*, and possibly *Myrmicocrypta* occupy this habitat, as do the nests of many species in the putative sister group of the Attini, the Blepharidattini (*Wasmannia* and *Blepharidatta*) (Schultz and Meier 1995; Diniz et al. 1998; R.C.F. Brandao and J. Delabie, pers. comm.). Like all ants, leaf-litter–nesting ants clean nest substrate surfaces in the vicinity of their brood, removing fungi, bacteria, and other debris and applying antibiotic secretions. Any leucocoprineaceous fungus capable of exploiting this competitor-free microhabitat, whether imported as infrabuccal pellets or invading independently, could become a protected camp follower. Such exploitation might require avoiding ant sensory, mechanical, and biochemical antifungal defenses, whether or not the fungus had an initially neutral effect on ant colony fitness. This would place a camp-following fungus in the same ecological guild as the many arthropod commensals that have successfully overcome the same obstacles to facultatively or obligately inhabit the protected and ecologically predictable microenvironments provided by ant nests (Schultz and McGlynn 2000). Obviously, by evolving ant-attractant properties, a fungus dispersed by ants via infrabuccal pellets would already possess a number of requisite preadaptations for moving into the nest environment and, eventually, becoming domesticated. A single observation of an extant (hence, obligately fungivorous) attine nest is at least consistent with one feature of this camp-follower scenario, that of continuity between adjacent litter-inhabiting and ant-associated fungal individuals: A veiled pterulaceous garden of a litter-nesting *Apterostigma* colony was observed connected to an extensive mycelial mat occupying the leaf litter beyond the nest (Mueller 2002).

Agricultural Evolution: Domestication

Archaeologists have dated domestication events by studying the remains of ancient human-associated plants and animals and identifying morphological traits that are clearly modified relative to the corresponding traits in wild populations. In some cases, the archaeological record preserves domestication sequences spanning thousands of years. Examples of sequential modifications include increasing seed size and decreasing seed coat thickness in a variety of domesticates, and increasingly apical position of seeds on stalks, reduced stalk branching, and seed indehiscence in domesticated grains and maize. If domestication is defined as the management of captive plants and animals, regardless of whether those plants and animals are modified due to domestication by humans, then the earliest detectable changes in the archaeological record provide only minimum dates of origin. In other words, the historical origin of a particular human–domesticate symbiosis is necessarily older than the earliest detectable morphological markers encountered in the archaeological record.

Modifications in domesticates are the result of selective forces exerted by humans, both unconsciously (especially during the earlier stages of association) and intentionally (especially during the later stages). This human-mediated selection is commonly referred to as "artificial selection" in order to distinguish it from "natural selection." Three features of attine fungiculture provide opportunities for the operation of the ant analogue of artificial selection (Mueller 2002). The first is garden founding. Foundress queens depart from their parent nests carrying pellets of garden mycelium that serve as the starting seeds for their new gardens (Weber 1972; Mueller 2002). If genetic variants coexist in the mycelium of the parent garden (e.g., due to somatic mutation), and if foundress queens discriminate between variants when incorporating mycelium into their infrabuccal pockets, then they exert selection on the domesticate population. The second feature is garden propagation. Garden-tending workers select hyphae of growing mycelium from healthy parts of the garden and plant them on newly added substrate (Weber 1972). Again, if domesticate variants coexist in attine gardens, and if workers preferentially propagate one of these variants while ignoring others, then garden-tending workers exert selection on the domesticate population. The third feature is garden reacquisition. Although attine fungi are usually transmitted clonally from parent to daughter nests, genetic evidence indicates that new garden strains are acquired occasionally from the wild and from the gardens of other ants (Mueller et al. 1998; Adams et al. 2000). If foragers are able to distinguish between candidate strains, then they can choose which strains to import into the nest and thus exert selection on the extended fungal population.

In support of the ant-imposed artificial selection hypothesis, Mueller et al. (2004) recently documented that the lower attine ant *Cyphomyrmex muelleri* has an acute ability to discriminate between domesticate strains. When presented with a range of domesticate choices, workers of *C. muelleri* invariably preferred their native garden domesticate, discriminating against even very close relatives of the native domesticate. A similar ability to differentiate between closely related cultivar strains has been described in leaf-cutting *Acromyrmex* species by Bot et al. (2001) and Viana et al. (2001). These observations suggest that attine ants may impose artificial selection against unwanted, presumably inferior domesticates or, alternatively, selectively favor desirable domesticate types that are more nutritious, more resistant to disease, easier to cultivate, or otherwise beneficial. It remains to be determined, however, whether attine ants have the ability to detect and artificially select for ant-benefiting traits in domesticate genotypes or, alternatively, whether naturally arising domesticate mutants spread to fixation in a garden due to competitive superiority over other strains independent of ant-subculturing biases.

Under the assumption that selection regimes on domesticates differ between human-mediated and wild-type environments, fixation of desirable traits in domesticates (favorable gene combinations and mutations) requires reduced gene flow between domesticate populations and their ancestral free-living populations. In the absence of such a reduction, genetic variants favorable for domestication can only become fixed in the domesticate population if selection within the symbiosis is adequately strong and/or gene flow between domesticate and free-living populations is asymmetrical, such that gene flow from the symbiosis into the free-living population adequately outweighs the reverse gene flow, and genes favorable for life within the symbiosis introgress into the population as a whole and are reimported into the symbiosis in subsequent domestication events.

Barriers to gene flow between domesticated and wild populations, both inten-tional and incidental, have obviously played an important role in the histories of a significant number of human domesticates. These barriers have included the isola-tion of domesticates in discrete garden plots and livestock pens, asexual propaga-tion (e.g., with cuttings and tubers), the domestication of self-fertilizing plants (e.g., barley, wheat, oats, rice, and sorghum), and the domestication of reproductively isolated polyploid and translocation races (e.g., some potato strains). Domesticated Mesoamerican beans (*Phaseolus vulgaris*), for example, appear to have arisen from a single small population; a second, separate domestication occurred in the south-ern Andes (Gepts et al. 1986; Gepts and Debouck 1991). Barriers to gene flow also have included the human dispersal of domesticates to areas well outside of their natural ranges, including cross-continental dispersal (e.g., North American and European strains of potatoes and tomatoes are descended from only a few individ-uals; Rick 1976; Quiros 2003). While under domestication by ants, attine fungi have been similarly isolated from wild populations; however, genetic evidence indicates that these periods of isolation through clonal propagation have been relatively brief when measured on evolutionary time scales (Mueller et al. 1998; Rehner et al., unpublished data). Judging by the distribution of the Attini, it is also likely that, while under domestication, attine fungi have been carried into regions relatively inhospitable to leucocoprineaceous fungi, including deserts and seasonally dry habitats, and it is possible, though by no means proven, that such habitats, removed from natural free-living populations, have served as crucibles for the morphologi-cal modifications encountered in some of the attine domesticates (Fowler 1982).

Alternatively, the histories of many human domesticates suggest that barriers to gene flow may have been relatively permeable and that modifications of domesti-cates nonetheless occurred due to strong, human-mediated selection. In a prolonged domestication process that may have lasted for hundreds of years, maize (*Zea mays* ssp. *mays*) arose, probably in southern Mexico, from *Zea mays* ssp. *parviglumis* (Wang et al. 1999; Matsuoka et al. 2002). In spite of its extreme morphological modifications, domesticated maize retains the majority of the variability present in its progenitor subspecies, as well as variability acquired through subsequent intro-gression from another free-living subspecies, *Zea mays* ssp. *mexicana* (Eyre-Walker et al. 1998; Matsuoka et al. 2002; Vigouroux et al. 2002). Similar genetic variabil-ity due to persistent outbreeding with wild populations has been demonstrated for rice (Morishima and Oka 1979; Second 1982), barley (Brown and Munday 1982), dogs (Vilà et al. 1997, 1999; Leonard et al. 2002), horses (Vilà et al. 2001), and potatoes (discussed below). So common is this pattern of genetic continuity between human domesticates and free-living ("weed") populations that interactive crop–companion weed reciprocal evolution has been hypothesized as the prevailing norm in human agriculture (Harlan 1965; Wilson 1990), and such reciprocal systems have been demonstrated for pairs of domesticated and free-living populations in *Cucurbita* (Wilson 1989, 1990) and *Chenopodium* (Wilson 1981, 1990).

A relatively unexplored area of inquiry is modifications of humans due to selec-tion pressures exerted by their domesticates and serving the domesticates' evolu-tionary interests. Whether they serve the evolutionary interests of the domesticates, however, domesticate-related modifications of humans are well documented and include the varying frequencies, in some human populations relative to others, of such traits as lactose intolerance, wheat allergies, resistance to livestock-borne diseases, susceptibility to morning sickness as a function of diet, and ability to detect and/or tolerate a variety of toxic plant secondary compounds (Johns 1990; Jackson 1991; Haig 1993; Diamond 1997). Aside from such population-level genetic adaptations, humans as a species have remained genetically unmodified by their associations with domesticated plants and animals. This paucity of genetic change in humans contrasts sharply with the major changes that have occurred in their domesticates and tends to support the widely held, rather common-sense view that human plant and animal husbandry are symbioses of asymmetrical control, in which one symbiont, humans, has effectively enslaved the other symbionts and adapted them to human needs (discussed in detail below).

Unlike humans, attine ants are extensively modified for their agricultural symbiosis, and all such modifications are obviously the result of genetic (rather than cultural) evolution. Known modifications are largely behavioral and include the detection and removal ("weeding") of *Escovopsis* and other garden parasites (Currie 2001a,b); the cultivation of antibiotic-producing actinomycete bacteria on their exoskeletons (Currie et al. 1999b, 2003b); specialized foraging behaviors to select substrates suitable for their garden fungi; the transport of cultivar by virgin queens from parent to daughter nests (von Ihering 1898; Huber 1905a,b); the weaving of aerial hyphae by Pterulaceae-cultivating *Apterostigma* species into the protective tentlike veils that surround their gardens; and, in *Atta*, the division of workers into ethological and morphological castes specialized for garden tending, foraging, and refuse disposal. In addition, there are a series of suspected, but so far unstudied, nonbehavioral modifications that include physiological adaptations for fungivory; biochemical adaptations that enable attine ant species to specialize on narrowly defined domesticate groups; sensory adaptations for distinguishing between symbiont strains and between suitable and unsuitable fungi (e.g., parasitic fungi like *Escovopsis*); sensory adaptations for evaluating the health or growth rate of the garden in order to adjust foraging or weeding activities (Ridley et al. 1996; Currie and Stuart 2001); and, in the yeast-cultivating *Cyphomyrmex* species, the induction of the yeast morph in their garden fungi.

In contrast to the many adaptations present in the attine ants, modifications associated with domestication in the attine fungi are surprisingly difficult to document. One clear example is gongylidia (fig. 7.9), which serve as food for higher attine ants and which are so distinctive that the vegetative dikaryotic hyphae of the leaf-cutter fungus was described by Kreisel (1972) as a separate species, *Attamyces bromatificus*. When higher attine fungi are cultured in the laboratory in the absence of ants, the production of gongylidia frequently declines over time and may cease altogether, suggesting that gongylidia production is in some way linked to life with attine ants. Whether gongylidia are induced by specific nutritional or environmental conditions present within the symbiosis or whether they are the product of continuous positive selection by the ants during garden propagation remains an open question. Stradling (1978) considered the higher attine fungi to constitute a "rich and complete diet" for leaf-cutter ants, and Bass and Cherrett (1995) found that gongylidia prolonged the lives of *Atta* workers compared to an exclusively hyphal diet. Curiously, the scant data available suggest that, judged by crude protein, lipid, and carbohydrate proportions, the nutritional content of both the hyphae and gongylidia of attine fungi (so far analyzed only for the fungi of *Atta colombica* and *Atta sexdens*) are not obviously modified relative to the nutritional content of free-living leucocoprineaceous fungi, except that gongylidia appear to contain less protein and more lipids and carboydrates than do hyphae (Mueller et al. 2001). It remains possible that the higher attine fungi (or, indeed, all attine fungi) are modified in terms of (1) the production of particular amino acids, lipids, and/or carbohydrates; (2) the production of trace nutrients (e.g., vitamins, minerals, or steroids); (3) the loss of toxins present in the ancestral forms; (4) the production of ant-attractant allomones (Mueller et al. 2001); or (5) the sequestering of ant-produced colony-recognition hydrocarbons (Viana et al. 2001).

A second likely modification occurs in the *Cyphomyrmex* yeast fungi, cultivated by a probably monophyletic subset of ant species within the *Cyphomyrmex rimosus* group (Kempf 1966; Snelling and Longino 1992; Schultz and Meier 1995). Compared to conspicuous attine mycelial gardens, yeast gardens (fig. 7.6) are so easily overlooked that some early researchers concluded that *Cyphomyrmex rimosus*-group ants did not practice fungiculture (Forel 1893; Emery 1895; Urich 1895). Once discovered, yeast gardens proved so unusual that these domesticates are among the few attine fungi to have been specifically assigned a formal taxonomic name (*Tyridiomyces formicarium;* Wheeler 1907). Yeast gardens consist of small, irregularly shaped nodules about 0.5 mm in diameter that are composed of a fungus growing in the yeast phase (i.e., as separate, single cells; fig. 7.6) rather than in the typical mycelial phase, in which cells are connected in linear filaments (fig. 7.4). Ants nourish yeast gardens with insect frass and nectar collected while foraging. Nectar is transported to the nest in the crops of workers and regurgitated directly onto the garden; it is also shared with nestmates via trophallaxis (i.e., regurgitative feeding; Murakami and Higashi 1997).

Yeast-phase growth in the order Agaricales is entirely unexpected. Outside of the attine fungi, yeast morphology is known among the Basidiomycota (basidium-forming fungi, including the true mushrooms) only in two distantly related orders: the Tremellales (the jelly fungi), the basal lineage of the hymenomycetes (the true mushrooms), and the even more distantly related Ustilaginales (smut fungi) (Fell et al. 2001). In these two groups, the yeast phase occurs only in the uninucleate haploid state, whereas the attine *Cyphomyrmex* yeasts appear to be dikaryotic. The attine yeasts were derived independently of these other yeast groups. Although yeast-phase growth has a genetic basis, it appears to be induced in mycelium by the presence of *Cyphomyrmex* ants. This hypothesis is supported by four facts: (1) a free-living fruiting body of a feral yeast domesticate has been collected, produced by typical mycelial growth on leaf litter (fig. 7.5; Mueller et al. 1998); (2) phylogenetic analyses reconstruct the attine yeast fungi as a derived, monophyletic group nested within the leucocoprineaceous "clade 1" subclade of the lower attine fungi (fig. 7.1; Mueller et al. 1998); (3) the mycelial morph is also present in gardens, growing on the integuments of ant larvae (Schultz and Meier 1995); and (4) in culture, the yeast morph eventually reverts to mycelial growth (Mueller et al. 1998).

Because the attine yeast fungi all belong to a compact monophyletic group within the lower attine fungi (fig. 7.1), it is possible that, for reasons unknown and unrelated to ant fungiculture, they share a derived tendency to convert to yeast-phase growth under certain conditions. Under this hypothesis, the ants take advantage of this preexisting tendency to induce the yeast morph (perhaps by unusual gardening behaviors), and the lower attine yeast fungi are not necessarily modified for life with ants. The data, however, favor an alternative hypothesis. The independent origin of the attine yeasts, the complete absence of the yeast phase in other Homobasidiomycetidae (mushroom fungi), and the tight association of the yeast fungi with a probably monophyletic group of *Cyphomyrmex* species (Schultz 1995) all suggest that yeast growth is a derived modification for life with *Cyphomyrmex* ants. The adaptive function of the yeast morph is unknown, but at least two explanations are plausible. First, yeast nodules are easily transportable, allowing for a seminomadic existence and/or rapid escape from predators like army ants (LaPolla et al. 2002) and *Megalomyrmex* "agropredators" (Adams et al. 2000). Second, yeast gardens may be less susceptible to *Escovopsis* infection. So far, *Escovopsis* has not been isolated from yeast gardens (Currie, unpublished data), but this remains a largely uninvestigated question. It is interesting to note that dimorphic ascomycetes in several distinct clades have yeastlike growth phases in association with insects such as ambrosia and bark beetles. The derived yeastlike state occurs in mycangia, and hyphal conversion occurs in the beetle galleries.

Agricultural Pathogens

Human-domesticated plants and animals are infected by a range of pathogens, including fungi, bacteria, viruses, arthropods, and nematodes (Maloy 1993; Agrios 1997). Pathogens have devastated human agricultural societies throughout recorded history. Agricultural diseases are listed in the Old Testament, along with human diseases and war, as one of the great scourges of mankind. The study of crop diseases dates back to the Greek philosopher Theophrastus (c. 370–286 BC). More recently, the Irish potato famine of the 1840s, caused by the late blight of potato agent (*Phytophthora infestans*), resulted in the deaths of more than 2 million people (Lang 2001). This disaster demonstrates the potential of agricultural diseases to devastate human populations.

The gardens of attine ants are also devastated by pathogens. Although other garden pathogens and pests probably await discovery, the only currently known attine garden disease is caused by microfungi in the genus *Escovopsis* (Ascomycota: Hypocreales) (Currie et al. 1999a, 2003a; Currie 2001a), necrotrophic parasites that grow in contact with and extract nutrients from the attine fungal domesticates (Reynolds and Currie, in press). *Escovopsis* infections of fungus gardens are typically chronic, resulting in significantly decreased rates of garden growth and substantially depressed rates of worker production (Currie 2001b). Less typically, *Escovopsis* can rapidly overwhelm gardens, completely overgrowing them and leading ants to abandon the infected gardens, sometimes resulting in colony death (Currie et al. 1999a; Currie 2001a). *Escovopsis* is specialized on the attine symbiosis and has been found only in the nest habitats of both leucocoprineaceous and pterulaceous fungus-growing ants (Currie et al. 1999a; Bot et al. 2001; Currie 2001a,b). Molecular phylogenetic analyses indicate that *Escovopsis* was an early participant in the attine ant–microbe symbiosis and that it shares a long history of coevolution with the ants and their fungal domesticates (both leucocoprineaceous and pterulaceous; Currie et al. 2003a). Thus, like human agriculture, ant agriculture has a long history of crop disease.

Agricultural Pathogens: Disease Susceptibility and Control

In both human and ant agriculture, domesticates face increased susceptibility to disease for two reasons. First, cultivation involves growing domesticated organisms at greater population densities than those of their free-living counterparts. Higher densities facilitate the spread of pathogens between individuals, contributing to the evolution of increased virulence in the pathogens (Anderson and May 1981, 1982; Ewald 1994). Second, artificial selection, inbreeding, and clonal propagation limit the genetic diversity of agricultural crops compared to their free-living counterparts, and genetic diversity is believed to facilitate the evolution of resistance to pathogens (Jaenike 1978; Hamilton 1980).

The success of agriculture depends on the control of domesticate pathogens. Human agriculture employs dozens of methods to prevent and suppress pathogens. These methods can be assigned to four general categories: exclusion, eradication, protection, and immunization (resistance) (Whetzel 1929; Maloy 1993; Agrios 1997). Exclusion prevents pathogens from entering and establishing themselves in a new area and is typically achieved in human agriculture through quarantines and embargoes. Eradication is accomplished by the removal, elimination, or destruction of pathogens from areas or individuals. Protection requires the separation of infected from uninfected individuals to prevent the spread of pathogens; it is primarily achieved by manipulating the environment, applying protectants, or erecting barriers. Immunization (resistance) uses breeding, medication, vaccination, and nutrition management to modify the domesticates or their growth conditions to make them less susceptible to or more tolerant of pathogens.

Although a full comparison of agricultural disease-control methods used by humans and attine ants is beyond the scope of this review, it is worth considering how the four mechanisms of crop defense in human agriculture parallel those in ant agriculture. First, attine ants practice exclusion by preventing inoculum of potential pathogens from coming into contact with the garden. This is achieved by cleaning nest surfaces and new substrate before it is added to the garden (Stahel and Geijskes 1939; Autuori 1941; Quinlan and Cherrett 1977, 1979) and by excluding the refuse-tending worker caste from physical contact with the fungus garden and with garden-tending castes (Hart and Ratnieks 2002). Second, ants practice eradication by removing pathogen inoculum that comes into contact with the garden before infection can be established. This is primarily accomplished through a behavior called fungus grooming, in which workers use their mouthparts to separate pathogen inoculum from domesticate mycelium (Currie and Stuart 2001). Attine ants also weed out and discard infected garden material (Currie and Stuart 2001). Attine research has so far neglected the category of protection, but relevant features of attine agriculture include: (1) the allocation of colony resources to the production of worker castes dedicated to monitoring gardens and detecting infections; (2) the architectural separation, in the nests of some attine species, of multiple fungus gardens into different chambers, which may prevent infections present in one garden chamber from spreading to other, uninfected fungus gardens; and (3) the permanent quarantine of gardens with advanced *Escovopsis* outbreaks by sealing them off with soil plugs (Currie and Mueller, pers. obs.). The final category of human agricultural disease control is immunization (resistance), and at least one resistance defense mechanism has been established in attine ant agriculture: the use of antibiotics produced by mutualistic filamentous bacteria (Actinomycetes) in the family Pseudonocardiaceae (Currie et al. 1999b, 2003b). Antibiotic compounds are also produced by the ants (Bot et al. 2002) and by the fungal domesticates (Nair and Hervey 1978; Hervey and Nair 1979; Angeli-Papa 1984; Kermarrec et al. 1986; Wang et al. 1999), although the role of the domesticates in disease control remains poorly understood. Resistance is a promising area of future research on the attine agricultural symbiosis, particularly with respect to the selection and spread of domesticate strains that are resistant to *Escovopsis* infection, including, possibly, the *Cyphomyrmex* yeast domesticates mentioned above.

Agricultural Pathogens: Origins

The pathogens that infect human domesticates are typically, but not always, closely related to the pathogens that infect free-living populations of the same or closely related species. Some of these domesticate diseases may have originated subsequent to domestication and then switched hosts from nondomesticated to domesticated plants and animals, whereas other diseases may already have been established before domestication and may have been introduced into human agricultural systems at the same time as, or shortly after, the domestication event. Recent molecular phylogenetic analyses of *Escovopsis* indicate the same pattern in attine agriculture. The sister group to *Escovopsis* is the ascomycete family Hypocreaceae (Currie et al. 2003a,b), which includes a large number of fungi that are pathogens of free-living mushrooms. Thus, it is likely that the ancestor of *Escovopsis* was an established pathogen of the ancestral attine domesticate and that it invaded the attine agricultural symbiosis at the time of its origin.

Alternatively, and perhaps less likely, because the Hypocreales also includes fungi that are parasites of insects, and because some hypocrealean pathogens can even facultatively switch between fungal and arthropod hosts, the ancestor of *Escovopsis* may have been a parasite of attine ants that switched to the fungal domesticates after fungus-growing behavior arose. Under this scenario, *Escovopsis* is analogous to the many diseases that humans have acquired from their domesticated animals, including measles, tuberculosis, smallpox, influenza, pertussis, and malaria (Diamond 1997, 1998).

Agricultural Pathogens: Conclusions

Because the study of natural ecosystems holds great promise for improving both human agriculture (Denison et al. 2003) and medicine (Williams and Nesse 1991), a better understanding of attine disease ecology may generate new ideas for controlling pathogens of human domesticates and perhaps even for controlling the agents of human disease. Attine ants have been using antibiotics derived from mutualistic actinomycete bacteria to suppress *Escovopsis* for millions of years (Currie et al. 1999b, 2003a,b). In addition, attine ants use antibiotics derived from their metapleural and mandibular glands (Bot et al. 2001), and some attine fungal domesticates also produce defensive antibiotics (Nair and Hervey 1978; Hervey and Nair 1979; Wang et al. 1999). Given the long history of this strategy, it is surprising, judging by our short human experience with antibiotics (approximately 60 years) and with agricultural pesticides (approximately 140 years), that *Escovopsis* has not yet evolved a generalized resistance to the actinomycete or other attine antibiotics. The most likely explanation for the continuing effectiveness of anti-*Escovopsis* antibiotics is that *Escovopsis* may be continually coevolving new resistance to particular actinomycete and other attine antibiotics, which are likewise evolving. Under this scenario, attine disease control may have proceeded as an ancient coevolutionary arms race, in which the actinomycete, ant, and domesticated fungal lineages continually evolve new antibiotics, and in which associated *Escovopsis* lineages continually evolve new forms of resistance to those antibiotics. Future research must characterize the antibiotic chemical or chemicals produced by the actinomycetes, ants, and fungi as well as their physiological effects on *Escovopsis*. Future research should also characterize the selection pressures, if any, on the actinomycete symbiont that affect antibiotic evolution.

Attine agricultural disease management incorporates a number of effective features, some of which may be applicable to human agriculture and medicine. First, lower attine domesticates are genetically linked to free-living fungal populations; attine fungi thus retain a large pool of genetic variability that likely serves as a source of pathogen-resistant strains (as well as a source of strains with other desirable features). Second, attine ants use antibiotics produced by evolving populations of bacteria. Again, the genetic variability in these populations probably serves as a source of new antibiotic variants and facilitates rapid response to newly evolved pathogen strains. Third, attine ants police their gardens intensively. Worker castes solely dedicated to gardening constantly patrol gardens, rapidly weeding out and discarding infected mycelium. It is interesting that the leaf-cutting higher attines, which cultivate domesticates that may be inbred and thus less resistant to new pathogen strains, possess physical gardening worker castes of minute ants that appear to be present in greater numbers than the gardening castes of the lower attines and that may generally be better at garden sanitation than are the morphologically unspecialized castes of the lower attines.

The Issue of Control: Enslavement of Domesticates by Agriculturalists versus Manipulation of Agriculturalists by Domesticates

Agricultural evolution—human or ant—is traditionally interpreted from the perspective of the agriculturalist, who appears to act with active intent, rather than from the perspective of the domesticate, which appears to be behaviorally inert and sessile in the case of plants and fungi. Thus, research programs have historically focused on such issues as how the quality of life has improved or worsened for the agriculturalist after the transition from hunting-gathering to agriculture, how the agriculturalist has imposed artificial selection and prevented domesticate escape, or what specific evolutionary modifications have arisen in the domesticate to better serve the agriculturalist (e.g., Sauer 1952; Flannery 1973; MacNeish 1991; Cowan and Watson 1992; Harlan 1992; Diamond 1997; Smith 1998a). These research questions take a one-sided perspective, that of the agriculturalist, and ignore the evolutionary interests of and leverages exerted by the proto-domesticate during the origin and subsequent evolution of the domesticate–agriculturalist association. This biased perspective seems to be intuitively justified because the agriculturalists appear to be in total control: Agriculturalists seem to manipulate critical life-history stages of the domesticate (e.g., timing of growth and reproduction); they dictate the fitnesses of different domesticate types (e.g., through artificial selection, whether intentional or incidental); and they can terminate an existing association either by switching from one domesticate to another or, in the case of humans, even abandon agriculture entirely and return to hunting and gathering. At first glance, then, agriculturalists—humans or ants—seem to direct the fates of domesticates, suggesting that the domesticates are completely enslaved.

An alternative perspective holds that domesticates have partial or even complete control over their evolutionary fates, if not in the present then at least at the origin of domesticate–agriculturalist associations, and that the proto-domesticates were initially acted upon by natural selection in ways that favored increased participation in symbioses with proto-agriculturalists who had yet to evolve the ability to dictate or direct the evolutionary fates of their domesticates (Rindos 1984). Under this perspective, domesticates do not become enslaved, if ever, until the later stages of a coevolutionary process. Precisely when in that process the transition from domesticate participation (complete control) to domesticate enslavement (reduced control) occurs is difficult to discern.

Taking an extremist domesticate-control perspective, one can even postulate that, before the origin of domestication, (1) the proto-domesticates exploited the proto-agriculturalists for their own reproductive purposes; (2) natural selection favored proto-domesticates that associated with proto-agriculturalists in symbiotic relationships that may have decreased the agriculturalists' fitness relative to a domesticate-free (hunting-gathering) strategy; and (3) domesticates ultimately ensnared agriculturalists in relationships that the agriculturalists found difficult to terminate. This radical view of agricultural evolution naturally conflicts with our intuition (and delusion?) that we humans were and are in charge of our past and present agricultural decisions. In contrast to this intuitive agriculturalist-control perspective of human agriculture, a domesticate-control perspective underlies the infrabuccal-pellet dispersal hypothesis proposed for the origin of attine ant fungiculture (Mueller et al. 2001).

Which of these alternative perspectives – the traditional agriculturalist-control perspective, the domesticate-control perspective, or a perspective that recognizes an intermediate tug-of-war–like coevolutionary interplay (Reeve et al. in press)—is the appropriate one depends on the extent to which each participant held control over its evolutionary fate during the initial formation of the domesticate–agriculturalist interaction (evolutionary origin) and retained this control during subsequent evolution to a more derived agricultural state (subsequent evolutionary modification). A thorough comparison of agricultural evolution in humans and ants therefore must consider both origin and subsequent evolution, first comparing human and ant preagricultural states and, in a separate, second analysis, comparing the derived human and ant agricultural systems that arose from those antecedent states.

The term "control" subsumes a set of factors, all of which help empower a symbiotic partner to elude the domination and exploitation of a coevolving partner. These factors include a partner's ability to (1) facultatively leave or escape from a symbiosis to lead an independent existence; (2) facultatively switch between partner species; (3) choose between genetic variants of the other partner (and thus influence or even dictate selective processes operating on the other partner); and (4) manipulate behavior or life-history parameters (e.g., growth and reproduction) of the other partner to modify it for the manipulator's benefit, sometimes even to the detriment of the manipulated partner. A partner that scores high in all these abilities is least likely to be exploited by the other partner, whereas a partner that scores low in all of these abilities is more likely to be exploited and enslaved (i.e., domesticated). Humans undoubtedly score high in the listed abilities in their recent agricultural systems, but, for understanding agricultural origins, it is necessary both to assess these abilities in the preagricultural states and to consider the perspectives of both the proto-agriculturalist and the proto-domesticate.

Domesticate control is easier to assess for the preagricultural coevolutionary interactions between the attine ants and their fungi than it is for human preagricultural interactions. For example, as discussed above, leucocoprineaceous fungi may use ants for dispersal via infrabuccal pellets, and they may have done so for millions of years preceding the origin of attine agriculture. Once dependent on vectoring by ants, such fungi may have evolved the ability to manipulate ant behavior by presenting ants with food rewards, a process convergent with the evolution of similar ant-reward structures (elaiosomes) that have originated many times in plants (Serenander 1906; Handel et al. 1981; Beattie 1985; Handel and Beattie 1990a,b). If so, then the fungi would have evolved to track the nutritional requirements and sensory preferences of the ant proto-agriculturalists before the advent of fungiculture, and the ants would have been engaged as reactive participants (passive respondents) in a coevolutionary process dictated by the evolutionarily "proactive" fungi.

Did such a stage, in which "reactive" humans coevolved with "proactive" plants, exist before the advent of human agriculture 10,000 years ago? It is possible, although this process is unlikely to have left any archaeological evidence that could conclusively document such a stage. As discussed previously in this chapter, ancestral humans probably dispersed plants in a number of ways, including as seeds (e.g., seeds that remained viable after passing through the human gut), as cuttings that were accidentally discarded at human campsites, or even as living stakes that were thrust into the ground during the building of fences, shelters, and other structures (Flannery 1973; Bye 1981; Rindos 1984; Harlan 1992; Diamond 1997; Smith 1998a). The associations between human-dispersed plants and humans are analogous to the association between infrabuccally dispersed fungi and ants, and it is such incidental associations that are expected to provide the raw material for further evolution. This further evolution includes derived agricultural behavioral repertoires that enhance the interaction for the benefit of the agriculturalist and that may eventually lead to intentional, planned domestications of the same or of other species.

Learning through trial, error, observation, and imitation no doubt played a major role in the development of the agricultural behavioral repertoires of humans, whereas the behavioral repertoires of ants were gradually modified through the prolonged interaction of mutation and selection—the evolutionary analogs of trial and error. This distinction between learning versus mutation-induced behavioral change is critical because learning can greatly accelerate adaptive modification in a species. Through learning, human agriculturalists rapidly modified the behavioral repertoires they used in their coevolutionary interactions with plants. Thus, in the case of a change in the human–domesticate relationship that benefited the human at the expense of the domesticate, the rapid pace of human behavioral change could preclude a corresponding evolutionary response in the domesticate. Through learned behaviors humans could prevent the facultative escape of a domesticate or prevent a domesticate from evolving toxicity or some other defense against human control. This rapid response on the part of humans leads to the rapid loss of control on the part of the domesticate and results in eventual enslavement (see criteria for control above). Learning thus enabled humans to take the role of the proactive partner during preagricultural and agricultural evolution and relegated the domesticate to the role of reactive partner. We can only speculate about the sophisticated agricultural systems the attine ants might have achieved during their 50 million years of evolution if ants were capable of human-scale learning and transmission of cultural information.

We suspect that most readers will resist our suggestion that human agriculturalists were once under the partial control of their proto-domesticates during the early evolutionary process that ultimately led to human agriculture. Human intuition suggests that we are not under the control of the cabbages and tomatoes that we plant in our backyards, that cabbages cannot facultatively escape from our gardens and from their inevitable destinies of death in our kitchens, and that cabbages have not enslaved us to labor on their evolutionary behalf. Human intuition can be misleading, however. We know, for example, that human symbionts can sometimes induce profound behavioral changes in humans that benefit the symbiont. The rabies virus induces drastic aggressive behavior to facilitate its spread to new potential hosts, and coca plants induce in humans a physical addiction and a craving for more coca, which requires the cultivation of more coca plants. Though seemingly farfetched and in conflict with our intuitions, we cannot at this point rule out similar manipulations during the preagricultural evolution of humans, a stage when humans began to assemble the behavioral repertoires that ultimately led to agricultural systems guided by human planning and intentional experimentation.

A properly unbiased evolutionary analysis of human agriculture (conducted, for example, by a Martian evolutionary biologist), neither anthropocentric nor domesticate-centric, needs to address what the separate, selective advantages were to both humans and their domesticates during the long preagricultural process that eventually led to more derived agricultural systems. Many recent domesticates were clearly imported into the human agricultural symbiosis in a process of instantaneous enslavement guided by human foresight (e.g., grocery-store "button mushrooms" and cherry trees). The domestication of other organisms, however, including those that were domesticated earliest, was preceded by a long coevolutionary process, the dynamics and outcomes of which may well have been determined by an interplay of control exerted by the proto-domesticates and the proto-agriculturalists. It is during this ancient time period, perhaps 50,000–100,000 years ago and occurring well before the recognized origin of true agriculture 10,000 years ago, that the incipient states of human-domesticate coevolutionary associations may be most directly comparable to the coevolutionary ant–fungus associations that led to attine agriculture.

The Issue of Control: Conclusions

Did ants domesticate fungi or did fungi domesticate ants? We have already explained why, before the origin of attine agriculture, the fitness of nondomesticated leucocoprineaceous fungi may have been increased through a symbiotic association in which the fungi used ants as dispersal agents. Once this association evolved into an agricultural symbiosis, the attine fungi could have retained some measure of control, manipulating the relationship in their continuing self-interests. As already pointed out, compared to free-living fungi, attine-cultivated Leucocoprineae are better dispersed and distributed, better protected from parasites and pathogens, and possibly better represented in terms of sheer abundance due to the husbanding activities of their ant hosts. With the possible exception of the higher attine fungi, the attine domesticates retain the ability to leave the symbiosis and to become feral. The same cannot be said for the ants, which are highly modified for and obligately dependent on fungiculture, and which are generally faithful to particular domesticate clades (although not to single domesticate genotypes). This asymmetry in terms of commitment to and modification for the symbiosis might superficially seem to support the notion that the fungi retain more control than do the ants.

A few studies provide weak evidence that attine fungi may exert some measure of control over their ant hosts. Bot et al. (2001) described incompatibility interactions within experimentally created ant–fungus associations involving two sympatric *Acromyrmex* leaf-cutter ant species and their fungal symbionts. The degree to which ants from a particular colony were motivated to remove and destroy an unfamiliar domesticate strain (ant–fungus incompatibility) was uncorrelated with the ant species from which the strain was taken but was correlated with the degree of genetic difference between the unfamiliar strain and the ants' resident domesticate strain; this genetic difference precisely paralleled observed patterns of somatic incompatibility between fungi, characterized by antagonistic interactions between fungal strains. Significantly, ant–fungus incompatibility disappeared when the ants were deprived of their resident domesticate and force-fed an unfamiliar domesticate for at least several days; at that point, the new strain assumed the role of resident strain with regard to ant–fungus incompatibility. Because the ants' incompatibility with unfamiliar fungi was due to recognition cues produced by the resident fungus, one interpretation of the results (not favored by Bot et al. 2001) is that the resident fungus manipulates its ant hosts' behavior as a means to guarantee its monopoly. Alternatively, the ants may simply take advantage of the preexisting fungal incompatibility system to maintain fungal monocultures. In either case, the fungi have retained the ability to interact antagonistically with other conspecific fungi, and Bot et al. (2001) suggest that the fungal domesticates also may have retained the ability to escape from a particular ant association and to move laterally to a new ant nest (e.g., when their current ant hosts are threatened by disease or senescence).

Ridley et al. (1996) and North et al. (1997, 1999) suggested that the cultivated fungus of the leaf-cutter *Atta sexdens rubropilosa* regulates the selection of plant material by foragers by chemically signaling the ants regarding the suitability or toxicity of substrates, and that it uses chemical manipulation to compel a colony of ants to provide it with a healthy diet. Alternatively, the ants may be judging the health of the garden and the suitability of the substrates via indirect cues, communicating these judgments to other ant nestmates and adjusting their behaviors accordingly. Obviously, the issue of fungus versus ant control in the attine agricultural symbiosis is not resolved, but it remains a promising area for future research (Mueller 2002). One obvious line of inquiry is whether there is variability in the ability of fungal domesticate strains to attract new ant hosts or to move in and replace resident domesticates, independent of the strains' ant-beneficial traits.

Ant and Human Agriculture: Synthesis

Links between domesticated fungi and free-living populations are known for two of the four attine agricultural systems (table 7.1), specifically, for the lower attine and yeast domesticates (figs. 7.3, 7.5; Mueller et al. 1998; Vo and Mueller, unpublished data). Links to free-living populations cannot be ruled out for the pterulaceous *Apterostigma* domesticates (Munkacsi and McLaughlin 2001; Villesen et al. in press). The remaining group, the higher attine fungi, represents a highly derived clade descended from a lower-attine–like leucocoprineaceous ancestor. Like some human domesticates, higher attine fungi appear to be inbred and possibly largely self-fertilizing (Rehner, unpublished data), a feature that may preserve gene combinations optimal for the requirements of their ant hosts. Although higher attine fungi are known to produce fruiting bodies, these mushrooms are known only within garden chambers or on the external surfaces of nest mounds, physically connected to and an extension of the garden mycelium (fig. 7.7; Möller 1893; Mueller 2002). Because free-living higher attine mushrooms are unknown, it remains possible that the higher attine fungi are not viable outside the symbiosis.

Attine ants may obtain carbohydrates and proteins from sources other than their fungus gardens. Some authors have asserted that some of the most derived higher attine cultivators, leaf-cutter ants in the genus *Atta*, obtain about 95% of their carbohydrates from foraging outside the nest (Littledyke and Cherrett 1976; Quinlan and Cherrett 1979; Bass and Cherrett 1995). This figure, however, may be a gross overestimate because it depends on largely uninvestigated assumptions about the total caloric requirements of the colony and about the proportion of that requirement furnished by the cultivated fungus (Turner 1974), and it neglects the possible contribution due to larva-worker anal trophallaxis (Schneider 2000). Foraging for nectar and other sugary liquids and the redistribution of these resources by trophallaxis to other adult nestmates (but not larvae) has been observed in the yeast-cultivating ant *Cyphomyrmex rimosus* (Murakami and Higashi 1997) and in the leucocoprineaceous-cultivating lower attine ants *Myrmicocrypta ednaella* (Murakami and Higashi 1997) and *Mycocepurus goeldii* (Oliveira et al. 1995; Leal and Oliveira 2000). Some anecdotal evidence suggests that *Apterostigma* species may forage for sources of protein. For example, captive colonies of *Apterostigma* species have been maintained with an ant diet prepared from a mixture of eggs, honey, vitamins, and agarose (L. Alonso, pers. comm.); a forager of *Apterostigma collare* was observed carrying a dead mosquito into its nest (Schultz, unpublished obs.); and a recent study of nitrogen cycling in ants indicates that *Apterostigma* species are relatively high on the food chain, with nitrogen isotope ratios more similar to those of predators than to those of attine leaf-cutter ant species (Davidson et al. 2003).

Thus, judged on the human agriculture-based continuum, the Attini practice a mixed food-acquisition strategy, exemplified in humans by the North American Hopewell civilization (2100–1600 years ago), which combined sophisticated farming with hunting and fishing (Smith 2001a), and by early lowland Neotropical agriculturalists (table 7.2), who combined agriculture with hunting and gathering (Piperno and Pearsall 1998). There are two important differences between the ant and human systems, however. First, nutritionally, attine ants are obligately dependent on agriculture and facultatively dependent on foraging for food; whereas they are obligately dependent on foraging for nutritional substrate for their fungi. Humans have so far retained all nutritional options, including potential reversion to hunting and gathering, and human agriculture is therefore facultative rather than obligate as in the Attini. Agriculture is necessary, however, for the maintenance of current human population levels. Second, in most Attini, foraging apparently provides an additional source of carbohydrates, whereas in many human mixed-strategy systems with an agricultural component, such as the Hopewell example, foraging provides additional protein. Foraging for carbohydrate and cultivating protein may be more common in human agriculture than is generally recognized, however (Bray 2000). Such a system of protein production is exemplified by the environmental management strategy of the pre-Columbian savannah people of the Bolivian Amazon, who, through the construction of earthworks forming large weirs and artificial ponds, harvested fish on a massive scale (Erickson 2000).

Among the most ancient human domesticates, perhaps the closest analogues of the attine fungi are the root crops such as potatoes, yams, arrowroot, taro, and manioc, which, like the attine fungi, are clonally propagated. The simplicity of vegetative cultivation makes tubers ideal proto-domesticates, leading some authors to argue that these and other root crops may be more ancient than seed crops such as wheat, corn, maize, and barley (Sauer 1952; Johns 1990; Harlan 1992; Piperno and Pearsall 1998; Bray 2000; Piperno et al. 2000; Lang 2001). For example, Australian aborigines regularly cut off and replant the stems and tops of gathered wild yams (Gregory 1886, cited by Harlan 1992), and natives of the Ubangui-Chari region of equatorial Africa use some gathered yams immediately and plant the surplus near their camps for future use (Chevalier 1936). Because tubers are propagated clonally, artificial selection is also a straightforward process (Rindos 1984). "Strikingly superior types can be found by screening large natural populations" and, once under cultivation, "if clones are found that are better tasting, less poisonous, more poisonous, more productive, etc., they can be propagated and cultivars are developed immediately" (Harlan 1992, p. 131).

One root crop, the potato (*Solanum* spp.), has become the fourth most intensively cultivated food crop in the world. Potatoes were first domesticated between 7,000 to 10,000 years ago in the Lake Titicaca basin of the central Andes, and this region remains the center of potato genetic diversity (Hawkes 1990). Here farmers make use of eight domesticated potato species, representing 2000 to 3000 known varieties, in a stable, traditional system of subsistence agriculture that has persisted since the origin of this human–plant association (Brush et al. 1981; Hawkes 1990; Lang 2001; Quiros 2003). Human management of this extended multi-species base, which contains both free-living and domesticated populations, provides many striking parallels with attine ant management of their associated leucocoprineaceous and pterulaceous fungi.

Traditional Andean potato farmers propagate their crops clonally by replanting tubers with desirable traits. This allows for the persistence, over many years, of selected clonal lineages, and, paralleling the exchange of fungal strains between ant nests (Mueller et al. 1998; Green et al. 2002), potato farmers share and distribute the most favorable domesticate strains via extensive intervillage "seed" (i.e., vegetative clone) networks. The particular beneficiaries of these networks are farmers living in mostly lowland climates who must frequently replace their domesticates, which, in those climates, are more prone to blight, aphid-borne viruses, and other potato pathogens (Brush et al. 1981). It has likewise been suggested that particular strains of attine fungi may be adapted to particular ecological conditions (Mueller et al. 1998; Green et al. 2002). Even though some subsistence farmers may grow separate plots of commercially improved potato varieties as a cash crop, they continue to grow the more genetically variable native varieties for subsistence because they taste better, they store better, and they remain viable year after year, whereas the commercially improved varieties can only be clonally propagated for 1–3 years before they lose their vigor, possibly due to viral infections (Brush et al. 1981).

Paralleling the ability of both lower and higher attine ants to discriminate between fungal strains (Bot et al. 2001; Viana et al. 2001; Mueller et al. 2004), Andean potato farmers recognize about 100 different phenotypic domesticates with colorful names like "cat's nose" and "eyes of a jungle native" (Quiros 2003), but these phenotypes may be produced by a variety of genotypes. Thus, native phenotype-based classification and selection maintains genetic diversity (Brush et al. 1981; Rindos 1984). Growing inside and around the periphery of Andean potato fields are feral and wild potatoes, which interbreed with domesticates and produce volunteer seedlings that also grow in or near the fields (Brush et al. 1981; Quiros 2003). These are generally tolerated by farmers, and individuals with desirable properties are occasionally recruited into domesticate pools (Brush et al. 1981). Free-living populations of lower attine fungal domesticates are likewise present in the vicinity of attine nests and readily available for domestication (figs. 7.3, 7.5; Mueller et al. 1998; Vo and Mueller, unpublished data).

As noted by Rindos (1984), this diversity base allows potatoes to better respond to the evolution of pathogens and pests, including a variety of viruses, bacteria, fungi, nematodes, and insects. Brush et al. (1981) observed no organized strategies to control pests and pathogens in the cultivated fields of Andean subsistence potato farmers. In contrast, the most intensively cultivated potatoes in Europe and North America are descended from a few individuals purchased in a market in Panama after the Irish potato blight disaster of the 1840s (Quiros 2003). This genetically monotonous domesticate lineage is highly susceptible to disease, including the A1 strain of the late blight disease (*Phytophthora infestans,* the cause of the Irish potato famine) and the recently widespread A2 strain. The predominant disease-management strategy in intensive potato agricultural systems is a cycle of developing new chemical fungicides, leading to selection on the pathogen for resistance against the fungicides, leading to the development of another generation of fungicides, and so on. An alternative strategy, developed by the International Potato Center (CIP), relies on developing resistant strains of potatoes by drawing on the genetic diversity of the potatoes of the Andean highlands. When the CIP developed a potato variety incorporating a single, major gene with highly specific resistance against a single, dominant pathogen strain, they found that, initially, the potato was immune to the pathogen. The pathogen rapidly evolved to overcome that resistance, however, and became virulent. When the CIP developed a variety that incorporated multiple, minor resistance genes, the potato was not entirely immune, but the pathogen persisted under less drastic selection that did not generate the rapid spread of major resistance. This latter, preferred strategy, modeled on the traditional approach, achieves a mutual coexistence in which hosts and pathogens coexist at levels that are acceptable to farmers (Lang 2001).

This successful, long-term strategy of mutual pathogen-domesticate coexistence in traditional human potato agriculture parallels the pattern found in attine fungiculture. Attine fungi are clonally propagated from strains obtained both from free-living sources and from the nests of other fungus-growing ants. Free-living populations of Leucocoprineae are presumably genetically diverse. They may or may not be targeted by *Escovopsis* pathogens (this is an entirely unstudied phenomenon), but if they are, then they represent, like Andean feral and wild potatoes, a diverse source of resistant domesticate strains. Like humans, ants may select new domesticates based on various desirable traits; in any case, if *Escovopsis* is capable of infecting free-living fungi (both leucocoprineaceous and pterulaceous) and if *Escovopsis* is abundant, then the mere presence of a successfully growing mycelium or fruiting body (figs. 7.3, 7.5) within the foraging area of the nest is potential proof of its ability to resist locally dominant pathogens. The modified higher attine (figs. 7.7, 7.8) and yeast (figs. 7.5, 7.6) domesticates may represent the analogues of our more highly domesticated potatoes; the putatively inbred higher attine domesticates (Rehner et al. unpublished data) may even be analogous to the potato varieties cultivated in North America and Europe, which are descended from only a few individuals but which retain sexual competency and are routinely crossed to produce botanical seed and new varieties (C. Quiros, pers. comm.).

Andean subsistence potato agriculture appears to strike a balance between the selection of desirable domesticate strains on the one hand and constant, low-level outcrossing with feral and wild strains on the other. The decisive factor optimizing this balance appears to be pathogen pressure. A similar set of forces may be at work in the attine agricultural symbiosis. If so, then the insights gained from an examination of human potato agriculture may help explain the continuing existence of genetically linked free-living and domesticated populations of lower attine fungi (Mueller et al. 1998; Vo and Mueller, unpublished data) and the persistence of occasional sexual recombination in the apparently inbred and self-fertilizing higher attine fungi (Rehner et al., unpublished data).

Conclusion

There are clearly many differences between ant and human agriculture. Humans are a single species perhaps 100,000 years old. Attine ants represent a clade of more than 210 known extant species that is 50 million years old. Humans mostly domesticate plants and animals. Attine ants domesticate fungi. Humans are omnivorous and are facultative agriculturalists (i.e., they can choose to return to nonagricultural hunting and gathering). The attine ants, at least at the colony level, are obligate fungivores and obligate fungiculturalists, and they perish when deprived of their fungus gardens. Perhaps most important, modern humans act with conscious intent, and they can thus improve their agricultural systems quite rapidly through learning. Ant behavior is largely genetically determined and evolves through the much slower processes of mutation and selection. While the earliest stages of human domestication probably involved unconscious, incidental associations with plants and animals, these associations were later subjected to conscious planning and experimentation. Certainly, learned behaviors and the cultural transmission of information dominate the last 10,000 years of human agriculture and animal breeding. Just as important, human agriculture is a single facet of larger societal systems replete with social hierarchies, traditions, religions, and other factors that are not clearly comparable to anything in ants. Although these features of humans have proven to be an effective short-term evolutionary strategy, it remains to be seen whether they will remain effective over as long a time period as the one characterizing the success of attine ant agriculture.

In spite of these differences, many significant similarities remain between ant and human agriculture. These similarities strongly suggest that both systems represent convergent solutions to similar ecological problems, including the management of coevolving agricultural pests and pathogens. If so, then many seemingly commonsense notions about human agriculture deserve to be reexamined through a comparison with ant agriculture. Ultimately, both systems are biological and are thus subject to the rules of natural selection. By comparing the parallel agricultural systems of ants and humans, perhaps these rules may be brought more sharply into focus.

Acknowledgments We are particularly grateful to R. Ford Denison, Dolores Piperno, Bruce Smith, and Melinda Zeder for stimulating discussions and feedback on various drafts of the manuscript. These colleagues do not necessarily agree with our conclusions and are, of course, by no means responsible for the shortcomings in our background knowledge of human agricultural evolution. We are also grateful to Timothy J. Baroni, Faridah Dahlan, Jennifer Leonard, Maureen Mello, Christopher Marshall, David J. McLaughlin, Esther G. McLaughlin, Andy Munkacsi, Eugenia Okonski, Carlos Quiros, and Rebecca Wilson. We gratefully acknowledge the continuing support of the National Science Foundation for research on fungus-growing ants (DEB-9707209, IRCEB DEB-0110073). U.G.M. was supported by CAREER (DEB-9983879) and IRCEB (DEB-0110073) grants from the National Science Foundation during the writing of this chapter.

Literature Cited

Adams, R. M. M., U. G. Mueller, T. R. Schultz, and B. Norden. 2000. Agro-predation: Usurpation of attine fungus gardens by *Megalomyrmex* ants. *Naturwissenschaften* 87:549–554.

Agrios, G. N. 1997. *Plant pathology*, 4th ed. San Diego, CA: Academic Press.

Anderson, R. M., and R. M. May. 1981. The population dynamics of microparasites and their invertebrate hosts. *Philosophical Transactions of the Royal Society of London, Series B* 291:451–524.

Anderson, R. M., and R. M. May. 1982. Coevolution of hosts and parasites. *Parasitology* 85:411–426.

Angeli-Papa, J. 1984. La culture d'un champignon par les fourmis attines; mise en evidence de pheromones d'antibioses dans le nid. *Cryptogamie Mycologie* 5:147–154.

Angeli-Papa J., and J. Eymé. 1985. Les champignons cultivés par les fourmis Attinae. *Annales des Sciences Naturelles, Botanique, Paris* 13:103–129.

Autuori, M. 1940. Algumas observações sobre formigas cultivadoras de fungo (Hym. Formicidae). *Revista de Entomologica* 11:215–226.

Autuori, M. 1941. Contribuiçao para o conhecimento da saúva (*Atta* spp.). I. Evolução do saúveiro (*Atta sexdens rubropilosa* Forel, 1908). *Arquivos do Instituto Biologico São Paulo* 12:197–228.

Bailey, I. W. 1920. Some relations between ants and fungi. *Ecology* 1:174–189.

Bass, M., and J. M. Cherrett. 1995. Fungal hyphae as a source of nutrients for the leaf-cutting ant *Atta sexdens*. *Physiological Entomology* 20:1–6.

Bates, H. W. 1863. *The naturalist on the river Amazons.* London: John Murray.

Beattie, A. J. 1985. *The evolutionary ecology of ant-plant mutualism.* Cambridge: Cambridge University Press.

Belt, T. 1874. *The naturalist in Nicaragua.* London: John Murray.

Boserup, E. 1965. *The conditions of agricultural growth.* London: Allen and Unwin.

Bot, A. N. M., S. A. Rehner, and J. J. Boomsma. 2001. Partial incompatibility between ants and symbiotic fungi in two sympatric species of *Acromyrmex* leaf-cutting ants. *Evolution* 55:1980–1991.

Bot, A. N. M., D. Ortius-Lechner, K. Finster, R. Maile, and J. J. Boomsma. 2002. Variable sensitivity of fungi and bacteria to compounds produced by the metapleural glands of leaf-cutting ants. *Insectes Sociaux* 49:363–370.

Bray, W. 2000. Ancient food for thought. *Nature* 408:145–146.

Brown, A. H. D., and J. Munday. 1982. Population genetic structure and optimal of landraces of barley from Iran. *Genetica* 58:85–96.

Brush, S. B., H. J. Carney, and Z. Huaman. 1981. Dynamics of Andean potato agriculture. *Economic Botany* 35:70–88.

Buckley, S. B. 1860. The cutting ant of Texas (*Oecodoma mexicana* Sm.) *Proceedings of the Academy of Natural Sciences of Philadelphia* 1860:233.

Bye, R. A., Jr. 1981. Quelites—ethnoecology of edible greens—past, present, and future. *Journal of Ethnobiology* 1:109–123.

Chapela, I. H., S. Rehner, T. R. Schultz, and U. Mueller. 1994. Evolutionary history of the symbiosis between fungus-growing ants and their fungi. *Science* 266:1691–1694.

Cherrett, J. M. 1986. History of the leaf-cutting ant problem. In *Fire ants and leaf-cutting ants: Biology and management*, ed. C. S. Lofgren and R. K. Vander Meer, pp. 10–17. Boulder, CO: Westview Press.

Cherrett, J. M., R. J. Powell, and D. D. Stradling. 1989. The mutualism between leaf-cutting ants and their fungus. In *Insect-fungus interactions*, ed. N. Wilding, N. M. Collins, P. M. Hammond, and J. F. Webber, pp. 93–120. London: Academic Press.

Chevalier, A. 1936. Contribution a l'étude de quelques espèces africaines du genre *Dioscorea*. *Bulletin du Muséum National d'Histoire Naturelle* Paris, 2e Série 8:520–551.

Cowan, C. W., and P. J. Watson. 1992. *The origins of agriculture: An international perspective*. Washington, DC: Smithsonian Institution Press.

Currie, C. R. 2001a. Prevalence and impact of a virulent parasite on a tripartite mutualism. *Oecologia* 128:99–106.

Currie, C. R. 2001b. A community of ants, fungi, and bacteria: A multilateral approach to studying symbiosis. *Annual Review of Microbiology* 55:357–380.

Currie, C. R., A. N. M. Bot, and J. J. Boomsma. 2003a. Experimental evidence of a tripartite mutualism: Bacteria protect ant fungus gardens from specialized parasites. *Oikos* 101:91–102.

Currie, C. R., U. G. Mueller, and D. Malloch. 1999a. The agricultural pathology of ant fungus gardens. *Proceedings of the National Academy of Sciences of the USA* 96:7998–8002.

Currie, C. R., J. A. Scott, R. C. Summerbell, and D. Malloch. 1999b. Fungus-growing ants use antibiotic-producing bacteria to control garden parasites. *Nature* 398:701–704.

Currie, C. R., and A. E. Stuart. 2001. Weeding and grooming of pathogens in agriculture by ants. *Proceedings of the Royal Society of London, Series B* 268:1033–1039.

Currie, C. R., B. Wong, A. E. Stuart, T. R. Schultz, S. A. Rehner, U. G. Mueller, G. H. Sung, J. W. Spatafora, and N. A. Straus. 2003b. Ancient tripartite coevolution in the attine ant-microbe symbiosis. *Science* 299:386–388.

Davidson, D. W., S. C. Cook, R. R. Snelling, and T. H. Chua. 2003. Explaining the abundance of ants in lowland tropical rainforest canopies. *Science* 300:969–972.

de Tapia, E. M. 1992. The origins of agriculture in Mesoamerica and Central America. In *The origins of agriculture: An international perspective*, ed. C. S. Cowan and P. J. Watson, pp. 143–171. Washington, DC: Smithsonian Institution Press.

Denham, T. P., S. G. Haberle, C. Lentfer, R. Fullagar, J. Field, M. Therin, N. Porch, and B. Winsborough. 2003. Origins of agriculture at Kuk Swamp in the highlands of New Guinea. *Science* 301:189–193.

Denison, R. F., E. T. Kiers, and S. A. West. 2003. Darwinian agriculture: When can humans find solutions beyond the reach of natural selection? *Quarterly Review of Biology* 78:145–168.

Diamond, J. 1997. *Guns, germs, and steel: The fates of human societies*. New York: Norton.

Diamond, J. 1998. Ants, crops, and history. *Science* 281:1974–1975.

Diniz, J. L. M., C. R. F. Brandão, and C. I. Yamamoto. 1998. Biology of *Blepharidatta* ants, the sister group of the Attini: A possible origin of fungus-ant symbiosis. *Naturwissenschaften* 85:270–274.

Emery, C. 1895. Die Gattung Dorylus Fab. und die systematische Eintheilung der Formiciden. *Zoologische Jahrbücher, Abtheilung für Systematik, Geographie und Biologie der Thiere* 8:685–778.

Erickson, C. L. 2000. An artificial landscape-scale fishery in the Bolivian Amazon. *Nature* 408:190–193.

Ewald, P. W. 1994. *The evolution of infectious disease.* Oxford: Oxford University Press.

Eyre-Walker, A., R. L. Gaut, H. Hilton, D. L. Feldman, and B. S. Gaut. 1998. Investigation of the bottleneck leading to the domestication of maize. *Proceedings of the National Academy of Sciences of the USA* 95:4441–4446.

Febvay, G., and A. Kermarrec. 1981. Morphologie et fonctionnement du filtre infrabuccal chez une attine *Acromyrmex octospinosus* (Reich) (Hymenoptera: Formicidae): Role de la poche infrabuccale. *International Journal of Insect Morphology and Embryology* 10:441–449.

Fell, J. W., T. Boekhout, A. Fonseca, and J. P. Sampaio. 2001. Basidiomycetous yeasts. In *The Mycota VII: Systematics and evolution*, Part B, ed. D. J. McLaughlin, E. G. McLaughlin, and P. A. Lempke, pp. 3–35. New York: Springer-Verlag.

Fisher, P. J., D. J. Stradling, and D. N. Pegler. 1994a. Leaf-cutting ants, their fungus gardens and the formation of basidiomata of *Leucoagaricus gongylophorus*. *Mycologist* 8:128–131.

Fisher, P. J., D. J. Stradling, and D. N. Pegler. 1994b. *Leucoagaricus* basidiomata from a live nest of the leaf-cutting ant *Atta cephalotes*. *Mycological Research* 98:884–888.

Flannery, K. V. 1973. The origins of agriculture. *Annual Review of Anthropology* 2:271–310.

Forel, A. 1893. Note sur les "Attini." *Annales de la Société Royale Belge d'Entomologie* 37:586–607.

Fowler, H. G. 1982. Evolution of the foraging behavior of leaf-cutting ants (*Atta* and *Acromyrmex*). In *The biology of social insects*, ed. M. D. Breed, C. D. Michener, and H. E. Evans, p. 33. Boulder, CO: Westview Press.

Fowler, H. G., L. C. Forti, V. Pereira-da-Silva, and N. B. Saes. 1986a. Economics of grass-cutting ants. In *Fire ants and leaf-cutting ants: Biology and management*, ed. C. S. Lofgren and R. K. Vander Meer, pp. 18–35. Boulder, CO: Westview Press.

Fowler, H. G., V. Pereira-da-Silva, L. C. Forti, and N. B. Saes. 1986b. Population dynamics of leaf-cutting ants: A brief review. In *Fire ants and leaf-cutting ants: Biology and management*, ed. C. S. Lofgren and R. K. Vander Meer, pp. 123–145. Boulder, CO: Westview Press.

Gepts, P., T. C. Osborn, K. Rashka, and F. A. Bliss. 1986. Phaseolin-protein variability in wild forms and landraces of the common bean (*Phaseolus vulgaris*): Evidence for multiple centers of domestication. *Economic Botany* 40:451–468.

Gepts, P., and D. Debouck. 1991. Origin, domestication, and evolution of the common bean *Phaseolus vulgaris.* In *Common beans: Research for crop improvement*, ed. O. Voysest and A. Ban Schoonhoven, pp. 7–53. Oxon, UK: CAB.

Green, A. M., U. G. Mueller, and R. M. M. Adams. 2002. Extensive exchange of fungal cultivars between sympatric species of fungus-growing ants. *Molecular Ecology* 11:191–195.

Gregory, A. C. 1886. Memoranda on the aborigines of Australia. *Journal of the Anthropological Institute* 16:131–133.

Haig, D. 1993. Genetic conflicts in human pregnancy. *Quarterly Review of Biology* 68:495–532.

Hamilton, W. D. 1980. Sex versus non-sex versus parasites. *Oikos* 35:282–290.

Handel, S. N., and A. J. Beattie. 1990a. La dispersion des graines par les fourmis. *Pour la Science* 156:54–61.

Handel, S. N., and A. J. Beattie. 1990b. Seed dispersal by ants. *Scientific American* 263:76–83.

Handel, S. N., S. B. Fisch, and G. E. Schatz. 1981. Ants disperse a majority of herbs in a mesic forest community in New York state. *Bulletin of the Torrey Botanical Club* 108:430–437.

Harlan, J. R. 1965. The possible role of weed races in the evolution of cultivated plants. *Euphytica* 14:173–176.

Harlan, J. R. 1992. *Crops and man*, 2nd ed. Madison, WI: American Society of Agronomy.

Hart, A. G., and F. L. W. Ratnieks. 2002. Waste management in the leaf-cutting ant *Atta colombica*. *Behavioral Ecology* 13:224–231.

Hawkes, J. G. 1990. *The potato: Evolution, biodiversity and genetic resources*. Washington, DC: Smithsonian Institution Press.

Hayden, B. 1995. A new overview of domestication. In *Last hunters—first farmers*, ed. T. D. Price and A. B. Gebauer, pp. 273–299. Santa Fe, NM: School of American Research Press.

Heim, R. 1957. A propos du *Rozites gongylophora* A. Möller. *Revue de Mycologie* 22:293–299.

Hervey, A., and M. S. R. Nair. 1979. Antibiotic metabolite of a fungus cultivated by gardening ants. *Mycologia* 71:1064–1066.

Hervey, A., C. T. Rogerson, and I. Leong. 1977. Studies on fungi cultivated by ants. *Brittonia* 29:226–236.

Hölldolber, B., and E. O. Wilson. 1990. *The ants*. Cambridge, MA: Belknap Press.

Huber, J, 1905a. Über die Koloniegründung bei *Atta sexdens*. *Biologisches Centralblatt* 25:606–619.

Huber, J, 1905b. Über die Koloniegründung bei *Atta sexdens*. *Biologisches Centralblatt* 25:625–635.

Jackson, F. L. C. 1991. Secondary compounds in plants (allelochemicals) as promoters of human biological variability. *Annual Review of Anthropology* 20:505–546.

Jaenike, J. 1978. A hypothesis to account for the maintenance of sex within populations. *Evolutionary Theory* 3:191–194.

Janzen, D. H. 1995. Who survived the Cretaceous? *Science* 268:785.

Johns, T. 1990. *The origins of human diet and medicine*. Tucson: The University of Arizona Press.

Johnson, J. 1999. Phylogenetic relationships within *Lepiota* sensu lato based on morphological and molecular data. *Mycologia* 91:443–458.

Kempf, W. W. 1966 ("1965"). A revision of the Neotropical ants of the genus *Cyphomyrmex* Mayr. Part II. Group of *rimosus* (Spinola) (Hym.: Formicidae). *Studia Entomologica* 8:161–200.

Kermarrec, A., G. Febvay, and M. Decharme. 1986. Protection of leaf-cutting ants from biohazards: Is there a future for microbiological control? In *Fire ants and leaf-cutting ants: Biology and management*, ed. C. S. Lofgren and R. K. Vander Meer, pp. 339–356. Boulder, CO: Westview Press.

Kreisel, H. 1972. Pilze aus Pilzgärten von *Atta insularis* in Kuba. *Zeitschrift für Allgemeine Mikrobiologie* 12:643–654.

Laden, G. 1992. Ethnoarchaeology and land use ecology of the Efe (Pygmies) of the Ituri rain forest, Zaire. PhD dissertation, Harvard University.

Lang, J. 2001. *Notes of a potato watcher.* College Station: Texas A&M University Press.

LaPolla, J. S., U. G. Mueller, M. Seid, and S. P. Cover. 2002. Predation by the army ant *Neivamyrmex rugulosus* on the fungus-growing ant *Trachymyrmex arizonensis. Insectes Sociaux* 49:251–256.

Leal, I. R. and P. S. Oliveira. 2000. Foraging ecology of attine ants in a Neotropical savanna: Seasonal use of fungal substrate in the cerrado vegetation of Brazil. *Insectes Sociaux* 47:376–382.

Lee, R. B. 1968. What hunters do for a living, or how to make out on scarce resources. In *Man the hunter*, ed. R. B. Lee and I. DeVore, pp. 30–48. Chicago: Aldine.

Lee, R. B., and I. DeVore, eds. 1968. *Man the hunter.* Chicago: Aldine.

Leonard, J. A., R. K. Wayne, J. Wheeler, R. Valadez, S. Guillén, and C. Vilà. 2002. Ancient DNA evidence for Old World origin of New World dogs. *Science* 298:1613–1616.

Letourneau, D. K. 1998. Ants, stem-borers, and fungal pathogens: experimental tests of a fitness advantage in *Piper* ant-plants. *Ecology* 79:593–603.

Little, A. E. F., T. Murakami, U. G. Mueller, and C. R. Currie. 2004. Construction, maintenance, and microbial ecology of fungus-growing ant infrabuccal pellet piles. *Naturwissenschaften* 90:558–562.

Littledyke, M., and J. M. Cherrett. 1976. Direct ingestion of plant sap from cut leaves by the leaf-cutting ants *Atta cephalotes* (L.) and *Acromyrmex octospinosus* (Reich) (Formicidae, Attini). *Bulletin of Entomological Research* 66:205–217.

MacNeish, R. S. 1991. *The origins of agriculture and settled life.* Norman: University of Oklahoma Press.

Maloy, O. C. 1993. *Plant disease control: Principles and practice.* New York: John Wiley & Sons.

Matsuoka, Y., Y. Vigoroux, M. M. Goodman, J. G. Sanchez, E. Buckler, and J. Doebley. 2002. A single domestication for maize shown by multilocus microsatellite genotyping. *Proceedings of the National Academy of Science of the USA* 99:6080–6084.

Meier, R., and T. R. Schultz. 1996. Pilzzucht und Blattschneiden bei Ameisen—Präadaptationen und evolutive Trends. *Sitzungsberichte der Gesellschaft Naturforschender Freunde zu Berlin* 35:57–76.

Möller, A. 1893. Die Pilzgärten einiger südamerikanischer Ameisen. *Botanische Mittheilungen aus den Tropen* 6:1–127.

Morishima, H., and H. I. Oka. 1979. Genetic diversity in rice populations of Nigeria: influence of community structure. *Agro-Ecosystems* 5:263–269.

Muchovej, J. J., T. M. Della Lucia, and R. M. C. Muchovej. 1991. *Leucoagaricus weberi* sp.nov. from a live nest of leaf-cutting ants. *Mycological Research* 95:1308–1311.

Mueller, U. G., S. A. Rehner, and T. R. Schultz. 1998. The evolution of agriculture in ants. *Science* 281:2034–2038.

Mueller, U. G., T. R. Schultz, C. R. Currie, R. M. M. Adams, and D. Malloch. 2001. The origin of the attine ant-fungus mutualism. *Quarterly Review of Biology* 76:169–197.

Mueller, U. G. 2002. Ant versus fungus versus mutualism: Ant-cultivar conflict and the deconstruction of the attine ant-fungus symbiosis. *American Naturalist* 160 (supplement): S67–S98.

Mueller, U. G., J. Poulin, and R. M. M. Adams. 2004. Symbiont choice in a fungus-growing ant (Attini, Formicidae). *Behavioral Ecology* 15:337–364.

Müller, F. 1874. The habits of various insects. *Nature* 10:102–103.

Munkacsi, A., and D. J. McLaughlin. 2001. Evolutionary relationships of *Pterula* and

Deflexula within Agaricales *sensu stricto* and their relationships with the tricholo-mataceous attine fungi. Abstract, 2001 Mycological Society of America Meeting, August 25–29, 2001, Salt Lake City, Utah.

Murakami, T., and S. Higashi. 1997. Social organization in two primitive attine ants, *Cyphomyrmex rimosus* and *Myrmicocrypta ednaella*, with reference to their fungus substrates and food sources. *Journal of Ethology* 15:17–25.

Nair, M. S. R., and A. Hervey. 1978. Structure of lepiochlorin, an antibiotic metabolite of a fungus cultivated by ants. *Phytochemistry* 18:326–327.

Neumann, K. 2003. New Guinea: A cradle of agriculture. *Science* 301:180–181.

North, R., C. W. Jackson, and P. E. Howse. 1997. Evolutionary aspects of ant-fungus inter-actions in leaf-cutting ants. *Trends in Ecology and Evolution* 12:386–389.

North, R., C. W. Jackson, and P. E. Howse. 1999. Communication between the fungus garden and workers of the leaf-cutting ant, *Atta sexdens rubropilosa*, regarding choice of sub-strate for the fungus. *Physiological Entomology* 24:127–133.

Oliveira, P. S., M. Galetti, F. Pedroni, and L. P. C. Morellato. 1995. Seed cleaning by *Mycocepurus goeldii* ants (Attini) facilitates germination in *Hymenaea courbaril* (Caesal-piniaceae). *Biotropica* 27:518–522.

Oppenheimer, J. R., and G. E. Lang. 1969. *Cebus* monkeys: Effects on tracking of *Gustavia* trees. *Science* 165:187–188.

Pimentel, D., and C. W. Hall. 1989. *Food and natural resources.* New York: Academic Press.

Piperno, D. R. The origins of plant cultivation and domestication in the Neotropics: A behav-ioral ecological perspective. In *Behavioral ecology and the transition to agriculture,* ed. D. J. Kennett and B. W. Winterhalder. Washington, DC: Smithsonian Institution Press, in press.

Piperno, D. R., and D. M. Pearsall. 1998. *The origins of agriculture in the lowland neotropics.* New York: Academic Press.

Piperno, D. R., A. J. Ranere, I. Holst, and P. Hansell. 2000. Starch grains reveal early root crop horticulture in the Panamanian tropical forest. *Science* 407:894–897.

Price, S. L., T. Murakami, U. G. Mueller, T. R. Schultz, and C. R. Currie. 2003. Recent findings in fungus-growing ants: Evolution, ecology, and behavior of a complex mi-crobial symbiosis. In: *Genes, behavior, and evolution in social insects*, ed. M. Kikuchi and S. Higashi, pp. 255–280. Sapporo: Hokkaido University Press.

Pringle, H. 1998. The original blended economies. *Science* 282:1447.

Quinlan, R. J., and J. M. Cherrett. 1977. The role of substrate preparation in the symbiosis between the leaf-cutting ant *Acromyrmex octospinosus* (Reich) and its food fungus. *Ecological Entomology* 2:161–170.

Quinlan, R. J., and J. M. Cherrett. 1978a. Studies on the role of the infrabuccal pocket of the leaf-cutting ant *Acromyrmex octospinosus* (Reich) (Hym., Formicidae). *Insectes Sociaux* 25:237–245.

Quinlan, R. J., and J. M. Cherrett. 1978b. Aspects of the symbiosis of the leafcutting ant *Acromyrmex octospinosus* (Reich) and its food fungus. *Ecological Entomology* 3:221–230.

Quinlan, R. J., and J. M. Cherrett. 1979. The role of the fungus in the diet of the leaf-cutting ant *Atta cephalotes* (L.). *Ecological Entomology* 4:151–160.

Quiros, C. 2003. "Solanacea: POTATO: *Solanum tuberosum.*" University of California, Davis, Vegetable Crops 221, spring 2003. Available: http://veghome.ucdavis.edu/classes/vc221/potato/potsum1.html

Reeve, H. K., N. J. Mehdiabadi, and U. G. Mueller. Conflicts between hosts and symbionts as a tug-of-war. *American Naturalist* in press.

Reynolds, H. T., and C. R. Currie. Pathogenicity of *Escovopsis*: The parasite of the attine ant-microbe symbiosis directly consumes the ant cultivated fungus. Mycologia in press.

Rick, C. M. 1976. Tomato, *Lycopersicon esculentum* (Solanaceae). In *Evolution of crop plants*, ed. N. W. Simmonds, pp. 268–309. London: Longman.

Ridley, P., P. E. Howse, and C. W. Jackson. 1996. Control of the behavior of leaf-cutting ants by their "symbiotic" fungus. *Experientia* 52:631–635.

Rindos, D. 1984. *The origins of agriculture: An evolutionary perspective*. New York: Academic Press.

Sahlins, M. 1968. Notes on the original affluent society. In *Man the hunter*, ed. R. B. Lee and I. DeVore, pp. 85–89. Chicago: Aldine.

Sauer, C. O. 1952. *Agricultural origins and dispersals*. New York: American Geographical Society.

Schneider, M. 2000. Observations on brood care behavior of the leafcutting ant *Atta sexdens* L. (Hymenoptera: Formicidae) [abstract no. 3547]. In *Abstracts of the XXI International Congress of Entomology*, vol. 2, p. 895. XXI International Congress of Entomology, August 20–26, 2000, Foz do Iguassu. Londrina, Brasil:

Schultz, T. R. 1995. The evolutionary history of the attine ant-fungus symbiosis: Phylogenetic analyses of attine ants (Formicidae: Attini) and their fungi (Basidiomycotina: Lepiotaceae and Tricholomataceae) using morphological and molecular characters. PhD dissertation, Cornell University.

Schultz, T. R., and T. P. McGlynn. 2000. The interactions of ants with other organisms. In *Ants: Standard methods for measuring and monitoring biodiversity*, ed. D. Agosti, J. Majer, L. Alonso, and T. R. Schultz, pp. 35–44. Washington, DC: Smithsonian Institution Press.

Schultz, T. R., and R. Meier. 1995. A phylogenetic analysis of the fungus-growing ants (Hymenoptera: Formicidae: Attini) based on morphological characters of the larvae. *Systematic Entomology* 20:337–370.

Second, G. 1982. Origin of the genic diversity of cultivated rice (*Oryza* sp.): Study of the polymorphism scored at 40 isozyme loci. *Japanese Journal of Genetics* 57:25–57.

Serenander, R. 1906. Entwurf einer Monographie der europäischen Myrmekochoren. *Kungliga Svenska Vetenskapsakademiens Handlingar* 41:1–410.

Smith, B. D. 1997. The initial domestication of *Cucurbita pepo* in the Americas 10,000 years ago. *Science* 276:932–934.

Smith, B. D. 1998a. *The emergence of agriculture*. New York: Scientific American Library.

Smith, B. D. 1998b. Between foraging and farming. *Science* 279:1651–1652.

Smith, B. D. 2001a. Low-level food production. *Journal of Archaeological Research* 9:1–43.

Smith, B. D. 2001b. The transition to food production. In *Archaeology at the millenium: A sourcebook*, ed. G. M. Feinman and T. D. Price, pp. 199–229. New York: Kluwer Academic.

Snelling, R. R., and J. T. Longino. 1992. Revisionary notes on the fungus-growing ants of the genus *Cyphomyrmex, rimosus* group (Hymenoptera: Formicidae: Attini). In *Insects of Panama and Mesoamerica: Selected studies*, ed. D. Quintero and A. Aiello, pp. 479–494. New York: Oxford University Press.

Stahel, G., and D. C. Geijskes. 1939. Ueber den Bau der Nester von *Atta cephalotes* (L.) und *Atta sexdens* (L.) (Hym: Formicidae). *Revista Entomologica* 10:27–78.

Stradling, D. J. 1978. Food and feeding habits of ants. In *Production ecology of ants and termites*, ed. M. V. Brian, pp. 81–106. Cambridge: Cambridge University Press.

Tedlock, D. 1985. *Popol Vuh: The Mayan book of the dawn of life*. New York: Simon and Schuster.

Turner, J. A. 1974. The bioenergetics of leaf-cutting ants in Trinidad. MSc. dissertation, University of the West Indies.

Urich, F. W. 1895. Notes on some fungus-growing ants of Trinidad. *Journal of the Trinidad Field Naturalists' Club* 2:175–182.

Viana, A. M. M., A. Frézard, C. Malosse, T. M. C. Della Lucia, C. Errard, and A. Lenoire. 2001. Colonial recognition of fungus in the fungus-growing ant *Acromyrmex subterraneus subterraneus* (Hymenoptera: Formicidae). *Chemoecology* 11:29–36.

Vigouroux, Y., M. McMullen, C. T. Hittinger, K. Houchins, L. Schulz, S. Kresovich, Y. Matsuoka, and J. Doebley. 2002. Identifying genes of agronomic importance in maize by screening microsatellites for evidence of selection during domestication. *Proceedings of the National Academy of Sciences of the USA* 99:9650–9655.

Vilà, C., J. A. Leonard, A. Götherström, S. Marklund, K. Sandberg, K. Lidén, R. K. Wayne, and H. Ellegren. 2001. Widespread origins of domestic horse lineages. *Science* 291:474–477.

Vilà, C., J. E. Maldonado, and R. K. Wayne. 1999. Phylogenetic relationships, evolution, and genetic diversity of the domestic dog. *Journal of Heredity* 90:71–77.

Vilà, C., P. Savolainen, J. E. Maldonado, I. R. Amorim, J. E. Rice, R. L. Honeycutt, K. A. Crandall, J. Lundeberg, and R. K. Wayne. 1997. Multiple and ancient origins of the domestic dog. *Science* 276:1687–1689.

Villesen, P., U. G. Mueller, T. R. Schultz, R. M. M. Adams, and M. C. Bouck. Evolution of ant-cultivar specialization and cultivar switching in *Apterostigma* fungus-growing ants. *Evolution* in press.

von Ihering, R. 1898. Die Anlagen neuer Colonien und Pizgärten bei *Atta sexdens. Zoologischer Anzeiger* 21:238–245.

Wang, R., A. Stec, J. Hey, L. Lukens, and J. Doebley. 1999. The limits of selection during maize domestication. *Nature* 398:236–239.

Wang, Y., U. G. Mueller, and J. C. Clardy. 1999. Antifungal diketopiperazines from the symbiotic fungus of the fungus-growing ant *Cyphomyrmex minutus. Journal of Chemical Ecology* 25:935–941.

Weber, N. A. 1972. *Gardening ants: The attines.* Philadelphia: American Philosophical Society.

Weber, N. A. 1982. Fungus ants. In *Social insects*, vol. 4, ed. H. R. Hermann, pp. 255–363. New York: Academic Press.

Webber, J. F., and J. N. Gibbs. 1989. Insect dissemination of fungal pathogens of trees. In *Insect-fungus interactions*, ed. N. Wilding, N. M. Collins, P. M. Hammond, and J. F. Webber, pp. 161–193. London: Academic Press.

Wetterer, J. K., T. R. Schultz, and R. Meier. 1998. Phylogeny of fungus-growing ants (tribe Attini) based on mtDNA sequence and morphology. *Molecular Phylogenetics and Evolution* 9:42–47.

Wheeler, W. M. 1901. Notices Biologiques sur les fourmis Mexicaines. *Annales de la Société Entomologique de Belgique* 45:199–205.

Wheeler, W. M. 1907. The fungus-growing ants of North America. *Bulletin of the American Museum of Natural History* 23:669–807.

Wheeler, W. M., and I. W. Bailey. 1920. The feeding habits of pseudomyrmine and other ants. *Transactions of the American Philosophical Society* 22:235–279.

Whetzel, H. H. 1929. The terminology of plant pathology. In *Proceedings of the International Congress of Plant Sciences*, August 16–23, 1926, Ithaca, NY, ed. B. M. Duggar, pp. 1204–1215. Menasha, WI: George Banta Publishing Co.

Williams, G. C., and R. M. Nesse. 1991. The dawn of Darwinian medicine. *Quarterly Review of Biology* 66:1–22.

Wilson, H. D. 1981. Genetic variation among South American populations of tetraploid *Chenopodium* sect. *Chenopodium* subsect. *Cellulata*. *Systematic Botany* 6:380–398.

Wilson, H. D. 1989. Discordant patterns of allozyme and morphological variation in Mexican *Cucurbita*. *Systematic Botany* 14:612–623.

Wilson, H. D. 1990. Gene flow in squash species. *BioScience* 40:449–455.

Zeder, M. A., and B. Hesse. 2000. The initial domestication of goats (*Capra hircus*) in the Zagros Mountains 10,000 years ago. *Science* 287:2254–2257.

8

Evolutionary Dynamics of the Mutualistic Symbiosis between Fungus-Growing Termites and *Termitomyces* Fungi

Duur K. Aanen

Jacobus J. Boomsma

Colonies of fungus-growing termites (Isoptera: Termitidae) are among the most spectacular organismal phenomena in the world. One of the best-known species of fungus-growing termites is *Macrotermes bellicosus*. A queen of this species (fig. 8.1) can lay up to 40,000 eggs per day, and a mature colony, normally founded by a single queen and king, consists of millions of sterile individuals, the workers and soldiers. *Macrotermes bellicosus* builds mounds that can be up to 7 m tall (fig. 8.2; Korb 1997). This species and all fungus-growing termites live in an obligate symbiosis with basidiomycete fungi of the genus *Termitomyces*. The volume of the fungus garden of a live colony of *M. bellicosus* has been estimated to encompass several cubic meters. This chapter summarizes recent advances in our understanding of the major macroevolutionary developments that have shaped the symbiosis between the fungus-growing termites and their fungal symbionts and places these changes in an ecological context.

Not all termite species have such a spectacular life history as *M. bellicosus*. All termites, however, are eusocial as defined by cooperative brood care, overlapping adult generations, and a reproductive division of labor (Wilson 1971). They live in societies consisting of a pair of long-lived reproductive adults, one or several worker castes, and, in many cases, a soldier caste. Both sexes are normally represented in all worker castes and workers are often immatures (nymphs). Adult sexuals (alates) are winged but remove their wings after a nuptial flight and before initiating a new nest, where they will function as king and queen. Although all termites are eusocial, the degree of eusociality ranges widely. Some species live in small colonies with reproductives being helped by temporary workers and soldiers that themselves might be future reproductives; species at the other end of the spectrum have large, complex colonies in which reproductives are helped by permanent workers (Shellman-Reeve

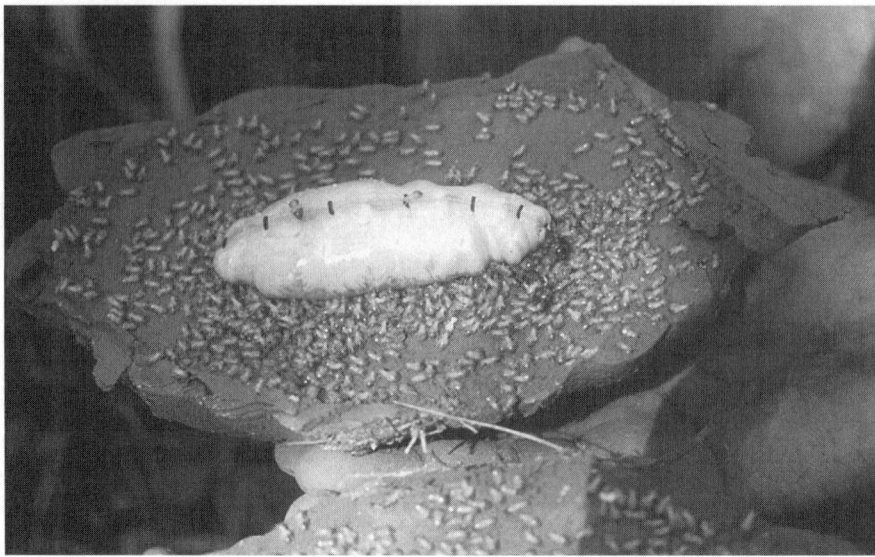

Figure 8.1. Opened royal chamber of the fungus-growing termite *Macrotermes bellicosus*. The large individual in the center is the queen, and the king can be seen just below the queen. The king has the same-sized head and thorax as the queen, but a much smaller abdomen. The other individuals are workers. (Photo by D. K. Aanen.)

1997). The distribution of termites is cosmopolitan, but it excludes the colder temperate regions and has its highest diversity in wet, lowland tropical forests (Eggleton 2000).

The major transitions in evolution (Maynard-Smith and Szathmáry 1995) are all characterized by the emergence of cooperation among lower level units into a new higher level unit (Michod 1999). One transition that has occurred many times is the permanent merger of mutualistic symbionts. Termites offer a prime example of multiple obligate mutualism, involving a range of microorganisms that belong to the three major groups of organisms: Archaea, Bacteria, and Eucarya (Bignell 2000). In a sense, termites can be visualized as nested Russian *Matreshka* dolls: some of their protist symbionts are in turn dependent on bacterial symbionts that they harbor either internally or attached to their surface (Breznak 2000). All termites have symbionts of at least two groups. There is, however, only one subfamily of termites that is obligately dependent on an ectosymbiont for food. Members of the Macrotermitinae live in a mutualistic symbiosis with fungi of the genus *Termitomyces*, close relatives of the wood-decomposing nonsymbiotic basidiomycete genus *Lyophyllum* (Moncalvo et al. 2000, 2002; Preslev et al. 2003).

Generally speaking, termites can be regarded as plant cell-wall feeders, in contrast to eusocial Hymenoptera (ants, some bees, some wasps), which are mainly cytoplasm consumers (Abe and Higashi 1991). It is probably because of the low food quality of cell walls (nitrogen poor, but rich in cellulose and lignin) that all termites are obligately dependent on mutualistic symbionts. Although it has fre-

Figure 8.2. Mound of *Macrotermes bellicosus*. Colonies of this species can build mounds of up to 7 m tall. (Photo by D. K. Aanen.)

quently been stated that the function of the symbionts for termites is cellulose degradation, this is not strictly true because all termite species have at least some ability to degrade cellulose without the help of symbionts (Bignell 2000). Thus, rather than cellulose degradation, the major benefit of symbiosis with bacteria may be the improved supply of nitrogen by lowering the carbon–nitrogen ratio during gut fermentations (Bignell 2000).

It is generally accepted that termites form a monophyletic group, closely related to mantids and cockroaches, and that these three groups are monophyletic (Eggleton 2001). The most recent evidence suggests that the subsocial wood-feeding cockroaches of the genus *Cryptocercus* form the sister group of termites (Lo et al. 2000). However, the exact relationship between Isoptera, mantids, and cockroaches is still debated (Shellman-Reeve 1997; Eggleton 2001). Termites have been divided into

7 families, 281 genera, and about 2600 species (Kambhampati and Eggleton 2000). Traditionally, termites also have been separated as "higher" and "lower" termites based on their symbionts All species of the paraphyletic grouping of lower termites harbor a dense and diverse population of prokaryotes and flagellate protists (single-celled eukaryotes) in their guts. Higher termites comprise only a single family (Termitidae with four subfamilies), but they represent more than 80% of all described species (Kambhampati and Eggleton 2000). Although higher termites also have a dense and diverse array of prokaryote associates, they typically lack flagellate protists. A recent hypothesis on the most likely phylogenetic relationships among the seven families of termites is given in figure 8.3. The phylogenetic relationships among the four subfamilies of the Termitidae have been studied recently using DNA sequences (Miura et al. 1998; fig. 8.3). However, the phylogenetic hypotheses obtained so far are equivocal [see Donovan et al. (2000) for an estimate based on morphological characters], and there is some evidence that not all the subfamilies are monophyletic (Donovan et al. 2000; Eggleton 2001).

Higher termites show considerable variation in their feeding behavior. Many feed exclusively on soil, probably deriving nutrition from humic compounds; others feed on wood, and the Macrotermitinae cultivate and consume cellulolytic fungi. The fungal symbionts of these termites have been placed in a single genus, *Termitomyces* (Termitomyceteae: Tricholomataceae: Basidiomycota). Incipient termite colonies that do not become established with a fungus do not survive (Sands 1969; Johnson 1981). The benefit of fungus cultivation for the termites is that food is predigested before the termites eat it. Fungus-growing termites have a great impact on most African and Asian ecosystems and play a major role in the degradation of plant material (Rouland-Lefèvre 2000). Some of them can be considered major pests of agriculture and wooden structures. On the other hand, the basidiocarps of many

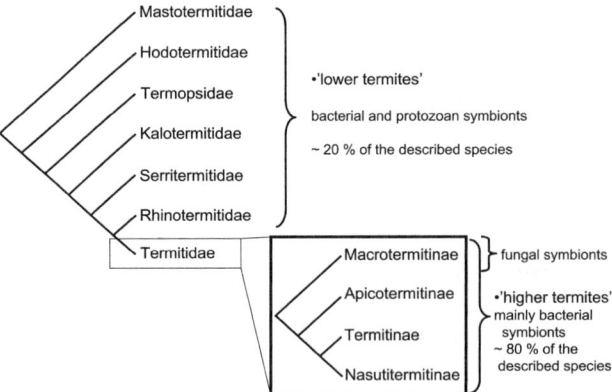

Figure 8.3. Phylogenetic relationships among the seven termite families (left; Eggleton 2001) and of the four subfamilies of the Termitidae (right). The family phylogeny given is a majority-rule consensus tree based on studies of Donovan et al. (2000), Kambhampati and Eggleton (2000), and Thompson et al. (2000). The phylogeny of Termitidae subfamilies is based on Miura et al. (1998).

Termitomyces species are appropriate for human consumption and have commercial value in some regions (Batra and Batra 1979).

The word "symbiosis" was popularized by de Bary (1879) and originally described a neutral coexistence of two or more distantly related organisms. Today, the term is often used for a relationship between two different kinds of organisms that interact in an obligately dependent, mutualistic, and physically intimate way (Bignell 2000). However, we prefer to use the term "mutualism" for symbioses of this kind (i.e., for the ones that provide a net benefit to both partners) and to reserve the term "symbiosis" for interactions in which the outcome can be both negative (parasitic) and positive (mutualistic). The distinction between mutualistic symbioses and parasitic symbioses may not be sharp: Mutualism can best be considered as a reciprocal exploitation that nonetheless provides net benefits to each partner (Herre et al. 1999). This view implies that parasitic tendencies can be present in partners of a mutualistic symbiosis.

In this chapter, we first review what is known about the natural history of the Macrotermitinae–*Termitomyces* symbiosis. Second, we highlight recent work that has shed light on the evolutionary developments that have shaped this symbiosis. Third, we discuss a number of key characteristics that have been identified as important for evolutionary innovations in this symbiosis. Fourth, we compare the main characteristics of the fungus-growing termites with analogous traits in other fungus-growing insects, particularly the fungus-growing ants (chapter 7). Finally, we discuss future prospects of the evolutionary study of fungus-growing termites. The ecology of the symbiosis between Macrotermitinae and *Termitomyces* has been reviewed in detail elsewhere (e.g., Batra and Batra 1979; Wood and Thomas 1989; Darlington 1994; Rouland-Lefèvre 2000), and we therefore only touch on these aspects in passing.

The Fungus-Growing Termites: Natural History

In Macrotermitinae, a single queen and king normally found a colony. They mate and seal themselves off in a copulatorium, a newly built cell of hard clay that becomes the "royal chamber." Here the first eggs are laid, and the first brood of sterile workers is reared entirely upon the bodily reserves of the parents. Most species rely on horizontal acquisition to establish the association with the fungal symbiont: the first workers apparently pick up *Termitomyces* spores from the environment (Johnson 1981; Sieber 1983). As the colony matures, the abdomen of the queen becomes enlarged and sausagelike (physiogastric), while the king remains much smaller (figs. 8.1, 8.4). Now food is continuously provisioned, and eggs are continuously produced and tended by the workers. Young termites (nymphs) remain in the fungus garden surrounding the royal chamber. Some species, especially in the genus *Macrotermes*, construct large mounds (fig. 8.2), whereas other species have nests that are entirely below ground. The Macrotermitinae often dominate the termite fauna in the tropical savannas and forests of the Old World (Rouland-Lefèvre 2000).

The fungus helps to degrade the plant-derived material (e.g., wood, dry grass, leaf litter) upon which the termites live (Johnson et al., 1981). The fungus comb,

Figure 8.4. The queen and king of a species of the genus *Microtermes*, surrounded by workers. The disparity in size between the castes is not so great as in *Macrotermes bellicosus*. (Photo by V. I. D. Ros.)

composed of fecal pellets inoculated with fungi, grows in a special structure in the nest (fig. 8.5). The combs are housed in specially constructed single or multiple chambers, either inside a mound or dispersed in the soil. Fecal pellets (primary feces) produced by workers are added continuously to the top of the comb, and fungal mycelium rapidly develops in the newly added substrate. After a few weeks, the fungus starts to produce vegetative structures, called nodules (also called mycotêtes), that are consumed by the termite workers. These nodules are a rich source of nitrogen, sugars, and enzymes, but an additional function may be to serve as inoculum of a new fungus comb with fungal spores (e.g., Leuthold et al. 1989). At a later stage, the older comb material permeated with mycelium is consumed (Darlington 1994), a behavior reminiscent of soil feeding, the feeding behavior of the sister clade of Macrotermitinae (Aanen et al. 2002). For some species of *Macrotermes* and *Odontotermes* there is a continuous turnover of comb by the addition of primary feces at the top and consumption of older comb material at the bottom. Other species, however, such as those in the genus *Pseudacanthotermes*, consume entire combs before building new ones in the empty chambers (Rouland-Lefèvre 2000).

The exact role of *Termitomyces* in the symbiosis with the Macrotermitinae is still being debated (Bignell 2000; Ohkuma 2003). Several possible functions have been proposed: (1) decomposition of lignin to improve palatability of the food and access to cellulose (Grasse and Noirot 1958; Ohkuma 2003); (2) acquisition of fungal

Figure 8.5. Detail of the fungus comb of *Macrotermes bellicosus*. (Photo by K. Machielsen.)

cellulases and xylanases to work synergistically and complementarily with the endogenous enzymes of the termites (for a review, see Rouland-Lefèvre 2000; but see Bignell 1994); (3) enrichment of food with nitrogen due to the intensive metabolism of carbohydrates (Collins 1983); and (4) the production of heat and metabolic water (Lüscher 1951). Some studies suggest that the role of the fungus may differ in different termites (Rouland-Lefèvre 2000; Hyodo et al. 2003). For example, Hyodo et al. (2003) reported that the role of the fungus in species of the genus *Macrotermes* was mainly to degrade lignin, whereas the fungus in three other genera served primarily as food.

The Macrotermitinae have been subdivided into 11 well-supported genera (Kambhampati and Eggleton 2000) and about 330 species. A recent study, however, has shown that the exclusively Asiatic genus *Hypotermes* is derived from within the genus *Odontotermes*, reducing the number to 10 monophyletic genera (Aanen et al. 2002). Most of the macrotermitine diversity occurs in Africa, where all genera (with the exception of *Hypotermes*) are found. Four genera also are native to Asia, and two also occur in Madagascar (Kambhampati and Eggleton 2000).

The single species of the genus *Sphaerotermes*, *S. sphaerothorax*, builds combs that support growth of bacteria but not of fungi (Garnier-Sillam et al., 1989). This species has also been placed in the Macrotermitinae and has been considered by some authors as a missing link between non–fungus-growing termites and fungus-growing termites (Darlington 1994). However, this taxonomic treatment has been questioned (Eggleton and Kambhampati 2000), and the recent phylogenetic study by Aanen et al. (2002) has shown clearly that this species falls outside the clade of fungus-growing Macrotermitinae.

As mentioned before, the fungal symbionts of the Macrotermitinae are placed in a single genus, *Termitomyces*. Approximately 40 species of the *Termitomyces* symbiont have been described (Kirk et al. 2001). However, DNA-based phylogenetic studies (Aanen et al. 2002; Frøslev et al. 2003) suggest that many more species exist that are either morphologically similar to described species or that rarely if ever produce sexual fruiting bodies (required for traditional taxonomic identification). Molecular data have contributed greatly (see below) to our ability to identify the fungi and to hypothesize evolutionary patterns, not only for the fungi but for the termites as well.

How, Where, and When Did It All Start?

Compared to the other termite subfamilies, the geographic distribution of the Macrotermitinae is limited, occurring only in tropical Africa and parts of the Arabian and Indomalayan regions (fig. 8.6; Wood and Thomas 1989; Darlington 1994). The restricted distribution suggests a fairly late evolutionary origin in the Oligocene after the separation of the continents (Emerton 1955). The origin of the symbiosis has been hypothesized to be African because the highest taxon diversity is found in Africa (Darlington 1994). A DNA-sequencing study of both the termites and their fungal symbionts (Aanen et al. 2002) provided strong evidence for the African origin. In this study basal lineages of Macrotermitinae all proved to be African, while the Asian representatives appeared to belong to terminal clades. The termite-fungus samples analyzed contained two terminal clades of Asian termites within the genera *Macrotermes* and *Odontotermes* that cultivated five terminal clades of Asian *Termitomyces*. The most parsimonious explanation for this observation is that the fungus-growing termites have colonized Asia at least twice (but more likely four times, as some *Ancistrotermes* and *Microtermes* species known to occur in Asia were not sampled), and there were at least five intercontinental migrations of the associated fungi (Aanen et al. 2002). Given the limited size of the sample analyzed by Aanen et al. (2002), five intercontinental migrations is almost certainly an underestimate, suggesting that independent colonization of fungal symbionts have occurred after the various genera of Macrotermitinae colonized Asia.

Heim (1977) considered the symbiosis between fungus-growing termites and their fungal associates an antagonistic equilibrium in which the termites try to suppress the fungus. He believed that the fungus comb was simply part of the nest structure, built to provide a suitable nursery for the developing brood. He regarded it as a weakness of the system that the comb proved to be a favorable substrate for fungi. In his opinion the termite *Sphaerotermes sphaerothorax* had been particularly successful, because it had managed to rid itself of the fungus. However, Heim's theory did not gain a much support because other studies had already shown that species of *Termitomyces* play an essential role in the nutrition of the fungus-growing termites (Sands 1956; Grasse 1959). The phylogenetic analysis by Aanen et al. (2002) based on molecular characters provided a satisfactory close to this story by clearly showing that *S. sphaerothorax* falls outside the Macrotermitinae.

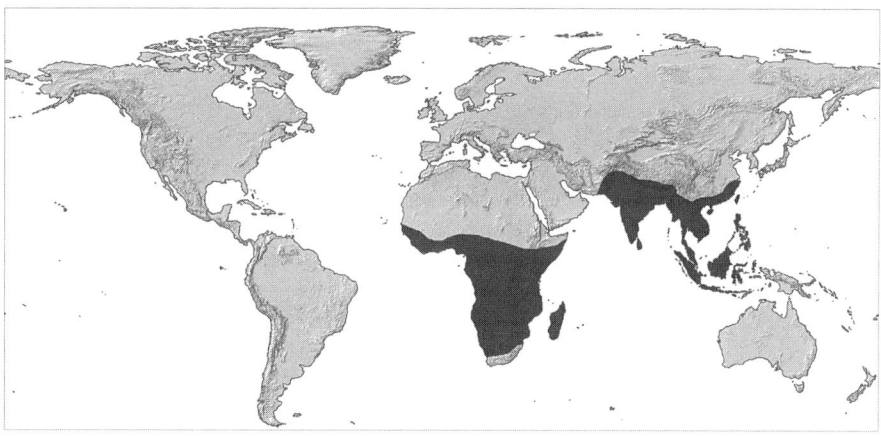

Figure 8.6. Current distribution of fungus-growing termites. (Modified from Batra and Batra 1979).

Although Heim's functional theory has now been decisively rejected, his hypothesis that the fungus comb began as an integral part of the nest structure is still widely accepted (e.g., Sands 1969; Wood and Thomas 1989; Darlington 1994). Sands (1969) proposed that the ancestors of the Macrotermitinae made nests that were built or lined with fecal carton (pulverized wood material), and saprotrophs subsequently invaded the carton, partially breaking it down into digestible products. In this scenario the termites began to exploit this additional, albeit minor, food source, and coevolution of termites and fungi began, ultimately leading to an obligate mutualistic relationship. More generally, it is known that many nonfarming termite species are attracted to feed on fungus-infested wood (Batra and Batra 1979; Rouland-Lefèvre 2000). It is therefore tempting to suggest that fungi were already an important food source before the origin of active fungus cultivation. The only additional step required would be for the termites to develop the ability to manipulate fungal growth in their nests (Sands 1969; Mueller and Gerardo 2002).

Evolution of Fungus-Growing Termites and Their Mutualistic Fungal Symbionts

DNA sequence data have been used to estimate phylogenies of the fungus-growing termites and their associated *Termitomyces* species (Aanen et al. 2002; Rouland-Lefèvre et al. 2002). Aanen et al. (2002) provided strong evidence for a single origin of the symbiosis because both the termites and their fungal symbionts were well-supported monophyletic groups (fig. 8.7). Therefore, secondary domestications of other fungi or reversals of one of the termites or fungi to a nonsymbiotic state would be highly unlikely.

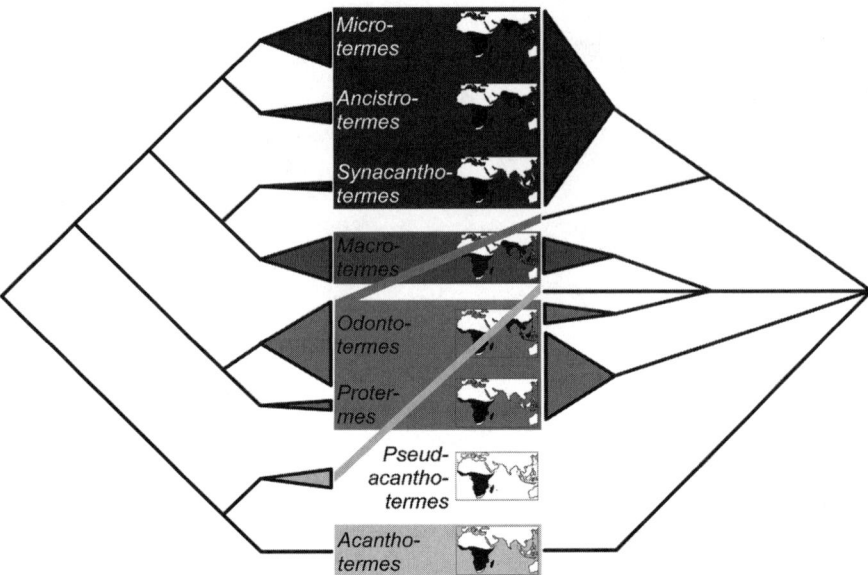

Figure 8.7. A simplified representation of the phylogenetic relationships of the five major clades of fungus-growing termites (left) and their mutualistic fungal symbionts (right), based on the more detailed results reported by Aanen et al. (2002). The size of the triangles on the left is roughly proportional to the number of termite species described within each genus (based on Kambhampati and Eggleton 2000). The main clades of termites and their corresponding clades of fungal symbionts are connected by rectangles. In general, termite clades (at the genus level or higher) tend to specialize on specific clades of *Termitomyces*, but strains of fungi within these termite clades are regularly exchanged between termite species. The distribution ranges of the different genera, African and/or Asian, are also indicated.

The high degree of specificity of the termite–fungus associations mainly at the level of termite genus clearly demonstrates that coevolution has occurred. Aanen et al. (2002) recognized five main clades of termites that are generally associated with particular clades of fungi. For example, the fungi associated with the monophyletic termite genus *Macrotermes* formed a monophyletic group. Likewise, the fungi of the closely related genera *Ancistrotermes* and *Microtermes* were monophyletic. However, there were also exceptions to the higher level pattern of specificity. For example, four termite species of the genus *Odontotermes* had symbionts that did not belong to the main clade of fungi associated with *Odontotermes*, and these exceptions indicate that host switching also took place between some of the main clades.

Within the main clades of the termite phylogeny, host switching has been frequent, especially at the lower taxonomic levels, and neighboring nests of a single termite species can have symbionts that are not closely related (Aanen et al. 2002; Katoh et al. 2002). Likewise, closely related fungal symbionts can be associated with different, sometimes distantly related termites. This view is consistent with horizontal transmission of fungal symbionts in most of the extant taxa. Because basal taxa are included in the analysis, the ancestral transmission mode in this mutualism was almost certainly horizontal. Clonal vertical transmission of fungi, previously shown to be common in the genus *Microtermes* (via females) and in *Macrotermes bellicosus* (via males) (Johnson et al. 1981), was derived independently on two occasions. We will return to further details of the evolution of transmission modes below.

Key Characteristics for the Evolution of the Symbiosis

Symbiosis, Mutualism, and Parasitism: Which Factors Are Important?

As stated in the introduction, mutualisms are best viewed as reciprocal exploitations that nonetheless provide net benefits to each partner (Herre et al. 1999). This view implies that the degree of mutualism in a symbiotic relationship can be considered a continuous variable and that there is no absolute distinction between truly parasitic and truly mutualistic interactions. Symbiont transmission mode has been proposed to be one of the major determinants for the ultimate beneficiality of a symbiosis (e.g., Frank 1996; Herre et al. 1999; i.e., vertical versus horizontal transmission and clonal versus sexual transmission). Vertical symbiont transmission proceeds from host parent to host offspring and implies that symbiont reproduction is aligned with host reproduction. Vertical transmission can further be uniparental (via one of the two sexes) or biparental (via both sexes). Horizontal symbiont transmission, in contrast, is independent of host reproduction and will often result in associations between symbionts and hosts outside the parental host lineage. Both horizontal transmission and vertical transmission can be either clonal (i.e., no genetic recombination occurs between strains of symbiont) or sexual (i.e., recombination occurs), but strict uniparental vertical transmission prohibits the opportunities for sex between symbionts of different hosts and is therefore usually clonal, whereas horizontal transmission allows ample opportunities for sexual exchange.

Symbiont fitness has two components: one arises from the overall success of the group of symbionts within a host, and the other originates from the competitive success and transmission of one symbiont strain relative to other strains within the same host. As relatedness among symbiont strains within a host decreases, the success of a symbiont genotype depends increasingly on its ability to outcompete its neighbors and less on the overall success of the group (Hamilton 1972). Thus, a lower relatedness among symbionts favors competition, which is likely to decrease the overall success of the group of symbionts and to have virulent side-effects on the host. It is therefore in the interest of hosts to keep their symbionts genetically homogeneous, and an obvious way to achieve this is to reduce mixing of symbionts by enforcing clonal and vertical uniparental transmission (Frank 1996). However, this is not in the ultimate reproductive interest of the symbiont because natural selection will never favor the complete loss of dispersal out of the host-imposed vertical lineage (Frank 1996). This conclusion is based on a general model by Hamilton and May (1977), which demonstrated that selection always favors some dispersal, even if the survival or success of propagules is very low. The reason for this outcome is that competition with relatives becomes an increasingly strong force when dispersal decreases. This implies that, unless hosts can completely suppress their symbionts, some horizontal mixing of symbionts should still be expected and that these rare events maintain selection for the conditional expression of virulent traits in symbionts. In other words, even in the most harmonious mutualisms, hosts and symbionts are in conflict over the mixing of symbionts, and this conflict may occasionally become evident (Frank 1996). The degree to which a host can control the transmission of its symbionts is therefore one of the determinants of the degree of mutual beneficiality to which a symbiosis can evolve (Frank 1996; Herre et al. 1999).

Observed Transmission Modes in the Macrotermitinae and Genetic Variation of Symbionts

As already mentioned, symbiont transmission in fungus-growing termites is predominantly horizontal (i.e., fungal symbionts normally disperse out of the vertical host lineage and potentially form novel combinations with new hosts every generation). Although studies of single taxa are still rare, a number of details about symbiont transmission in fungus-growing termites are known, based on studies by Sands (1960), Johnson (1981), Johnson et al. (1981), and Sieber (1983). When horizontal transmission occurs, sexual termites (alates) do not carry any symbiont fungus during their mating flights, and the first workers of the royal pair build an incipient comb before collecting *Termitomyces* spores from the environment during foraging trips and inoculating the comb. This mode of transmission has been demonstrated in species of the genera *Ancistrotermes*, *Macrotermes*, *Pseudacanthotermes*, and *Odontotermes* and is likely to represent the ancestral mode of transmission (Aanen et al. 2002). The *Termitomyces* symbionts of these macrotermitine genera regularly produce fruiting bodies (fig. 8.8), so that sexual basidiospores are readily available in the environment. In the two known cases of vertical transmission, alates carry fungal spores from their colony of origin with them on their mating flight to inoculate the fungus comb of the incipient colonies that they will found. This mode of transmission has been demonstrated for all five of the species of the genus *Microtermes* studied, and for a single species of the genus *Macrotermes*, *M. bellicosus*. Remarkably, vertical transmission is uniparental in both cases, but by different parents based on genus. As we mentioned earlier, in *Microtermes* the female reproductive transmits the fungus, whereas the male reproductive transmits in *M. bellicosus*. Recent phylogenetic studies have shown that these two cases of vertical symbiont transmission represent two independent evolutionary transitions (Aanen et al. 2002).

The two independent transitions from horizontal to uniparental vertical transmission strongly suggest that host control of symbiont transmission is adaptive (Korb and Aanen 2003), in line with the theoretical considerations given above. However, vertical transmission is apparently not a condition for the evolutionary stability of the mutualism, as the symbiosis evolved on the basis of horizontal transmission and has obviously been mutualistic from its start. Also, other mutualistic symbioses, such as those between mycorrhizal fungi and nitrogen-fixing rhizobia with their

Figure 8.8. An excavated fungus comb of *Ancistrotermes cavithorax* and a fruiting body arising from it. (Photo by K. Machielsen.)

respective host plants rely exclusively on horizontal transmission. This raises the question of whether termites have mechanisms other than controlling symbiont transmission to prevent the evolution and expression of virulent traits in their fungal symbionts that would damage the productivity of the interaction. One obvious possibility is that the termites limit the genetic variation of their symbionts by culturing only a single clone of fungus in a nest, after weeding out all competing strains. The evidence obtained to date, still based on a limited number of colony samples, indicates that there is indeed only a single fungal clone present per colony (Aanen et al. 2002; Katoh et al. 2002). The question is whether this purity of fungal culture is achieved actively, through selection of symbiont spores by the termites, or passively because the comb material is of such specific quality that only one or very few strains can grow there.

Comparison with Other Fungus-Growing Social Insects

In the other main symbiosis between social insects and fungi, the attine ants and their leucocoprine (Leucocoprineae) fungi (chapter 7), recent experimental work has provided evidence for the mechanism by which higher attine ants maintain genetic uniformity of the fungal symbionts in their nests (Bot et al. 2001). Using transplantation experiments, Bot et al. (2001) found that certain artificial combinations of ants and fungal strains were incompatible and that this incompatibility was to a large extent independent of the ant species (two sympatric *Acromyrmex* species were used). It was hypothesized that the incompatibility between ants and alien fungal strains was ultimately due to somatic incompatibility between genetically different fungal symbionts. The hypothesis also called for the ant hosts to reject strains that are genetically different from their own symbiont even in the absence of that symbiont. Two further observations supported this idea. First, the incompatibility was transient: When ants were force-fed the fungus with which they were incompatible, the incompatibility disappeared after about a week. Second, there was a positive correlation between the level of incompatibility and the genetic distances between the resident and the transplanted fungus. Somatic incompatibility reactions for many basidiomycete fungi have been shown to occur readily in interactions between genetically different dikaryotic individuals, and usually at least some correlation between the genetic distance and the strength of the reaction has been found (Worrall 1997). The occurrence of such somatic incompatibility reactions is normally accompanied by a reduction in growth of both dikaryons (Worrall 1997) and can therefore in the ant–fungus mutualism be considered a virulent trait for the host that is expressed as a direct response to the introduction of a nonrelated symbiont. The surprising aspect of this finding is that these incompatibility reactions have been maintained in leaf-cutting ants despite the fact that symbiont transmission is vertical and uniparental, which supposedly means the horizontal transmission of fungal symbionts does not happen very often.

The origin of the attine ant symbiosis with fungi is completely independent of the analogous termite symbiosis: Not only are termites and ants not sister groups, but also the fungi cultivated are only distantly related. Moreover, the distribution ranges of the two groups of social insects are completely separated, as the attine ants occur exclusively in North and South America. These two groups of fungus-growing social insects provide us, therefore, with a unique opportunity to compare two independent evolutionary developments of fungus growing, both of which have similarities with certain aspects of human agriculture (chapter 7).

In both ant and termite symbioses, the insects form monophyletic groups (i.e., only a single social insect ancestor domesticated fungi in each case). However, whereas the termite fungal symbionts form a monophyletic group, the extant fungal symbionts of the ants do not. In fact, there have been a minimum of at least three independent secondary domestications (Chapela et al. 1994; Mueller et al. 1998). The secondary domestications were all accomplished by lower attines that became specialized fungus growers but that still reared fungi closely related to free-living leucocoprine basidiomycetes (Chapela et al. 1994; Mueller et al. 1998). One ant clade, however, secondarily domesticated a fungus from an entirely different clade (Chapela et al. 1994). This finding implies that the symbiosis between the attine ants and their fungal symbionts has remained asymmetric (i.e., ants are obligately dependent on fungi, but fungi are not obligately dependent on the ants) for quite some evolutionary time, with little coevolution in the domesticated fungi, in spite of vertical transmission by the ants. The lack of fungal coevolution, therefore, did not necessitate adaptations in the ants to cultivate specific lineages of fungus, explaining that novel domestications remained possible. It was only in the terminal monophyletic group of the higher attine ants that the symbiosis became symmetrical and that coevolution with a single, gongylidia-producing clade of fungi began. *Gongylidia* are specialized bulbous hyphal tips upon which insects feed. This innovation ultimately gave rise to four extant ant genera, including the *Acromyrmex* and *Atta* leaf-cutter ants (Chapela et al. 1994). In contrast, and despite horizontal transmission, the mutualism between the termites and their fungi has been symmetrical (i.e., both insects and fungi are mutually obligately interdependent) from its beginning (Aanen et al. 2002).

Many of the intuitive ideas on domestication (with an active domesticator and a passive domesticate; but see Diamond 1997; Leach 2003) can therefore not be applied to the fungus-growing termites because the evolutionary modifications in this mutualism have been as much fungus-driven as insect-driven (Aanen et al. 2002; Mueller and Gerardo 2002). We hypothesize that, once the initial domestication innovation occurred, the macrotermitine termites were able to occupy a series of new food niches and that these diverging niches permitted adaptive radiation and selected for specific fungal adaptations to combs built from different plant-derived materials. This niche specialization scenario is consistent with the different roles of the fungal symbionts for termite nutrition across genera (see above; Rouland-Lefèvre 2000). However, the lower attine ants could cultivate many free-living strains of fungus, so that the details of the nutritional role of the fungal symbionts probably changed rather little until the symbiosis became symmetrical in the higher attine ants, the clade with fungal symbionts characterized by the synapomophic gongylidia. In conclusion, and as formulated by Aanen et al. (2002), the attine ants evolved specific adaptations to be farmers of rather unspecified fungal crops, and these fungi realized crucial adaptations only in the terminal clade of the higher attine ants. The obligate symmetrical interactions that followed allowed the symbiosis to become highly specialized and ultimately produced the leaf-cutter ants. The Macrotermitinae, in contrast, specialized on a single group of fungi that quickly became genetically isolated from its free-living sister group to cospeciate and specialize in response to the increasing diversity of fungus-comb substrates available across termite species and habitats.

As already indicated in passing, another important difference between ant and termite associations lies in the predominant mode of symbiont transmission. Throughout the attine ants, symbiont transmission happens vertically via the founding queen and is therefore automatically uniparental. In the fungus-growing termites, however, horizontal transmission is not only ancestral but has been maintained throughout with the exception of two independent developments toward vertical transmission in apical clades. Aanen et al. (2002) hypothesized that the evolution of vertical symbiont transmission in termites has been constrained because it produces single-strain fungus gardens only when restricted to a single termite sex. This is accomplished by default in ants, where males do not survive beyond mating, but not in the termites where colonies are founded by a female and male.

In spite of these differences, both termite and ant systems seem to have arrived at variable but intermediate levels of specificity in their host–symbiont interactions. The evolutionary pathways, however, have been very different. The termites domesticated a single group of fungi. These fungi retained sex, and the termites continued to acquire them horizontally; the fungi became fully dependent on termites. The fungus-growing ants may have had vertical transmission in place before becoming obligately dependent on fungal crops (Mueller et al. 2001) and were therefore successful throughout the clade in suppressing almost all sexual reproduction of their symbionts, but also realized a significant amount of secondary domestication of free-living relatives of their fungal symbionts. Secondary domestication apparently continues until the present day (Mueller et al. 1998). Both syndromes seem to allow sexual exchange and horizontal transmission at sufficient rates to prevent genetic deterioration (Muller's ratchet; Muller 1964; chapter 2, this volume) of their symbionts from long-term clonality, but the mechanisms by which this is achieved are as yet unknown.

Summary and Perspectives

The symbiosis between fungus-growing termites and their associated fungi is a prime example of mutualism. Both partners are completely interdependent, although the exact reciprocal benefits are still debated. Recent studies based on DNA characters greatly increased our insight into the evolution of this symbiosis. The two partners have been shown to form monophyletic groups, implying that the symbiosis can be considered symmetrical. Elaborations in the symbiosis, therefore, are as likely to be fungus driven as insect driven.

However, a series of major questions remain to be addressed. First, given the intermediate level of specificity that has been found, how does the selection of fungal symbionts occur and which partner is the active one? Do termites have behavioral mechanisms to select and discriminate between fungal spores, or do they provide a highly specific substrate for a variety of fungal spores on which only some strains can grow? Second, do the termites have mechanisms to prevent secondary infections of the fungus comb that prevent the evolution and expression of virulent traits? Have they also elaborated on existing somatic incompatibility mechanisms between genetically different strains in their fungi? Some evidence has been obtained recently that such mechanisms exist in the attine ants to prevent mixed colonizations of nests (Bot et al. 2001). We hypothesize that in termites, the first fungus that is selected will greatly limit the opportunities for additional colonizations of the comb by unrelated strains, and although these colonies may live for decades, changeovers of symbiont may be extremely rare if not impossible. Experiments similar to the ones that have been done with fungus-growing ants will be needed to test this hypothesis. Such tests will be more difficult, however, as mass rearing of fungus-growing termites in the laboratory is very difficult.

More detailed knowledge on the reproductive modes of *Termitomyces* is essential for improving laboratory cultivation and theoretical assessment. Most basidiomycetes are heterothallic, and mating between two different monokaryons depends on a difference in mating types between the two mycelia. The mating type is defined by one or two loci with multiple alleles within a population, so that outcrossing is usually promoted (e.g., Aanen and Kuyper, 1999). Based on the heterothallic condition of the majority of euagarics, it has been assumed that *Termitomyces* species are heterothallic, but the mating system has never been determined. If they are heterothallic, this implies that the fungus comb ultimately needs to be colonized by two genetically different fungal spores, whereas a single spore would suffice if the life cycle were not heterothallic. *Termitomyces* spores do not germinate readily under artificial conditions, and this factor has hindered the study of the life cycle.

Another way of ensuring a single fungal clone per nest is vertical uniparental symbiont transmission. The occurrence of two independent evolutionary transitions to uniparental transmission among macrotermitine termites indicates that this behavior is likely to be adaptive (Korb and Aanen 2003). Regular transmission modes have been studied in detail for only a few species, and a larger taxonomic spread of novel data of this kind is badly needed.

Finally, an entirely new direction of research could arise with the discovery that other organisms may have specific parasitic or mutualistic roles in the termite–*Termitomyces* symbiosis, something that has happened recently in the study of the fungus-growing ants (e.g., Currie et al. 1999a,b). Interesting candidates for such a role among termites are species of the ascomycete fungus *Xylaria*, an inhabitant of deserted nests of fungus-growing termites. Some workers have considered this fungus to be a parasite that normally is suppressed by the termites while they are active in a nest (e.g., Thomas 1987a,b,c; Wood and Thomas 1989; Darlington 1994); others have hypothesized that this fungus is yet another mutualist (e.g., Batra and Batra 1979). Experimental studies of the role of *Xylaria* will therefore be crucial.

Literature Cited

Aanen, D. K., and Th. W. Kuyper. 1999. Intercompatibility tests in the *Hebelomacrustulini-forme* complex in northwestern Europe. *Mycologia* 91:783–795.

Aanen D. K., P. Eggleton, C. Rouland-Lefèvre, T. Guldberg-Frøslev, S. Rosendahl, and J. J. Boomsma. 2002. The evolution of fungus-growing termites and their mutualistic fungal symbionts. *Proceedings of the National Academy of Sciences of the USA* 99: 14887–14892.

Abe, T., and M. Higashi. 1991. Cellulose centered perspective on terrestrial community structure. *Oikos* 60:127–133.

Batra, L. R., and W. T. Batra. 1979. Termite-fungus mutualism. In *Insect fungus symbiosis*, ed. L. R. Batra, pp. 117–163. Montclair, NJ: Allanheld, Osmun & Co.

Bignell, D. E. 2000. Introduction to symbiosis. In *Termites: evolution, sociality, symbioses, ecology,* ed. T. Abe, D. E. Bignell, and M. Higashi, pp. 189–208. Dordrecht: Kluwer Academic.

Bignell, D. E., M. Slaytor, and P. C. Veivers. 1994. Functions of symbiotic fungus gardens in higher termites of the genus *Macrotermes*: Evidence against the acquired enzyme hypothesis. *Acta microbiologica et immunologica Hungarica* 41:391–401.

Bot, A. N. M., S. A. Rehner, and J. J. Boomsma. 2001. Partial incompatibility between ants and symbiotic fungi in two sympatric species of *Acromyrmex* leaf-cutting ants. *Evolution* 10:1980–1991.

Breznak, J. A. 2000. Ecology of prokaryotic microbes in the guts of wood- and litter-feeding termites. In *Termites: evolution, sociality, symbioses, ecology,* ed. T. Abe, D. E. Bignell, and M. Higashi, pp. 209–232. Dordrecht: Kluwer Academic.

Chapela, I. H., S. A. Rehner, T. R. Schultz, and U. G. Mueller. 1994. Evolutionary history of the symbiosis between fungus-growing ants and their fungi. *Science* 266:1691–1694.

Collins, N. M. 1983. The utilization of nitrogen resources by termites (Isoptera). In *Nitrogen as an ecological factor*, ed. J. A. Lee, S. McNeil, and I. H. Rorison, pp. 381–412. Oxford: Blackwell Scientific.

Currie, C. R., U. G. Mueller, and D. Malloch. 1999a. The agricultural pathology of ant fungus gardens. *Proceedings of the National Academy of Sciences of the USA* 96:7998–8002.

Currie, C. R., J. A. Scott, R. C. Summerbell, and D. Malloch. 1999b. Fungus-growing ants use antibiotic-producing bacteria to control garden parasites. *Nature* 398:701–704.

Darlington, J. E. C. P. 1994. Nutrition and evolution in fungus-growing ants. In *Nourishment and evolution in insect societies*, ed. J. H. Hunt and C. A. Nalepa, pp. 105–130. Boulder, CO: Westview Press.

de Bary, A. 1879. Die erscheinung der symbiose. Straßburg: Verlag Trubner.

Diamond, J. 1997. *Guns, germs and steel*. London: Vintage.

Donovan, S. E., D. T. Jones, W. A. Sands, and P. Eggleton. 2000. Morphological phylogenetics of termites (Isoptera). *Biological Journal of the Linnean Society* 70:467–513.

Eggleton, P. 2000. Global patterns of termite biodiversity. In *Termites: evolution, sociality, symbioses, ecology*, ed. T. Abe, D. E. Bignell, and M. Higashi, pp. 25–52. Dordrecht: Kluwer Academic.

Eggleton, P. 2001. Termites and trees: A review of recent advance in termite phylogenetics. *Insectes Sociaux* 48:187–193.

Emerton, A. E. 1955. Geographical origins and dispersions of termite genera. *Fieldiana Zoology* 37:465–521.

Frank, S. A. 1996. Host symbiont conflict over the mixing of symbiotic lineages. *Proceedings of the Royal Society of London B* 263:339–344.

Frøslev, T. G., D. K. Aanen, S. Rosendahl, and T. Læssøe. 2003. Phylogenetic relationships in *Termitomyces* and related taxa. *Mycological Research* 107:1277–1286.

Garnier-Sillam, E., F. Toutain, G. Villemin, and J. Renoux.1989. Etudes préliminaires des meules originales du termite xylophage *Sphaerotermes sphaerothorax* (Sjöstedt). *Insectes Sociaux* 36:293–312.

Grasse, P.-P. 1959. Une nouveau type de symbiose: La meule alimentaire des termites champignonnistes. *Nature (Paris)* 3293:385–389.

Grasse, P.-P., and C. Noirot. 1958. La meule de termites champignonnistes et sa signification symbiotique. *Annales de la Société Zoologique et de Biologie Animale* 11:113–129.

Hamilton, W. D. 1972. Altruism and related phenomena, mainly in social insects. *Annual Review of Ecology and Systematics* 3:193–232.

Hamilton, W. D., and R. M. May. 1977. Dispersal in stable habitats. *Nature* 269:578–581.

Heim, R. 1977. *Termites et champignons. Les champignons termitophiles d'Afrique noire etd'Asie méridionale*. Paris: Société nouvelle des éditions Boubée.

Herre E. A., N. Knowlton, U. G. Mueller, and S. A. Rehner. 1999. The evolution of mutualisms: Exploring the paths between conflict and cooperation. *Trends in Ecology and Evolution* 14:49–53.

Hyodo, F., I. Tayasu, T. Inoue, J.-I. Azuma, T. Kudo, and T. Abe. 2003. Differential role of symbiotic fungi in lignin degradation and food provision for fungus-growing termites (Macrotermitinae, Isoptera). *Functional Ecology* 17:186–193.

Johnson, R. A. 1981. Colony development and establishment of the fungus comb in *Microtermes* sp. nr. *usambaricus* (Sjöst.) (Isoptera, Macrotermitinae) from Nigeria. *Insectes Sociaux* 28:3–12.

Johnson, R. A., R. J. Thomas, T. G. Wood, and M. J. Swift. 1981. The inoculation of the fungus comb in newly founded colonies of the Macrotermitinae (Isoptera) from Nigeria. *Journal of Natural History* 15:751–756.

Kambhampati, S., and P. Eggleton. 2000. Taxonomy and phylogeny of termites. In *Termites: evolution, sociality, symbioses, ecology*, ed. T. Abe, D. E. Bignell, and M. Higashi, pp. 1–24. Dordrecht: Kluwer Academic.

Katoh, H., T. Miura, K. Maekawi, N. Shinzato, and T. Matsumoto. 2002. Genetic variationof symbiotic fungi cultivated by the macrotermitine termite *Odontotermes formosanus* (Isoptera: Termitidae) in the Ryukyu Archipelago. *Molecular Ecology* 11:1565–1572.

Kirk, P. M., P. F. Cannon, J. C. David, and J. A. Stalpers. 2001. *Ainsworth & Bisby's dictionary of the fungi*. Wallingford, UK: CABI Publishing.

Korb, J. 1997. Lokale und regionale verbreitung von *Macrotermes bellicosus* (Isoptera; Macrotermitinae): stochastik oder determinik? Berlin: Wissenschaft und Technik Verlag.

Korb, J., and D. K. Aanen. 2003. The evolution of uniparental transmission of fungal symbionts in fungus-growing termites (Macrotermitinae). *Behavioural Ecology and Sociobiology* 53:65–71.

Leach, H. M. 2003. Human domestication reconsidered. *Current Anthropology* 44:349–368.

Leuthold, R. H., S. Badertscher, and H. Imboden. 1989. The inoculation of newly formed fungus comb with *Termitomyces* in *Macrotermes* colonies (Isoptera, Macrotermitinae). *Insectes Sociaux* 36:328–338.

Lo, N., G. Tokuda, H. Watanabe, H. Rose, M. Slaytor, K. Maekawa, C. Bandi, and H. Noda. 2000. Evidence from multiple gene sequences indicates that termites evolved from wood-feeding cockroaches. *Current Biology* 10:801–804.

Lüscher, M. 1951. Significance of 'fungus-gardens' in termite nests. *Nature* 167:34–35.

Maynard-Smith, J., and E. Szathmáry. 1995. *The major transitions in evolution.* Oxford: Oxford University Press.

Michod, R. E. 1999. Individuality, immortality and sex. In *Levels of selection in evolution,* ed. L. Keller, pp. 53–74. Princeton, NJ: Princeton University Press.

Miura, T., K. Maekawa, O. Kitade, T. Abe, and T. Matsumoto. 1998. Phylogenetic relationships among subfamilies in higher termites (Isoptera: Termitidae) based on mitochondrial COII gene sequences. *Annals of the Entomological Society of America* 91:516–521.

Moncalvo, J.-M., F. M. Lutzoni, S. A. Rehner, J. Johnson, and R. Vilgalys. 2000. Phylogenetic relationships of agaric fungi based on nuclear large subunit ribosomal DNA sequences. *Systematic Biology* 9:278–305.

Moncalvo, J. M., R. Vilgalys, S. A. Redhead, J. E. Johnson, T. Y. James, M. C. Aime, V. Hofstetter, S. J. W. Verduin, E. Larsson, T. J. Baroni, R.G. Thorn, S. Jacobsson, H. Clémençon, and O. K. Miller, Jr. 2002. One hundred and seventeen clades of euagarics. *Molecular Phylogenetics and Evolution* 23:357–400.

Mueller, U. G., and N. Gerardo. 2002. Fungus-farming insects: Multiple origins and diverse evolutionary histories. *Proceedings of the National Academy of Sciences of the USA* 99:15247–15249.

Mueller, U. G., S. A. Rehner, and T. R. Schultz. 1998. The evolution of agriculture in ants. *Science* 281:2034–2038.

Mueller, U. G., T. R. Schultz, C. R. Currie, R. M. M. Adams, and D. Malloch. 2001. The origin of the attine ant-fungus mutualism. *Quarterly Review of Biology* 76:169–197.

Muller, H. J. 1964. The relation of recombination to mutational advance. *Mutational Research* 1:2–9.

Ohkuma, M. 2003. Termite symbiotic systems: efficient bio-recycling of lignocellulose. *Applied Microbiology and Biotechnology* 61:1–9.

Rouland-Lefèvre, C. 2000. Symbiosis with fungi. In *Termites: evolution, sociality, symbioses, ecology,* ed. T. Abe, D. E. Bignell, and M. Higashi, pp. 289–306. Dordrecht: Kluwer Academic.

Rouland-Lefèvre C, M. N. Diouf, A. Brauman, and M. Neyra. 2002. Phylogenetic relationships in *Termitomyces* (family Agaricaceae) based on nucleotide sequences of ITS: a first approach to elucidate the evolutionary history of the symbiosis between fungus-growing termites and their fungi. *Molecular Phylogenetics and Evolution* 22:423–429.

Sands, W. A. 1956. Some factors affecting the survival of *Odontotermes badius*. *Insectes Sociaux* 3:531–536.

Sands, W. A. 1960. The initiation of fungus comb construction in laboratory colonies of *Ancistrotermes guineensis* (Silvestri). *Insectes Sociaux* 7:251–259.

Sands, W. A. 1969. The association of termites and fungi. In *Biology of termites*, vol. I, ed. K. Krishna and F. M. Weesner, pp. 495–524. London: Academic Press.

Shellman-Reeve, J. S. 1997. The spectrum of eusociality in termites. In *Social competition and cooperation in insects and arachnids*, ed. J. C. Choe and B. J. Crespi, pp. 52–93. Cambridge: Cambridge University Press.

Sieber, R. 1983. Establishment of fungus comb in laboratory colonies of *Macrotermes michaelseni* and *Odontotermes montanus* (Isoptera, Macrotermitinae). *Insectes Sociaux* 30:204–209.

Thomas, R. J. 1987a. Distribution of *Termitomyces* Heim and other fungi in the nests andmajor workers of *Macrotermes bellicosus* (Smeathman) in Nigeria. *Soil Biology and Biochemistry* 19:329–333.

Thomas, R. J. 1987b. Distribution of *Termitomyces* Heim and other fungi in the nests and major workers of several Nigerian Macrotermitinae. *Soil Biology and Biochemistry* 19:335–341.

Thomas, R. J. 1987c. Factors affecting the distribution and activity of fungi in the nests ofMacrotermitinae (Isoptera). *Soil Biology and Biochemistry* 19:343–349.

Thompson, G. J., O. Kitade, N. Lo, and R. H. Crozier. 2000. Phylogenetic evidence for a single ancestral origin of a true worker caste in termites. *Journal of Evolutionary Biology* 13:869–881.

Wilson, E. O. 1971. *The insect societies.* Cambridge: Harvard University Press.

Wood T. G., and R. J. Thomas. 1989. The mutualistic association between Macrotermitinae and *Termitomyces.* In *Insect-fungus interactions,* ed. N. Wilding, N. M. Collins, P. M. Hammond, and J. F. Webber, pp. 69–92. New York: Academic Press.

Worrall, J. J. 1997. Somatic incompatibility in basidiomycetes. *Mycologia* 89:24–36.

9

The Role of Yeasts
as Insect Endosymbionts

Fernando E. Vega
Patrick F. Dowd

Insect associations with fungi are common and may be casual or highly specific and obligate. For example, more than 40 fungal species are associated with the coffee berry borer (*Hypothenemus hampei*, Coleoptera: Curculionidae; Pérez et al. 2003) and about the same number with the subterranean termite *Reticulitermes flavipes* (Zoberi and Grace 1990; table 9.1). In one system 28 species of yeasts were isolated from the external parts of *Drosophila serido* and 18 species, including some not found on the external surfaces, from their crop (Morais et al. 1994; table 9.1).

In relatively few cases a specific role for the fungus has been identified, as is the case for associations with ants (chapter 7), termites (chapter 8), and bark beetles (Chapter 11; Six 2003). These associations imply that different species are living together, reinforced by specific interactions, a concept popularized as symbiosis by de Bary (1879). Symbiotic associations have been classified as ectosymbiotic when the symbiont occurs outside the body of the host or endosymbiotic when the symbiont occurs internally, either intra- or extracellularly (Steinhaus 1949; Nardon and Nardon 1998; Margulis and Chapman 1998). Several interesting symbiotic associations occur between insects and yeasts. In all cases that are well studied, the benefit that accrues for the insect is better understood than the benefit to the yeasts.

The term "yeast" is used to describe a particular fungal growth form (Steinhaus 1947; Alexopoulos et al. 1996). These predominantly unicellular ascomycetes divide by budding at some point in their life cycle (e.g., *Saccharomyces*). A surprising number of yeasts, however, also produce filamentous hyphae. At present, almost 700 species in 93 genera (Barnett et al. 2000) have been described in the ascomycete class Saccharomycetes, a group known informally as "true yeasts." True yeasts lack specialized sex organs, and sexual spores (ascospores) are produced in

Table 9.1. Yeasts internally isolated from insects.

Insect Species	Order: Family	Yeast Location (Species)[a]	Reference
Stegobium paniceum (= *Sitodrepa panicea*)	Coleoptera: Anobiidae	Mycetomes (*Saccharomyces*)[b]	Escherich 1900
			Buchner 1930
		Cecae (*Torulopsis buchnerii*)	Gräbner 1954
		Mycetome between foregut and midgut	Pant and Fraenkel 1954
		Mycetomes (*Symbiotaphrina buchnerii*)	Kühlwein and Jurzitza 1961
		Mycetomes and digestive tube (*Torulopsis buchnerii*)	Bismanis 1976
		Gut cecae (*Symbiotaphrina buchnerii*)	Noda and Kodama 1996
Lasioderma serricorne	Coleoptera: Anobiidae	Mycetome between foregut and midgut (*Symbiotaphrina kochii*)	van der Walt 1961; Jurzitza 1964
			Gams and von Arx 1980
			Noda and Kodama 1996
Ernobius abietis	Coleoptera: Anobiidae	Mycetomes (*Torulopsis karawaiewii*) (*Candida karawaiewii*)[c]	Jurzitza 1970
			Jones et al. 1999
Ernobius mollis	Coleoptera: Anobiidae	Mycetomes (*Torulopsis ernobii*) (*Candida ernobii*)	Jurzitza 1970
			Jones et al. 1999
Hemicoelus gibbicollis	Coleoptera: Anobiidae	Larval mycetomes	Suomi and Akre 1993
Xestobium plumbeum	Coleoptera: Anobiidae	Mycetomes (*Torulopsis xestobii*) (*Candida xestobii*)	Jurzitza 1970
			Jones et al. 1999
Criocephalus rusticus	Coleoptera: Cerambycidae	Mycetomes	Riba 1977
Phoracantha semipunctata	Coleoptera: Cerambycidae	Alimentary canal (*Candida guilliermondii*, *C. tenuis*)	Chararas and Pignal 1981
		Cecae around midgut (*Candida guilliermondii*)	Nardon and Grenier 1989
Harpium inquisitor	Coleoptera: Cerambycidae	Mycetomes (*Candida rhagii*)	Jurzitza 1960
Harpium mordax H.sycophanta	Coleoptera: Cerambycidae	Cecae around midgut (*Candida tenuis*)	Jurzitza 1960
Gaurotes virginea	Coleoptera: Cerambycidae	Cecae around midgut (*Candida rhagii*)	Jurzitza 1960
			Jones et al. 1999
Leptura rubra	Coleoptera: Cerambycidae	Cecae around midgut (*Candida tenuis*)	Jurzitza 1960
		Cecae around midgut (*Candida parapsilosis*)	Jurzitza 1959
			Jones et al. 1999

Species	Order: Family	Location (yeast)	Reference
Leptura maculicornis, L. cerambyciformis	Coleoptera: Cerambycidae	Cecae around midgut (Candida parapsilosis)	Jurzitza 1960
Leptura sanguinolenta	Coleoptera: Cerambycidae	Cecae around midgut (Candida sp.)	Jurzitza 1960
Rhagium bifasciatum	Coleoptera: Cerambycidae	Cecae around midgut (Candida tenuis)	Jurzitza 1960
Rhagium inquisitor	Coleoptera: Cerambycidae	Cecae around midgut (Candida guilliermondii)	Jurzitza 1959
Rhagium mordax	Coleoptera: Cerambycidae	Cecae around midgut (Candida)	Jurzitza 1959
Carpophilus hemipterus	Coleoptera: Nitidulidae	Intestinal tract (10 yeast species)	Miller and Mrak 1953
Odontotaenius disjunctus	Coleoptera: Passalidae	Hindgut (Enteroramus dimorphus)	Lichtwardt et al. 1999
Odontotaenius disjunctus Verres sternbergianus	Coleoptera: Passalidae	Gut (Pichia stipitis, P. segobiensis Candida shehatae, C. ergatensis)	Suh et al. 2003
Scarabaeus semipunctatus Chironitis furcifer	Coleoptera: Scarabaeidae	Digestive tract (10 yeast species)	Malan and Gandini 1966
Unknown species	Coleoptera: Scarabaeidae	Guts (Trichosporon cutaneum)	do Carmo-Sousa 1969
Dendroctonus and Ips spp.	Coleoptera: Scolytidae	Alimentary canal (13 yeast species)	Shifrine and Phaff 1956
Dendroctonus frontalis	Coleoptera: Scolytidae	Midgut (Candida sp.)	Moore 1972
Ips sexdentatus	Coleoptera: Scolytidae	Digestive tract (Pichia bovis, P. rhodanensis, Hansenula holstii, (Candida rhagii)	Pignal et al. 1987, 1988
		Digestive tract (Candida pulcherina)	Le Fay et al. 1969, 1970
Ips typographus	Coleoptera: Scolytidae	Alimentary canal	Grosmann 1930
		Alimentary tracts (Hansenula capsulata, Candida parapsilosis)	Lu et al. 1957
		Guts and beetle homogenates (Hansenula holstii, H. capsulata, Candida diddensii, C. mohschtana, C. nitratophila, Cryptococcus albidus, C. laurentii)	Leufvén et al. 1984 Leufvén and Nehls 1986
Trypodendron lineatum	Coleoptera: Scolytidae	Not specified	Kurtzman and Robnett 1998b
Xyloterinus politus	Coleoptera: Scolytidae	Head, thorax, abdomen (Candida, Pichia, Saccharomycopsis)	Haanstad and Norris 1985

(continued)

213

Table 9.1. (continued)

Insect Species	Order: Family	Yeast Location (Species)[a]	Reference
Periplaneta americana	Dictyoptera: Blattidae	Hemocoel (Candida sp. nov.)	Verrett et al. 1987
Blatta orientalis	Dictyoptera: Blattidae	Intestinal tract (Kluyveromyces blattae)	Henninger and Windisch 1976
Blatella germanica	Dictyoptera: Blattellidae	Hemocoel	Archbold et al. 1986
Cryptocercus punctulatus	Dictyoptera: Cryptocercidae	Hindgut (1 yeast species)	Prillinger et al. 1996
Philophylla heraclei	Diptera: Tephritidae	Hemocoel	Keilin and Tate 1943
Aedes (4 species)	Diptera: Culicidae	Internal microflora (9 yeast genera)	Frants and Mertvetsova 1986
Drosophila pseudoobscura	Diptera: Drosophilidae	Alimentary canal (24 yeast species)	Shihata and Mrak 1952
Drosophila (5 spp.)	Diptera: Drosophilidae	Crop (42 yeast species)	Phaff et al. 1956
Drosophila melanogaster	Diptera: Drosophilidae	Crop (8 yeast species)	de Camargo and Phaff 1957
Drosophila (4 spp.)	Diptera: Drosophilidae	Crop (7 yeast species)	Starmer et al. 1976
Drosophila (6 spp.)	Diptera: Drosophilidae	Larval gut (17 yeast species)	Fogleman et al. 1982
Drosophila (20 spp.)	Diptera: Drosophilidae	Crop (20 yeast species)	Lachaise et al. 1979
Drosophila (8 species groups)	Diptera: Drosophilidae	Crop (Kloeckera, Candida, Kluyveromyces)	Morais et al. 1992
Drosophila serido	Diptera: Drosophilidae	Crop (18 yeast species)	Morais et al. 1994
Drosophila (6 spp.)	Diptera: Drosophilidae	Intestinal epithelium (Coccidiascus legeri)	Lushbaugh et al. 1976 Ebbert et al. 2003
Protaxymia melanoptera	Diptera	Unknown (Candida, Cryptococcus, Sporoblomyces)	Bibikova et al. 1990
Astegopteryx styraci	Homoptera: Aphididae	Hemocoel and fat body	Fukatsu and Ishikawa 1992
Tuberaphis sp. Hamiltonaphis styraci Glyphinaphis bambusae Cerataphis sp.	Homoptera: Aphididae	Tissue sections	Fukatsu et al.1994
Hamiltonaphis styraci	Homoptera: Aphididae	Abdominal hemocoel	Fukatsu and Ishikawa 1996
Cofana unimaculata	Homoptera: Cicadellidae	Fat body	Shankar and Baskaran 1987

Leofa unicolor	Homoptera: Cicadellidae	Fat body	Shankar and Baskaran 1987
Lecaniines, etc.	Homoptera: Coccoidea[d]	Hemolymph, fatty tissue, etc.	Buchner 1965
Lecanium sp.	Homoptera: Coccidae	Hemolymph, adipose tissue	Tremblay 1989
Ceroplastes (4 sp.)	Homoptera: Coccidae	Blood smears	Mahdihassan 1928
Laodelphax striatellus	Homoptera: Delphacidae	Fat body	Noda 1974
			Mitsuhashi 1975
		Eggs	Kusumi et al. 1979
		Eggs (*Candida*)	Eya et al. 1989
Nilaparvata lugens	Homoptera: Delphacidae	Fat body	Chen et al. 1981a
		Eggs (2 unidentified yeast species)	Nasu et al. 1981
		Eggs, nymphs (*Candida*)	Shankar and Baskaran 1992
		Eggs (7 unidentified yeast species)	Kagayama et al. 1993
		Eggs (*Candida*)	Eya et al. 1989
Nisia nervosa	Homoptera: Delphacidae	Fat body	Shankar and Baskaran 1987
Nisia grandiceps			
Perkinsiella spp.			
Sardia rostrata			
Sogatella furcifera			
Sogatodes orizicola	Homoptera: Delphacidae	Fat body	Lienig 1993
Amrasca devastans	Homoptera: Jassidae	Eggs, mycetomes, hemolymph	Gupta and Pant 1985
Tachardina lobata	Homoptera: Kerriidae	Blood smears (*Torulopsis*)	Mahdihassan 1928
Laccifer (=*Lakshadia*) sp.	Homoptera: Kerriidae	Blood smears (*Torula variabilis*)	Pŕibram 1925; Mahdihassan 1929
			Tremblay 1989
Comperia merceti	Hymenoptera: Encyrtidae	Hemolymph, gut, poison gland	Lebeck 1989
Solenopsis invicta	Hymenoptera: Formicidae	Hemolymph (*Myrmecomyces annellisae*)	Jouvenaz and Kimbrough 1991
S. quinquecuspis			

(*continued*)

Table 9.1. (continued)

Insect Species	Order: Family	Yeast Location (Species)[a]	Reference
Solenopsis invicta	Hymenoptera: Formicidae	Fourth instar larvae (*Candida parapsilosis, Yarrowia lipolytica*)	Ba et al. 1995
		Gut and hemolymph (*Candida parapsilosis, C. lipolytica, C. guillermondii, C. rugosa, Debaryomyces hansenii*)	Ba and Phillips 1996
Apis mellifera	Hymenoptera: Apidae	Digestive tracts (*Torulopsis* sp.)	Cited in Gilliam et al. 1974
		Intestinal tract (*Torulopsis apicola*)	Hajsig 1958
		Digestive tracts (8 yeast species)	Cited in Gilliam et al. 1974
		Intestinal contents (12 yeast species)	Cited in Gilliam et al. 1974
		Intestinal contents (7 yeast species)	Cited in Gilliam et al. 1974
		Intestines (14 yeast species)	Cited in Gilliam et al. 1974
		Intestinal tract (*Pichia melissophila*)	van der Walt and van der Klift 1972
		Intestinal tracts (7 yeast species)	Gilliam et al. 1974
		Alimentary canal (*Hansenula silvicola*)	Gilliam and Prest 197
		Crop and gut (13 yeast species)	Grilione et al. 1981
Apis mellifera	Hymenoptera: Apidae	Midguts (9 yeast genera)	Batra et al. 1973
Anthophora occidentalis	Hymenoptera: Anthophoridae		
Nomia melanderi	Hymenoptera: Halictidae		
Halictus rubicundus	Hymenoptera: Halictidae		
Megachile rotundata	Hymenoptera: Megachilidae		

Bombus sp.	Hymenoptera: Apidae	Crop (*Hansenula anomala*, *Saccharomyces spp.*,	Batra et al. 1973
Lasioglossum sp.	Hymenoptera: Halictidae	*Schizoaccharomyces spp.*, *Rhodotorula spp.*)	
Adelura apii	Hymenoptera: Braconidae	Midgut	Keilin and Tate 1943
Comperia merceti	Hymenoptera: Encyrtidae	Hemolymph, gut, fat body, poison gland	Lebeck 1989
Pimpla turionellae	Hymenoptera: Ichneumonidae	Hemolymph, fat body, and most tissues	Middeldorf and Ruthman 1984
Termites (5 species)	Isoptera: Termitidae	Gut (*Candida*, *Pichia*, *Sporothrix*, *Debaryomyces*)	Schäfer et al. 1996
Termites (6 species)	Isoptera: Termitidae	Hindgut (12 yeast species)	Prillinger et al. 1996
Sigelgaita sp.	Lepidoptera: Phycitidae	Larvae (12 yeast species)	Rosa et al. 1992, 1994
Unknown species	Lepidoptera: Cossidae	Larval gut (*Endomycopsis wickerhamii*)	van der Walt 1959
Orgyia pseudotsugata	Lepidoptera: Lymantriidae	Alimentary tract (*Candida zeylanoides*)	Martignoni et al. 1969
Chrysopa (=*Chrysoperla*) *carnea*	Neuroptera: Chrysopidae	Crop (*Torulopsis* sp.)	Hagen et al. 1970
Chrysoperla ruﬁlabris	Neuroptera: Chrysopidae	Foregut, midgut, hindgut (*Metschnikowia pulcherrima*)	Woolfolk and Inglis 2003

[a]The insect organ where the yeast was isolated is given as presented in the original paper. In many cases it is not possible to ascertain the exact location from which the yeast was isolated due to authors using general terms, such as "intestinal tract," "alimentary canal," "intestines," and so on.

[b]Note progression in taxonomical identification: from *Saccharomyces* (Buchner 1930), to *Torulopsis buchnerii* (Gräebner 1954), to *Symbiotaphrina* (Kühlwein and Jurzitza 1961). The latter was disputed by Bismanis (1976), who revived the name *Torulopsis buchnerii*, but Gams and von Arx (1980) validated the species as *Symbiotaphrina buchnerii*. Noda and Kodama (1996) and Jones and Blackwell (1996) present evidence indicating that this classification is inappropriate.

[c]*Candida karawaiewii* is considered as a synonym of *C. ernobii* (Kurtzman and Robnett 1998a).

[d]Dozens of examples of yeast endosymbionts in the superfamily Coccoidea have been reported by Buchner (1965).

asci converted from somatic cells that are not produced in an ascocarp. In the scientific literature, some fungi that have been isolated from insects are referred to as yeastlike fungi or yeastlike symbionts (YLS), and they often are evolutionarily reduced, derived from the subphylum Pezizomycotina, the largest clade of ascomycetes, variously known as filamentous or ascocarpic ascomycetes (chapter 10). Some yeastlike fungi are dimorphic, alternating between a yeast phase and a hyphal phase according to environmental conditions (e.g., body temperature or CO_2). The dimorphic yeasts include human pathogens (e.g., *Coccidiomyces*) but also certain associates of insects (e.g., *Ophiostoma*). Other yeasts that will not be discussed further as insect associates are classified as basidiomycetes and zygomycetes.

Several roles have been determined for yeast and yeastlike fungi associated with insects, the most important being a nutritional role in which yeasts provide enzymes for digestion, improved nutritional quality, essential amino acids, vitamins, and sterols. Yeasts also play an important role in the detoxification of toxic plant metabolites in the host's diet. In this chapter we focus on examples of yeasts that are located both intra- and extracellularly in certain insects, the benefits yeasts provide for the insects and their role in insect ecology, and hypotheses on how these systems evolved. It is important to note that endosymbiosis seems to be much more significant to the insect than to the yeast, and as such it is hard to discern the mutualism (see Douglas and Smith 1989), but benefits to the yeasts are suggested to be dispersal and a protected and favorable environment rich in nutrients (Cooke 1977). Finally, we discuss the evidence for the way in which these systems evolved.

The Endosymbiotic Systems

Yeastlike Symbionts and Rice Planthoppers

One of the best known insect–YLS systems involves three rice planthoppers: the brown planthopper (*Nilaparvata lugens*), the small brown planthopper (*Laodelphax striatellus*), and *Sogatella furcifera* (table 9.1). These insects are of economic importance because they transmit several rice viruses. Vertical transmission of yeasts in planthoppers occurs by a mechanism involving symbiont movement from the fat body to the primary oocyte in the ovariole, where the yeasts form a symbiont ball from which eggs become infected (Nasu 1963; Noda 1974, 1977; Mitsuhashi 1975; Chen et al. 1981a; Cheng and Hou 2001).[1] Reduction of the yeast population via heat treatment in *N. lugens* results in aposymbiotic insects in which the following effects have been observed: (1) death of fifth-instar nymphs after failing to moult or complete ecdysis (Chen et al. 1981b); (2) reduction of egg hatch and increased nymphal duration (Bae et al. 1997; Zhongxian et al. 2001); (3) interference with normal embryonic and postembryonic development due to the absence of specific proteins involved in these processes (Lee and Hou 1987); and (4) reduction in weight, growth rate, and concentration of protein per unit of fresh weight (Wilkinson and Ishikawa 2001).

In heat-treated *L. striatellus*, adult molt is affected, indicating that the yeast is involved in sterol metabolism in the insect (Noda and Saito 1979a,b). Noda et al.

(1979) reported that in *L. striatellus*, the YLS is responsible for the production of 24-methylenecholesterol and that cholesterol concentrations were greatly reduced in heat-treated insects. Similarly, Eya et al. (1989) reported dramatically reduced levels of ergosterol and 24-methylenecholesterol in heat-treated *N. lugens* and *L. striatellus* and found that YLS isolated from eggs of *L. striatellus* produced lanosterol, 24-methylenelanosterol, dihydroergosterol, and ergosterol in culture broth. Wetzel et al. (1992) isolated another sterol, ergosta-5,7,24(28)-trien-3β-ol (trienol 6) from the *N. lugens* and *L. striatellus* YLS.

Eya et al. (1989) conducted the first study in which a planthopper symbiont was identified to genus, in this case, *Candida*, implying that it is a member of the true yeasts (Saccharomycetes). *Candida*, however, is a polyphyletic grouping of asexual yeasts from many clades and is largely without phylogenetic significance today. In addition, the modern phylogenetic placement of planthopper YLS outside of the true yeasts has bearing on hypotheses of the origin of symbioses and is discussed later.

N. lugens yeastlike symbionts also play a role in nitrogen recycling and, more specifically, in uric acid metabolism, whereby nitrogen waste products are converted into compounds that have nutritional value (Sasaki et al. 1996). In this process, uric acid is stored for subsequent transformation and use. In heat-treated (and therefore aposymbiotic) *N. lugens*, there was no uricase (urate oxidase) activity, and the concentrations of uric acid were much higher than in the control insects. Symbionts grown in artificial culture had uricase activity 15 times higher than that detected in control insects.

Yeasts and Drosophila

Associations between true yeasts with *Drosophila* are well known (table 9.1; Begon 1982), and all studies indicate that the fungi are necessary for optimal development due to their nutritional role. Yeasts provide essential nutrients, vitamins, and sterols (Sang 1978), and the associations secondarily may involve detoxification of plant metabolites and pheromone production (Starmer et al. 1986). Shihata and Mrak (1952) isolated 24 yeast species from the alimentary canal of *D. pseudoobscura*. Individual flies contained three or fewer different yeast species, and the taxonomic of makeup of the yeast flora varied depending on the time of year they were collected (table 9.1).

In most yeast–*Drosophila* associations, the yeasts clearly play a nutritional role at an extracellular level. For example, *Candida ingens* metabolized toxic fatty acids in cactus tissues, with positive consequences for *D. mojavensis* (Starmer et al. 1982). Similarly, *Candida sonorensis* and *Cryptococcus cereanus* have been shown to metabolize 2-propanol (toxic to *Drosophila* larvae and adults at moderate to high concentrations) in decaying cactus tissue, resulting in positive effects on three *Drosophila* species (Starmer et al. 1986). Yeasts, in turn, are reported to benefit by being transported by the fly to different habitats (Gilbert 1980; Starmer and Fogleman 1986, Morais et al. 1994) and by being provided with adequate moisture conditions (Gilbert 1980). In all the cases mentioned above, the presence of yeast within the insect appears to reflect what is present on the feeding substrate (Morais et al. 1992), and thus the yeasts are not intracellular. The only clear example of an intracellular yeast–*Drosophila* association is the yeastlike fungus *Coccidiascus legeri* (Lushbaugh et al. 1976), which accelerates development and improves eclosion rates (Ebbert et al. 2003).

Other YLS Associations with Insects

Other examples of yeasts associated with insects include members of the Coleoptera, Dictyoptera, Diptera, Homoptera, Hymenoptera, Isoptera, Lepidoptera, and Neuroptera (table 9.1). Of these families, Coleoptera is the best represented. One of the best studied Coleopteran–yeast associations is that of *Symbiotaphrina* and two anobiids, the cigarette beetle (*Lasioderma serricorne*) and the drugstore beetle (*Stegobium* [= *Sitodrepa*] *paniceum*). In these insects the YLS live intracellularly in enlarged cells associated with grape-bunch–like proliferations of tissues around the midgut (Buchner 1965). Yeasts are transmitted vertically when the female smears them on the eggshell, which is consumed by the hatching larva (Buchner 1965). These YLS are important in providing nutrients for the host insects and also in detoxifying plant toxins (see below).

Benefits of Yeast Associations to Insects

Parasitoids and Yeasts

Various examples of parasitoids vertically transmitting yeasts at the time of oviposition have been reported: (1) *Comperia merceti*, an encyrtid parasitoid of *Supella longipalpa* (Blattodea: Blatellidae) ootheca (Lebeck 1989); (2) *Adelura apii*, a braconid parasitoid of the celery fly (*Philophylla* (*Acidia*) *heraclei;* Diptera: Agromyzidae) (Keilin and Tate 1943); and (3) *Pimpla turionellae*, an ichneumonid parasitoid (Middeldorf and Ruthmann 1984). Keilin and Tate (1943) reported that *A. apii* larvae developing within the host feeds on the yeast-filled hemolymph, suggesting a nutritional role for the yeast, about which little is known. These studies did not determine whether yeasts become established in nonkilled hosts. It is possible that yeast transmission during oviposition might have a role in egg protection (Lebeck 1989), as has been reported for hymenopteran polydnaviruses that interfere with hemocytes involved in egg encapsulation (Schmidt et al. 2001).

Yeasts and Plant Allelochemicals

Responses of endosymbiotic yeasts to plant allelochemicals have not received much attention. In evolutionary terms it would be advantageous for a plant to produce chemicals that inhibit yeast growth in cases when the yeast is providing essential nutrients to the insect pest. This would serve as an indirect pest control method. It is obvious, then, that endosymbiotic yeasts are exposed to the chemistry of the plant on which the insect is feeding, as evidenced by the detoxification mechanisms presented in the next section. But what happens when yeastlike symbionts are not prepared to handle plant chemistry? Milne (1961) showed that nicotine inhibited growth of *Lasioderma serricorne* YLS, and further showed that sorbic acid at 0.25% and higher inhibited growth of the symbiont. This work resulted in a proposal to use sorbic acid as a pest control mechanism due to its effect on the endosymbiont (Milne 1963). Similarly, Vega et al. (2003) found that increased concentrations of caffeine in *in vitro* studies resulted in reduced levels of *Pichia burtonii*, a yeast isolated from the coffee berry borer (*H. hampei*). Although not working with yeastlike symbionts, Jones (1981) examined the effects of 2-furaldehyde, a bald cypress (*Taxodium distichum*) allelochemical, and found a significant reduction in seven bacterial and two fungal (*Mucor*, *Curvularia*) enteric isolates of *Bombyx mori* larvae. This kind of research is of value not only for developing novel insect pest management strategies, but also in formulating evolutionary hypotheses aimed at understanding YLS associations with insects.

Yeasts and Pheromones

Ectosymbiotic yeasts are involved in the production of pheromones in bark beetle systems. Hunt and Borden (1990) showed that *Hansenula capsulata* and *Pichia pinus* associated with *Dendroctonus ponderosae* are responsible for the conversion of *cis*- and *trans*-verbenol to verbenone in the beetle galleries. Verbenone serves as an antiaggregation pheromone. Leufvén et al. (1984) also reported a similar yeast-mediated transformation in *Ips typographus*.

Yeasts and Insect Nutrition

Most insects have an obligate requirement for 10–14 amino acids, including aromatic and sulfur-containing types, sterols, several B vitamins, and specific fatty acids, usually linolenic and/or linoleic (Dadd 1985). Only sterols with certain functional groups are acceptable, depending on the insect (Dadd 1985; Nes et al. 1997). Many nutritional substrates of insects (e.g., plant sap and wood) provide very low levels of these nutrients or, in some cases, none. Because the biochemical constituents of certain yeasts and other fungi contain these essential requirements (Brues 1946; Southwood 1973), simple digestion of them will help provide the nutrients. Essential nutrient provision through digestion of yeast associates may be occurring in cases where yeasts are present in localized areas of the insect gut where absorption of nutrients occurs. For example, the yeasts associated with *Carpophilus* sap beetles were rapidly digested (Miller and Mrak 1953). What was described as an intracellular yeast (but which may be a filamentous fungus based on mycelium production in old cultures) symbiont of the soft scale *Pulvinaria innumerabilis* (Homoptera: Coccidae) produced lipase, diatase (amylase), and protease (Brues and Glaser 1921).

Many studies suggesting that yeast associates provide nutrients have inferred this role based on determination of the growth rate of the insects with and without the microbes. Removal of the microbes to provide aposymbiotic insects was accomplished through rearing under aseptic conditions or by interfering with the transmission process of the yeast from one generation to the next (such as surface sterilizing the eggs of *L. serricorne* or *S. paniceum*). Although these methods should lead to valid comparisons, studies in which antibiotics have been used to remove associates should be viewed with caution because the antibiotics may also affect processes in the insect itself.

Several studies have shown that insect survival, growth, or reproduction is deterred in the absence of their yeast associates, without identifying the nutritional reason. In several instances the biomass of colonies of *Solenopsis invicta* was significantly greater in colonies containing symbiotic yeasts than those that did not (Ba and Phillips 1996). In another study of unspecified benefits, the intracellular yeastlike fungus of *Drosophila*, *Coccidiascus legeri* (Lushbaugh et al. 1976), accelerated development and improved eclosion rates (Ebbert et al. 2003).

In other studies the nutrients provided by the yeast associates have been identified, mainly through studies in which nutritional supplements were added to the diets of insects in the absence of their yeast associates. For example, nitrogen appears to be provided by yeasts in the scale insect *Pseudococcus citri* (Koch 1954). Vitamins synthesized by cultures of the yeasts appear to have benefitted two wood-boring cerambycids (*Leptura* and *Rhagium*; Gräbner 1954; Jurzitza 1959).

Some of the best information on the role of symbionts has been provided in studies involving the intracellularly derived yeasts of the genus *Symbiotaphrina* of two anobiid beetles (*L. serricorne* and *S. paniceum*), in which it is relatively easy to eliminate the symbionts by surface-sterilizing eggs. These yeasts have been demonstrated to provide nitrogen (Pant et al. 1959; Jurzitza 1969a; Bismanis 1976), sterols (Pant and Fraenkel 1954; Pant and Kapoor 1963), and vitamins (Fraenkel and Blewett 1943a,b; Gräbner 1954; Pant and Fraenkel 1954; Jurzitza 1964, 1969a,b,c, 1972, 1976; Bismanis 1976) to their hosts. However, nutritional studies with defined diets indicated the symbionts could not supply all of the B vitamins necessary for optimal growth of *L. serricorne* (Pant and Anand 1985).

Other studies have demonstrated that the yeasts provide essential amino acids. In studies with *S. paniceum* involving defined diets with and without amino acids and with and without symbionts, the symbionts appeared to provide various amino acids, although the restoration of growth to levels noted with the symbiont-containing hosts varied (Pant et al. 1959). Tryptophan was clearly provided by the symbionts, because in its absence survival was close to that for a casein diet; there were no survivors among the symbiont-free insects (Pant et al. 1959). Survival of symbiont-containing insects in the absence of histidine was about 50% that of the casein diet, and for symbiont-free insects the survival was about half again that many (Pant et al. 1959). Overall, at least some survivors were present for all individual amino acid deficiencies in symbiont-containing insects, but there were no survivors in symbiont-free insects when arginine, isoleucine, leucine, lysine, phenylalanine, threonine, tryptophane, or valine were absent (Pant et al. 1959). Similar results were obtained in studies with *L. serricorne* (Jurzitza 1969a,b). Later Jurzitza (1972) also showed that with *L. serricorne*, the symbionts appeared to be recycling uric acid, as its addition to protein-deficient diets helped growth of symbiont-containing larvae, but not symbiont-free ones.

The symbionts also appear to provide *L. serricorne* with vitamin B_1, riboflavin, nicotinic acid, pyridoxine, pantothenic acid, and choline (Fraenkel and Blewitt 1943). Similar results in a study using symbiont elimination, followed by nutrient supplementation, indicated that the yeasts did in fact provid thiamine, folic acid, biotin, riboflavin, and nicotinic acid, in addition to pyridoxine, pantothenic acid, and choline (Pant and Fraenkel 1954). These results conflict somewhat with another study of similar design in which a different species of anobiid, *S. paniceum*, was not supplied with thiamine by the yeast (Pant and Fraenkel 1954). Additional studies with *L. serricorne* reared on totally artificial diets indicated insects were able to develop through four generations on a vitamin-free diet, although insects grew more slowly (Jurzitza 1969c). Through addition of vitamins to these diets, it appeared that the symbionts supplied choline, lactoflavin, nicotinic acid, pantothenic acid, pyrodoxine, thiamine, and a sterol (Jurzitza 1969c). Studies with different plant materials and symbiont-free *L. serricorne* indicated tobacco leaves supplied enough vitamins for normal development of the beetle; deciduous wood, however, had lower amounts of vitamins (resulting in slower development), and coniferous wood had almost no vitamins, allowing no development (Jurzitza 1976).

Studies using powders from different plant sources indicated growth of symbiont-free larvae of *L. serricorne* was improved with addition of casein and/or vitamins, except for material from *Angelica archangelica* (Umbelliferae), which contains toxic isopsoralens, suggesting the symbionts were needed to detoxify these toxic compounds (Jurzitza 1969d).

Symbiont exchange experiments indicated *S. paniceum* with *L. serricorne* yeasts could better tolerate thiamine deficiency (Pant and Fraenkel 1954). Presence of the symbionts allowed larvae to develop in the absence of vitamins, although growth was slower than if vitamins had been present (Jurzitza 1969c). Although some differences in nutrient requirements for *S. buchnerii* compared to prior studies were noted, the yeasts were capable of providing amino acids and vitamins to the host as they were able to use inorganic nitrogen and required only biotin as a vitamin (Bismanis 1976). Pantothenic acid, pyridoxine, and choline chloride were essential to *L. serricorne* in artificial diet studies, even in the presence of their symbionts (Blewett and Fraenkel 1944; Pant and Anand 1985).

In studies using different proteins and amino acids, gelatin-based diets supplemented with tryptophan permitted some growth of symbiont-free *L. serricorne*, while histidine and methionine did not (Jurzitza 1969b). Symbiotic yeasts were able to use inorganic sulfur to synthesize methionine (Jurzitza 1969b). Symbionts appeared to be more important in protein metabolism than vitamin provision for foliage feeders: Foliage provides an adequate supply of vitamins but not essential amino acids, whereas wood is both deficient in essential amino acids and vitamins (Jurzitza 1969d).

Sterol was provided to the anobiids by their respective symbionts, as indicated by similar types of nutrient supplement studies where cholesterol restored normal survival and growth of symbiont-free insects (Pant and Fraenkel 1950, 1954). Similar results with cholesterol and *L. serricorne* were reported in similarly designed studies (Pant and Fraenkel 1950; Pant and Kapoor 1963). When cholesterol was excluded from artificial diets, 89% of larvae of *L. serricorne* matured into adults in the presence of symbionts, but none matured in the absence of symbionts (Pant and Kapoor 1963). When 1% cholesterol was added to the diets, 88% of *L. serricorne* larvae without symbionts matured to adults (Pant and Kapoor 1963).

Yeast-mediated Digestive–Detoxifying Reactions

The next section provides a discussion of symbiotic associations that are generally catabolic in nature, resulting in digestive or detoxifying reactions. In several cases, it appears that the same enzyme or enzyme group can catalyze reactions that are both digestive and detoxifying, sometimes involving the same substrate. For example, removal of glucose molecules from tannic acid to form gallic acid detoxifies the tannic acid (Dowd 1990) and at the same time releases glucose molecules that can be absorbed as nutrients. Other detoxifying reactions involve conjugation and will also be considered.

The kinds of information available to support the symbiont roles in digestion or detoxification are variable. New techniques such as gene cloning or genome sequencing similar to that done with the bacterial symbionts of aphids (Clark et al. 1998; Shigenobu et al. 2000), followed by specific gene knockout or replacement will be necessary to definitively establish which enzymes are involved in the degradative reactions of yeast endosymbionts of insects.

Studies demonstrating digestive/detoxifying capabilities of yeast endosymbionts include those involving symbionts cultured apart from the host and studies involving symbionts as they naturally occur in insect tissues (see below). In some cases standard assimilation studies used to describe new species have used substrates that may be relevant in considering digestive capabilities that may be contributed to the host (e.g., starch, cellulose, cellobiose, inulin) or detoxifying capabilities (e.g., of salicin, a phenolic glycoside from willow).

Studies using cultures of symbionts can be useful in determining what the capabilities of the symbionts are but may not reflect what actually occurs in the natural host state (either under- or overrepresenting what actually occurs). In studying the capabilities of insect symbionts in culture substrate utilization studies, enzyme type or presence (using indicator substrates involving colorimetric detection, for example) or enzyme induction studies have been used. Certainly, using biologically relevant substrates is most appropriate, but indicator substrates are often useful in determining the spectrum of enzyme activity.

In studies involving the presence or absence of symbionts in the host, the same strategies are used that have been used to study enzyme activity of symbionts in culture. Some studies have indicated the symbionts (or perhaps their hosts) produce different enzymes in association with the host insect than apart from the insect. In some cases, it has been possible to obtain aposymbiotic strains of insects to compare performance in the presence of different nutrient sources or toxins. This type of study can be particularly useful if symbiont-free insects can be obtained without use of antibiotics because of their potentially confounding effects. The sections below cover digestive and detoxification reactions separately, organized on the basis of nutrient resource degraded by the different yeasts. Some associations have been studied more than others; wood ingesting insects have been of paramount interest.

Digestive Reactions

Digestive reactions include (1) protein/peptide degradation (catalyzed by a variety of endo- and exoproteases and peptidases), (2) polysaccharide/starch/sugar degradation (catalyzed by myriad hydrolytic enzymes), and (3) fat/fatty acid degradation (catalyzed by lipases). In the case of starches/sugars, substrate specificity is of particular importance. Protein degradation theoretically would occur with nearly all proteins, but forms resistant to degradation (such as allergens) may be more highly N- or O-glycosylated and require additional enzymes. Binding of different compounds (e.g., phenolics) to proteins also may inhibit their degradation (Dowd 1990).

Typically, insects do not produce many of the enzymes necessary to degrade polysaccharides or even simpler sugar molecules. They are limited even in their ability to break α-glucose bonds, principally with amylases (Applebaum 1985). Many starches or other plant polysaccharides (including cellulose, hemicellulose, and pectins) and lignins are linked by other types of bonds, such as β forms. Thus, the availability of many of these potential sources of sugars depends on having the appropriate enzymes. Complete cellulose degradation to simple sugars typically requires endo-β-glucanases (Cx-cellulases), exo-β-glucanases (cellobiohydrolases), and a β-glucosidase (cellobiose) (Klemm et al. 2002). A different enzyme, β-xylanase, will break down both cellulose and hemicellulose (Klemm et al. 2002). Binding of lignins to cellulose requires further degradative enzymes, which many fungi possess (Klemm et al. 2002).

Inulin, a β-fructose polymer, is found in high concentrations in some plant species, especially the tubers of plants such as *Dahlia*, *Helianthus*, and *Cichorium* (Franck and De Leenheer 2002). Exo- and endoinulanases produced by yeasts and molds degrade inulin to fructose monomers (Franck and De Leenheer 2002). Fructose can be absorbed and assimilated by insects (Turunen 1985). Pectins are highly complex polysaccharides that require multiple enzymes for effective degradation, including endo- and exo-polygalacturonase, methyl and acetyl esterases, endopectin lyase and endopectate lyase, endoarabinase, arabinofuranosidase, feruloyl esterase, endogalactanase, and rhamnogalacturonan dimer and galacturonohydrolases (Ralet et al. 2002). Again, interconversion of various sugar monomers to glucose, fructose, or other monomers may be necessary for insects to use them.

Even though cellulose degradation in insects is traditionally associated with the presence in the termite gut of flagellates that can break down cellulose (Bignell 2000; but see chapter 8, this volume), yeasts and their fungal relatives can degrade a variety of recalcitrant sugar sources through the activity of such enzymes as β-glucosidases, xylases, cellulases, and so on. (Klemm et al. 2002). Many studies on insect yeast endosymbionts have identified these enzymatic reactions using various substrates in indicator reactions generating a chromophore (e.g., Shen and Dowd 1989, 1991a,b). Supplying a compound of interest as a sole carbon (or nitrogen) source can also be useful in identifying enzymes capable of degrading various complex sugars, polysaccharides, or proteins.

Although sometimes producing toxic aglycones (Applebaum 1985), glycoside removal from some toxic compounds can also be interpreted as a digestive reaction because the liberated glucose or other sugar can then be absorbed as a nutrient (see "Detoxifucation Reactions" below). Most of the information available on polysaccharide or complex polymer degradation by yeast endosymbionts has come from beetle endosymbionts. For example, in studies with anobiid beetles, *Symbiotaphrina kochii* from the cigarette beetle *L. serricorne* (van der Walt 1961; Jurzitza 1964) and *Symbiotaphrina buchnerii* from the drugstore beetle *Sitodrepa panicea* (Kuhlwein and Jurzitza 1961; Bismanis 1976) assimilated cellobiose. Breakdown of indicator substrates suggested that cultures of *S. kochii* produced lipases, α- and β-glucosidase, phosphatase, and trypsin (Shen and Dowd 1991a). Cellobiose assimilation by yeast species isolated from the wood-boring anobiids *Ernobius mollis* and *E. abietis* (e.g., *Candida ernobii*), and *Xestobium plumbeum* (e.g., *C. xestobii*) was positive, negative, and limited, respectively (Meyer et al. 1998).

The degradative abilities of the *Candida* yeasts of cerambycid beetles have been studied to some extent. Some symbiont species have free-living strains (e.g., *C. guilliermondii* of *Phoracantha semipunctata*), so interpretations are difficult without the use of markers that resolve at infraspecific levels. However, *C. guilliermondii* is closely related to *C. xestobii*, an obligate symbiont of *X. plumbeum* (Jones et al. 1999). Previous studies primarily have investigated the ability of the microbes to assimilate different sugars or to degrade cellulose. Larval stages of most Cerambycidae are wood boring, so the capability to degrade wood polymers would be important. Thus far, studies of the microbial enzyme systems have involved only pure cultures of the yeast. *Candida* spp. isolated from *P. semipunctata* assimilated pectin (Chararas et al. 1972) and exhibited glycosidase (Chararas and Pingal 1981) and β-glucosidase activity (Chararas et al. 1983). Symbionts from *Stromatium barbatum* (Mishra and Singh 1978) and *Homochambyx spinicornis* (Mishra and Sen-Sarma 1985) produced several polysaccharide-hydrolyzing enzymes, including xylanase, which is important in cellulose degradation. *Candida rhagii* (endosymbiont of *Rhagium inquisitor* and *Gaurotes virginea*), and *C. tenuis* (endosymbiont of *Leptura rubra*) were tested in culture for their ability to assimilate cellobiose; they did not, however, utilize starch as a carbon source (Meyer et al. 1998).

Other yeast endosymbionts of beetles also have been studied. Yeasts closely related to *Pichia stipitis* have been isolated from the passalid beetle *Odontotaenius disjunctus* from a wide region in eastern North America (chapter 10). It is of interest that all isolates were identical based on 600 bp of the internal transcribed spacer (ITS) region as well as large subunit rDNA (Suh et al. 2003). Another passalid (*Verres sternbergianus*) is associated with a yeast of very similar genotype. Cultural studies indicated that the yeast degrades xylose, and it may assist in the digestion of the wood into which these insects bore (Suh et al. 2003). Cultures of *Pichia burtonii* and *Candida fermentati* isolated from the coffee berry borer (*H. hampei*) produced trypsinlike protease, α- and β-glucosidase, phosphatase, and lipase activity (Vega and Dowd, unpublished data).

Other species of insects with yeast endosymbionts also have been examined, including several species of yeasts discovered in termites. Cellobiose was assimilated by previously undescribed species of yeast from the termites *Zootermopsis nevadensis, Z. angusticollis, Reticulitermes santonensis, Heterotermes indicola, Mastotermes darwiniensis, Neotermes jouteli,* and the wood cockroach, *Cryptocercus punctulatus* (Prillinger et al. 1996). Inulin was used by a yeast isolated from the termites *Z. nevadensis* and *M. darwiniensis,* and starch was used by another yeast isolated from *Z. nevadensis, Z. angusticolis, H. indicola,* and *M. darwiniensis* (Prillinger et al. 1996). Hemicellulose degradation by termite yeasts also has been reported (Schäfer et al. 1996). The strains of *Candida guilliermondii* and *Debaromyces hansenii* isolated from the red imported fire ant *Solenopsis invicta* could use inulin and cellobiose (Ba and Phillips 1996).

Comparative symbiont removal studies demonstrating digestive contributions of yeasts to their hosts (similar to those already described for nutrition) to determine the relative importance of the symbiont to the health of the insect host under conditions of varying recalcitrant nutrient sources have not been performed. However, some interesting possibilities exist. For example, the major polysaccharide in the coffee berry is composed of an unusual complex of sugars: arabinogalactan, mannan, and an unsubstituted glucan (Bradbury 2001). β-glucosidase activity demonstrated using naphthol glucoside may be an indicator that the yeasts associated with the coffee berry borer are capable of breaking down this polysaccharide to a form that can be used by its insect host (Vega and Dowd, unpublished data).

Detoxification Reactions

Detoxification reactions consist primarily of hydrolytic reactions (performed by esterases or proteases), oxidative reactions (performed by monooxygenases), or conjugating reactions (performed by such enzymes as glutathione transferases; Brattsten and Ahmad 1986). The result of these reactions is either to degrade a complex polymer into a relatively smaller molecular weight compound that can be used or excreted (for protein or carbohydrate toxins) or to convert the toxins into more polar forms, which instead of being absorbed into tissues are excreted through the digestive system. Although there are some exceptions, yeast and fungi do not commonly produce monooxygenases or other oxygenases capable of detoxification (laccases are an exception). However, these organisms have potent hydrolytic capabilities. Conjugating reactions are also less common compared with arthropod or vertebrate sources. Thus, as for digestive reactions, it appears that the host insect contributes detoxifying enzymes (e.g., monooxygenases) while symbionts provide enzymes characteristic to yeasts or fungi. This is consistent with evolutionary gain and loss of function reported for genome studies of bacterial symbionts of aphids, whereby the bacterial symbionts have many amplified copies of some amino acid biosynthetic enzymes but have lost the enzymes responsible for biosynthesizing cell-surface lipids and defensive genes, unnecessary defenses due to host provisions (Shigenobu et al. 2000).

Detoxification reactions have been demonstrated for a variety of insect symbionts, including both those that are involved in farming associations (e.g., cultivated fungi of ants and termites) and those that are intracellular (e.g., bacteria and yeasts). These reactions have degraded compounds such as aromatic hydrocarbons, insecticides, aromatic esters, benzoxazolinones, and phenolic acids (Dowd 1991, 1992). Most of the information on yeast or yeastlike symbionts came from studies done with the YLS *Symbiotaphrina kochii* associated with the cigarette beetle *Lasioderma serricorne*. Because the symbiont is culturable and can be eliminated without including antibiotics in the insect's diet, it has been a particularly useful system for investigating insect–symbiont interactions. A variety of studies have explored the detoxification capabilities of this organism.

Earlier studies indicated salicin could be assimilated by *S. kochii* (van der Walt 1961; Jurzitza 1964) but not by *S. buchnerii* (Kuhlwein and Jurzitza 1961; Bismanis 1976). More recently, studies have shown that cultures of *S. kochii* used a variety of plant flavonoids, plant aromatic acids, and plant meal toxins, as well as aromatic alcohols, mycotoxins, insecticides, and herbicides as sole carbon sources; not all compounds tested, however, could be used (Shen and Dowd 1991a). Enzymes produced in culture were consistent with compounds used, and these included indicator substrates for esterase, α- and β-glucosidase, phosphatase, glutathione transferase, trypsin, and specifically parathion hydrolase; oxidative O-demethylation and laccase activity were not detected (Shen and Dowd 1991b). Esterase activity was induced significantly by flavone, griseofulvin (a *Penicillium* mycotoxin), *cis*-β-pinene, and malathion, with flavone causing almost two times higher levels of induction (Shen and Dowd 1989). Isozymes were induced with griseofulvin, malathion, and β-pinene (Shen and Dowd 1989). Thus, in culture the symbionts appeared to produce enzymes capable of detoxifying a variety of compounds, which were in some cases apparently inducible. It is interesting that compounds that were not targeted substrates induced enzymes; these may serve as general triggers for inducing detoxifying enzyme complexes.

Some comparative studies were also performed with intact insect–symbiont systems of *S. kochiii* and *L. serricorne*. Histological assays indicated the mycetomes were a concentrated source of the total gut enzymes capable of hydrolyzing the esterase substrate 1-naphthyl acetate and tannic acid (Dowd 1989). The yeasts appeared to be the source of the enzyme activity when examined using histochemical tests; the symbiont-free mycetomes showed only low enzyme levels (Dowd 1989). Symbiont-free insects took three to four times as long to emerge as adults when flavone or tannic acid was present in the diet, compared to insects that contained symbionts (Dowd and Shen 1990). A major band of esterase activity also was absent from the gut of symbiont-free insects compared to those with symbionts (Dowd and Shen 1990). The yeasts *C. ernobii* (as *C. karawaiewii*) and *C. xestobii*, isolated from the wood-boring anobiids, *Ernobius mollis*, *E. abietis*, and *Xestobium plumbeum*, differed in their ability to assimilate salicin; the reactions were positive, negative, and limited, respectively (Meyer et al. 1998). *Candida guilliermondii* and *Debaromyces hansenii* isolated from the red fire ant *Solenopsis invicta* have been shown to assimilate salicin (Ba and Phillips 1996).

The information available for cerambycid detoxification of certain substrates has been discussed previously in regard to digestive enzymes. Cultures of *Candida rhagii* (endosymbionts of *Rhagium inquisitor* and *Gaurotes virginea*) and *C. tenuis* (endosymbionts of *Leptura rubra*) assimilated salicin (Meyer et al. 1998), indicating production of β-glucosidases. These enzymes may be important in degrading hydrolyzable tannins or other phenolic glycosides, although increased toxicity may result when aglycones of certain chemicals are produced, as previously discussed (Dowd 1992).

Pichia burtonii and *Candida fermentati* were isolated from the coffee berry borer, *H. hampei* (Vega et al. 2003). Another *Pichia* species, *P. guilliermondii*, has also been implicated as a symbiont in cerambycids. Yeast-derived esterase isozymes were produced only by these yeast strains while in association with the insect, but wild strains produced them in artificial culture (Vega and Dowd, unpublished data). Enzymes were not induced in the symbiotic isolates even when coffee berry material was added to culture media, whereas Dopa was oxidized by yeast culture homogenates (Vega and Dowd, unpublished data). Caffeine was not detoxified by *P. burtonii*, however (Vega et al. 2003).

Origin of Endosymbiotic Associations

How did endosymbiotic yeast–insect associations come about? Two hypotheses have been proposed. The first states that symbionts were originally pathogenic parasites or nonpathogenic commensals (Steinhaus 1949), while the second presents them as descendants of phytopathogenic or saprophytic fungi (Dowd 1991). A third hypothesis involving feeding habits has been suggested on the basis of very recent work (Vega, unpublished data).

Taming the Insect Pathogen

Phylogenetic analysis of yeastlike fungi associated with planthoppers suggests that the endosymbionts are derived from within a clade of insect pathogens. Fukatsu and Ishikawa (1995, 1996) conducted a phylogenetic analysis of a YLS from the hemolymph of the aphid *Tuberaphis* (*Hamiltonaphis*) *styraci* (Tribe Cerataphidini; Stern et al. 1997). The symbiont was not a true yeast, but it was closely related to three other symbionts present in the planthoppers *N. lugens*, *L. striatellus*, and *S. furcifera* (Noda et al. 1995; Fukatsu and Ishikawa 1996). These YLS were all placed among the filamentous ascomycetes (Ascomycota: Pezizomycotina). Fukatsu and Ishikawa (1996) suggested that because filamentous ascomycetes include the most common entomopathogenic fungi (e.g., *Beauveria*, *Cordyceps*, and *Metarhizium*), it was possible that the ancestor of the present endosymbiont in *T. styraci* might have been an entomopathogen. Fukatsu et al. (1994) also used this hypothesis to explain the presence of yeasts in some aphid genera in the Cerataphidini.

Fukatsu and Ishikawa (1996) presented three possible explanations for the close phylogenetic relation between the aphid and planthopper symbionts: (1) horizontal symbiont transfer; (2) a common ancestor for both aphids and planthoppers having acquired the symbiont; and (3) independent acquisition of the symbiont in both aphids and planthoppers. A phylogenetic analysis of the uricase gene in aphids and planthoppers by Hongoh and Ishikawa (2000) supports the first explanation, (i.e., horizontal transfer from the aphids to the planthoppers).

The taxonomic status of the aphid and planthopper YLS (Fukatsu et al. 1994; Noda and Kodama 1995; Noda et al. 1995; Fukatsu and Ishikawa 1996) was not determined to a taxonomic level specific enough to place it among known insect pathogens. However, a study using the small and large subunit rRNA genes (Suh et al. 2001) concluded that the YLS of *N. lugens*, *L. striatellus*, and *S. furcifera* should be placed within one of several clades of the polyphyletic genus *Cordyceps* (Pezizomycotina: Hypocreales: Clavicipitaceae), all well known as insect pathogens.

Plant Pathogen Progenitor

It has been suggested that endosymbionts might have originated as plant pathogens because the detoxifying ability for the symbiont would have to be similar when acting as a plant pathogen or saprophyte or when detoxifying components of the insect diet that originate in the plant (Dowd 1991). The relationships of *Symbiotaphrina* spp. are not well resolved, but these yeastlike fungi appear to have arisen from filamentous ascomycetes (Pezizomycotina) in a notoriously poorly resolved area including discomycetes and loculoascomycetes (Jones and Blackwell 1996). The plant pathogen progenitor hypothesis implies that insects acquired the YLS from the plant, suggesting a feeding-habit component as a possible mechanism for the origin of yeasts as endosymbionts (discussed below).

Feeding Habits and Endophytes

Another possibility for the origin of yeasts as endosymbionts involves insect feeding habits and the presence of endophytic yeasts. Leafhoppers and planthoppers feed on phloem, xylem, or mesophyll tissue (depending on the species), whereas most aphids feed only on phloem (Backus 1985). Feeding behavior would bring insects in contact with yeasts occurring endophytically or on the phylloplane. Endophytic yeasts have been reported in wheat (*Triticum aestivum*; Larran et al. 2002), *Eucalyptus* (de Sá Peixoto Neto et al. 2002), and pines (Zhao et al. 2002), and yeasts on flowers, fruits, and the phylloplane are quite common (Phaff and Starmer 1987). *Drosophila* is a typical example of an insect that has close associations with yeasts present on food substrates, although these yeasts are not intracellular. Thus, insect feeding behavior presents the opportunity either for acquiring yeasts from the plant vascular tissue and/or the phylloplane and, if already acquired, for inoculating them into the plants in a manner similar to the transmission of plant pathogens. Various piercing-sucking insects can serve as vectors for plant pathogenic yeasts in the genera *Ashbya* and *Nematospora* (Phaff and Starmer 1987).

The possible role of endophytes as eventual YLS remains unexplored. Vega et al. (unpublished data) recently isolated various unidentified endophytic yeasts from coffee plants, as well as two yeast species isolated internally from the coffee berry borer (*P. burtonii*, *C. fermentati*; Vega et al. 2003). Pérez et al. (2003) isolated various yeasts (e.g., *Candida diddensiae*, *C. fermentati*, *P. burtonii*, *Hanseniaspora* sp.) from coffee berry borer's guts, feces, and cuticles, as well as from the galleries bored by the insect inside the coffee berry. Fungal endophytes also have been reported on the feces of grasshoppers, indicating that the insects are not only consuming them, but serving as a mechanism for their dispersal (Monk and Samuels 1990).

It is worth noting that one of the possibilities presented by Fukatsu and Ishikawa (1996) and Hongoh and Ishikawa (2000) to explain the close phylogenetic relation between the aphid and planthopper symbionts was horizontal symbiont transfer. Placement of the aphid and planthopper YLS within a clade of insect pathogens implies that a possible mechanism for horizontal transfer was sharing an entomopathogen. Another possible mechanism for horizontal transfer is the plant via the feeding habits of the insect. It would be worthwhile to search for endophytic yeasts in the plants shared by *T. styraci* and the planthoppers *N. lugens*, *L. striatellus*, and *S. furcifera* to determine whether these, if present, are the same as the YLS in the insects. By definition, endophytes are asymptomatic in the plant, and consequently endophytic yeasts have not been a cause of concern for plant pathologists, mycologists, or ecologists. Thus, the presence of yeasts as endophytes is likely to have been undersampled based on the low number of reports in the literature compared to reports of other endophytic fungi.

Future Directions and Practical Concerns

Despite the amount of work that has been done on insects and their yeast endosymbionts, considerable work will be necessary to comprehend the complexities of the endosymbiotic relationship. Certainly, based on the preceding discussion, it appears that symbionts are playing a vital role in the life strategy of insects, in a relatively unsubtle manner. Molecular techniques have indicated the association between *Symbiotaphrina* spp. and their anobiid hosts is ancient compared to that between other anobiids and *Candida* yeasts, which may have been perpetuated due to the detoxifying capabilities of the *Symbiotaphrina* (Jones et al. 1996).

Recent studies with vertebrate gut microbes may give clues about additional roles played by insect endosymbionts. Research on the human gut symbiont *Bacteroides thetaiotamicron* indicates that is capable of providing a wide variety of polysaccharide-degrading enzymes on the cell surface that are not produced in the human digestive system, suggesting that this organism can provide additional sugars for its host (Xu et al. 2003). Vertebrates without their normal complement of gastrointestinal tract microflora require significantly more calories to maintain weight, presumably because gut microflora are responsible for liberating additional sugars or other nutrients from undigested materials (Gilmore and Ferretti 2003). Similar roles almost certainly are played by insect symbionts, but more exacting studies analogous to vertebrate studies are necessary. Considering the numbers of microorganisms insects are exposed to and the numbers they may harbor, we need to determine how these organisms are interacting. Cascade-type metabolism has been reported in the termite and its symbionts (Breznak 2000; Slaytor 2000). Does this also occur in other insects? Can symbionts act as biocompetitors for other organisms that are potentially neutral or pathogenic?

Molecular biology has clarified the relationships between symbionts and their relatives, resolving long-standing questions about nature and identity (Noda and

Kodama 1995, 1996; Noda et al. 1995; Jones and Blackwell 1996; Jones et al. 1999; Suh et al. 2001). As discussed previously, genome studies of the bacterial symbionts of aphids indicated many similarities but some differences among the three species of aphids. Knowledge of phylogenetic relationships is helpful in evaluating hypotheses about the origin of such symbioses. The available information also has helped clarify some relationships of yeasts with digestive and detoxifying capabilities. Histological work with aphid symbionts of *Myzus essigi* indicated the potential ability to detoxify aryl esters and benzoxazolinones, tannic acid, and the insecticide diazinone (Dowd 1991). If the genome of the *M. essigi* symbiont is closely related to the genome of other aphid symbionts (e.g., *Baizongia pistaciae, Acyrthosiphon pisum, and Schizaphis graminum*), whose genomes are very similar, it might be possible to identify some strong candidate genes for detoxification. A variety of proteases were reported, including serine proteases (e.g., GenBank AAO26942, NP 660570, NP 777837) and aminopeptidases (e.g., GenBank AAO27053, NP 777948, BAB13071). Other than simple protein degradation/housekeeping in the cells, it does not make sense for these symbionts to produce so many protein-degrading enzymes. However, some aminopeptidases have been reported to hydrolyze toxins, including pyrethroid insecticides (Dowd and Sparks 1988), so it is likely that the hydrolytic enzymes involved in detoxification in *M. essigi* are the proteases sequenced in genomic studies of the other aphid species symbionts. Although mutability studies of specific enzymes from these unculturable aphid bacterial symbionts are impractical, once similar genomic information becomes available on intracellular yeast/fungal symbionts that can be cultured apart from the insect, more definitive studies can be undertaken to determine which genes code for degradative enzymes contributing to host welfare by using transformation techniques already available for yeasts or fungi.

The enzymatic capabilities of yeast endosymbionts could be exploited in a number of areas. Imbalances in amino acids, B vitamins, or other nutrients that can be provided by these eukaryotic symbionts could be corrected in crop plants by using their genes, in a manner similar to that reported for carotenoid biosynthesis in golden rice or other such nutrient-fortified plants (e.g., Potrykus 2001). But will these nutrient-fortified plants also have more problems with diseases and insect pests because they provide them with a more balanced diet? As has been suggested (for review, see Dowd 1991), the essential nature of the symbiont makes it a potentially unique target for control of insect pests that contain symbionts.

Acknowledgments   We thank Francisco Posada, Meredith Blackwell, Wendy S. Higgins, and Don Weber for comments on a previous version of this chapter and Regina Kleespies and Don Weber for their help in translating articles from German.

Note

1. In contrast to planthoppers, in Anobiidae and Cerambycidae there are distinct structures associated with the reproductive system that result in the yeast being smeared on the egg surface; these are consumed by the emerging larvae upon hatching (Buchner 1965).

Literature Cited

Alexopoulos, C. J., C. W. Mims, and M. Blackwell. 1996. *Introductory mycology*, 4th ed. New York: John Wiley & Sons.

Applebaum, S. W. 1985. Biochemistry of digestion. In *Regulation, digestion, nutrition and excretion*, vol. 4, *Comprehensive insect physiology, biochemistry and pharmacology*, ed. G. A. Kerkut and L. I. Gilbert, pp. 279–311. New York: Pergamon Press.

Archbold, E. F., M. K. Rust, D. A. Reierson, and K. D. Atkinson. 1986. Characterization of a yeast infection in the German cockroach (Dictyoptera: Blatellidae). *Environmental Entomology* 15:221–226.

Ba, A. S., D.-A. Guo, R. A. Norton, S. A. Phillips, Jr., and W. D. Nes. 1995. Developmental differences in the sterol composition of *Solenopsis invicta*. *Archives of Insect Biochemistry and Physiology* 29:1–9.

Ba, A. S., and S. A. Phillips, Jr. 1996. Yeast biota of the red imported fire ant. *Mycological Research* 100:740–746.

Backus, E. A. 1985. Anatomical and sensory mechanisms of leafhopper and planthopper feeding behavior. In *The leafhoppers and planthoppers*, ed. L. R. Nault and J. G. Rodriguez, pp. 163–184. New York: John Wiley & Sons.

Bae, S.-D., Y.-H. Song, and K. L. Heong. 1997. Effect of temperature on the yeast-like symbiotes in the brown plant hopper, *Nilaparvata lugens* Stål. *Rural Development Administration Journal of Crop Protection* 39:34–42.

Barnett, J. A., R. W. Payne, and D. Yarrow. 2000. *Yeasts—characteristics and identification*. Cambridge: Cambridge University Press.

Batra, L. R., S. W. T. Batra, and G. E. Bohart. 1973. The mycoflora of domesticated and wild bees (Apoidea). *Mycopathologia et Mycologia Applicata* 49:13–44.

Begon, M. 1982. Yeasts and *Drosophila*. In *The genetics and biology of Drosophila*, vol. 3B, ed. M. Ashburner, H. L. Carson, and J. N. Thompson, Jr., pp. 345–384. New York: Academic Press.

Bibikova, I. I., M. V. Fateeva, and B. M. Mamaev. 1990. Yeast in the microflora of the Far East Diptera *Protaxymia melanoptera* [in Russian]. *Izvestiya Akademii Nauk SSSR, Seriya Biologicheskaya* 3:474–476.

Bignell, D. E. 2000. Introduction to symbiosis. In *Termites: Evolution, sociality, symbiosis, ecology*, ed. T. Abe, D. E. Bignell and M. Higashi, pp. 189–208. Boston: Kluwer Academic.

Bismanis, J. E. 1976. Endosymbionts of *Sitodrepa panicea*. *Canadian Journal of Microbiology* 22:1415–1424.

Blewett, M., and G. Fraenkel. 1944. Intracellular symbiosis and vitamin requirements of two insect, *Lasioderma serricorne* and *Sitodrepa panicea*. *Proceedings of the Royal Society B* 132:212–221.

Bradbury, A. G. W. 2001. Chemistry I: Non-volatile compounds. 1A. Carbohydrates. In *Coffee: recent developments*, ed. R. J. Clarke and O. G. Vitzthum, pp. 1–17. Oxford: Blackwell.

Brattsten, L. B., and S. Ahmad, eds. 1986. *Molecular aspects of insect-plant interactions*. New York: Plenum Press.

Breznak, J. A. 2000. Ecology of prokaryotic microbes in the guts of wood- and litter-feeding termites. In *Termites: Evolution, sociality, symbiosis, ecology*, ed. T. Abe, D. E. Bignell and M. Higashi, pp. 209–231. Boston: Kluwer Academic.

Brues, C. T. 1946. *Insect dietary. An account of food habits of insects*. Cambridge: Harvard University Press.

Brues, C. T., and R. W. Glaser. 1921. A symbiotic fungus occurring in the fat-body of *Pulvinaria innumerabilis*. *Biological Bulletin* 40:299–324.

Buchner, P. 1930. *Tier und pflanze in symbiose*, 2nd ed. Berlin: Borntraeger.

Buchner, P. 1965. *Endosymbiosis of animals with plant microorganisms*. New York: Interscience Publishers.

Chararas, C., J.-E. Courtois, A. Thuillier, A. Le Fay, and H. Laurent-Hube. 1972. Nutrition de *Phoracantha semipunctata* F. (Coleoptère: Cerambicidae): étude des osidases du tube digestif et de la flore intestinale. *Comptes Rendus des Seances de la Societe de Biologie* 166:304–308.

Chararas, C., and M.-C. Pignal. 1981. Étude du rôle de deux levures isolées dans le tube digestif de *Phoracantha semipunctata*, Coléoptère Cerambicidae xylophage spécifique des *Eucalyptus*. *Comptes Rendus de l'Académie des Sciences de Paris* 292:109–112.

Chararas, C., M. C. Pignal, G. Vodjdana, and M. Bourgeay-Causse. 1983. Glycosidases and B-group vitamins produced by six yeast strains from the digestive tract of *Phoracantha semipunctata* larvae and their role in the insect development. *Mycopathologia* 83:9–15.

Chen, C.-C., L.-L. Cheng, C.-C. Kuan, and R. F. Hou. 1981a. Studies on intracellular yeast-like symbiote in the brown planthopper, *Nilaparvata lugens* Stål. I. Histological observations and population changes of the symbiote. *Zeitschrift für Angewandte Entomologie* 91:321–327.

Chen, C.-C., L.-L. Cheng, and R. F. Hou. 1981b. Studies on intracellular yeast-like symbiote in the brown planthopper, *Nilaparvata lugens* Stål. II. Effects of antibiotics and elevated temperature on the symbiotes and their host. *Zeitschrift für Angewandte Entomologie* 92:440–449.

Cheng, D. J., and R. F. Hou. 2001. Histological observations on transovarial transmission of a yeast-like symbiote in *Nilaparvata lugens* Stål (Homoptera: Delphacidae). *Tissue & Cell* 33:273–279.

Clark, M. A., L. Baumann, and P. Baumann. 1998. Sequence analysis of a 34.7-kb DNA segment from the genome of *Buchnera aphidicola* (endosymbiont of aphids) containing *groEL*, *dnaA*, the *atp* operon, *gidA*, and *rho*. *Current Microbiology* 36:158–163.

Cooke, R. 1977. *The biology of symbiotic fungi*. New York: John Wiley & Sons.

Dadd, R. H. 1985. Nutrition: organisms. In *Regulation, digestion, nutrition and excretion*, vol. 4, *Comprehensive insect physiology, biochemistry and pharmacology*, ed. G. A. Kerkut and L. I. Gilbert, pp. 313–390. New York: Pergamon Press.

de Bary, A. 1879. Die Erscheinung der Symbiose. Straßburg: Verlag Trubner.

de Camargo, R., and H. J. Phaff. 1957. Yeasts occurring in *Drosophila* flies and in fermenting tomato fruits in Northern California. *Food Research* 22:367–372.

de Sá Peixoto Neto, P. A., J. L. Azevedo, and W. L. Araú. 2002. Microrganismos endofíticos. *Biotecnología Ciência & Desenvolvimiento* 29:62–76.

do Carmo-Sousa, L. 1969. Distribution of yeasts in nature. In *Biology of yeasts*, vol. 1, *The yeasts*, ed. A. H. Rose and J. S. Harrison, pp. 79–105. London: Academic Press.

Douglas, A. E., and D. C. Smith. 1989. Are endosymbioses mutualistic? *Trends in Ecology and Evolution* 4:350–352.

Dowd, P. F. 1989. *In situ* production of hydrolytic detoxifying enzymes by symbiotic yeasts in the cigarette beetle (Coleoptera: Anobiidae). *Journal of Economic Entomology* 82:396–400.

Dowd, P. F. 1990. Detoxification of plant substances by insects. In *Insect attractants and repellents*, vol. VI, *CRC handbook of natural pesticides*, ed. E. D. Morgan and N. B. Mandava, pp. 181–225. Boca Raton, FL: CRC Press.

Dowd, P. F. 1991. Symbiont-mediated detoxification in insect herbivores. In *Microbial mediation of plant-herbivore interactions*, ed. P. Barbosa, V. A. Krischik and C. G. Jones, pp. 411–440. New York: John Wiley & Sons.

Dowd, P. F. 1992. Insect fungal symbionts: A promising source of detoxifying enzymes. *Journal of Industrial Microbiology* 9:149–161.

Dowd, P. F., and S. K. Shen. 1990. The contribution of symbiotic yeast to toxin resistance of the cigarette beetle (*Lasioderma serricorne*). *Entomologia Experimentalis et Applicata* 56:241–248.

Dowd, P. F., and T. C. Sparks. 1988. A comparison of properties of trans-permethrin hydrolase and leucine aminopeptidase from the midgut of *Pseudoplusia includens*. *Pesticide Biochemistry and Physiology* 31:195–202.

Ebbert, M. A., J. L. Marlowe, and J. K. Burkholder. 2003. Protozoan and intracellular fungi gut endosymbionts in *Drosophila*: prevalence and fitness effects of single and dual infections. *Journal of Invertebrate Pathology* 83:37–45.

Escherich, K. 1900. Über das regelmäßige vorkommen von sproßpilzen in dem darmepithel eines käfers. *Biologisches Zentralblatt* 20:350–358.

Eya, B. K., P. T. M. Kenny, S. Y. Tamura, M. Ohnishi, Y. Naya, K. Nakanishi, and M. Sugiura. 1989. Chemical association in symbiosis: sterol donor in planthoppers. *Journal of Chemical Ecology* 15:373–380.

Fogelman, J. C., W. T. Starmer, and W. B. Heed. 1982. Comparisons of yeast florae from natural substrates and larval guts of Southwestern *Drosophila*. *Oecologia* 52:187–191.

Fraenkel, G., and M. Blewett. 1943a. Vitamins of the B-group required by insects. *Nature* 150:703–704.

Fraenkel, G., and M. Blewett. 1943b. Intracellular symbionts of insects as sources of vitamins. *Nature* 152:506–507.

Franck, A., and L. De Leenheer. 2002. Inulin. In *Biopolymers*, vol. 6, *Polysaccharides II. Polysaccharides from Eukaryotes*, ed. E. J. Vandamme, S. De Baets and A. Steinbuchel, pp. 439–479. New York: John Wiley & Sons.

Frants, T. G., and O. A. Mertvetsova. 1986. Yeast associations with mosquitoes of the genus *Aedes* mg. (Diptera, Culicidae) in the Tom-Ob river region [in Russian]. *Nauchnye Doki Vyss Shkoly Biol. Nauki* 4:94–98.

Fukatsu, T., S. Aoki, U. Kurosu, and H. Ishikawa. 1994. Phylogeny of Cerataphidini aphids revealed by their symbiotic microorganisms and basic structure of their galls: implications for host-symbiont coevolution and evolution of sterile soldier castes. *Zoological Science* 11:613–623.

Fukatsu, T., and H. Ishikawa. 1992. A novel eukaryotic extracellular symbiont in an aphid, *Astegopteryx styraci* (Homoptera, Aphididae, Hormaphidinae). *Journal of Insect Physiology* 38:765–773.

Fukatsu, T., and H. Ishikawa. 1995. Molecular phylogenetic analyses on evolutionary origin of yeast-like symbionts in Cerataphidini. *Zoological Science 12* (suppl.):34.

Fukatsu, T., and H. Ishikawa. 1996. Phylogenetic position of yeast-like symbiont of *Hamiltonaphis styraci* (Homoptera, Aphididae) based on 18S rDNA sequence. *Insect Biochemistry and Molecular Biology* 26:383–388.

Gams, W., and J. A. von Arx. 1980. Validation of *Symbiotaphrina* (Imperfect Yeasts). *Persoonia* 10:542–543.

Gilbert, D. G. 1980. Dispersal of yeasts and bacteria by *Drosophila* in a temperate forest. *Oecologia* 46:135–137.

Gilliam, M., and D. B. Prest. 1977. The mycoflora of selected organs of queen honey bees, *Apis mellifera*. *Journal of Invertebrate Pathology* 29:235–237.

Gilliam, M., J. L. Wickerham, H. L. Morton, and R. D. Martin. 1974. Yeasts isolated from honey bees, *Apis mellifera*, fed 2,4-D and antibiotics. *Journal of Invertebrate Pathology* 24:349–356.

Gilmore, M. S., and J. J. Ferretti. 2003. The thin line between gut commensal and pathogen. *Science* 299:1999–2002.

Gräbner, K.-E. 1954. Vergleichend morphologische und physiologische Studien an Anobiiden- und Cerambyciden-symbionten. *Zeitschrift für Morphologie und Ökologie der Tiere* 41:471–528.

Grilione, P., F. Federici, and M. W. Miller. 1981. Yeasts from honey bees (*Apis mellifera* L.). In *Current developments in yeast research*, ed. G. G. Stewart and I. Russell, pp. 599–605. Toronto: Pergamon Press.

Grosmann, H. 1930. Beiträge zur Kenntnis der Lebensgemeinschaft zwischen Borkenkäfern und Pilzen. *Zeitschrift für Parasitenkunde* 3:56–102.

Gupta, M., and J. C. Pant. 1985. Some studies on the association of symbiotes in *Amrasca devastans* (Distant). *Indian Journal of Entomology* 47:3101–316.

Haanstad, J. O., and D. M. Norris. 1985. Microbial symbiotes of the ambrosia beetle *Xyloterinus politus*. *Microbial Ecology* 11:267–276.

Hagen, K. S., R. L. Tassan, and E. F. Sawall, Jr. 1970. Some ecophysiological relationships between certain *Chrysopa*, honeydews and yeasts. *Bollettino del Laboratorio di Entomologia Agraria Filippo Silvestri di Portici* 28:113–134.

Hajsig, M. 1958. *Torulopsis apicola* nov. spec., new isolates from bees. *Antonie van Leeuwenhoek* 24:18–22.

Henninger, W., and S. Windisch. 1976. *Kluyveromyces blattae* sp. n., eine neue vielsporige Hefe aus *Blatta orientalis*. *Archiv für Mikrobiologie* 109:153–156.

Hongoh, Y., and H. Ishikawa. 2000. Evolutionary studies on uricases of fungal endosymbionts of aphids and planthoppers. *Journal of Molecular Evolution* 51:265–277.

Hunt, D. W. A., and J. H. Borden. 1990. Conversion of verbenols to verbenone by yeasts isolated from *Dendroctonus ponderosae* (Coleoptera: Scoytidae). *Journal of Chemical Ecology* 16:1385–1397.

Jones, K. G. 1981. Baldcypress allelochemics and the inhibition of silkworm enteric microorganisms: some ecological considerations. *Journal of Chemical Ecology* 7:103–114.

Jones, K. G., and M. Blackwell. 1996. Ribosomal DNA sequence analysis places the yeast-like genus *Symbiotaphrina* within filamentous ascomycetes. *Mycologia* 88:212–218.

Jones, K. G., P. F. Dowd, and M. Blackwell 1999. Polyphyletic origins of yeast-like endocytobionts from anobiid and cerambycid beetles. *Mycological Research* 103: 542–546.

Jouvenaz, D. P., and J. W. Kimbrough. 1991. *Myrmecomyces annellisae* gen. nov., sp. nov. (Deuteromycotina: Hyphomycetes), an endoparasitic fungus of fire ants, *Solenopsis* spp. (Hymenoptera: Formicidae). *Mycological Research* 95:1395–1401.

Jurzitza, G. 1959. Physiologische Untersuchungen an Cerambycidensymbionten. *Archiv für Mikrobiologie* 33:305–332.

Jurzitza, G. 1960. Zur Systematik einiger Cerambycidensymbionten. *Archiv für Mikrobiologie* 36:229–243.

Jurzitza, G. 1964. Studien an der Symbiose der Anobiiden. II. Physiologische Studien an Symbionten von *Lasioderma serricorne* F. *Archiv für Mikrobiologie* 49:331–340.

Jurzitza, G. 1969a. Untersuchungen über die Wirkung sekundärer Pflanzeninhaltsstoffe auf die Pilzsymbiose des Tabakkäfers *Lasioderma serricorne* F. 2. Die Entwicklung normaler und aposymbiontischer Larven in tabak mit verschiedenem Nikotingehalt. *Zeitschrift für Angewandte Entomologie* 63:233–236.

Jurzitza, G. 1969b. Die Rolle der hefeartigen Symbionten von *Lasioderma serricorne* F. (Coleoptera, Anobiidae) im Proteinmetabolismus ihrer Wirte. 1. Das Wachstum normaler und aspsymbiontischer Larven in Diäten mit Proteinen, Proteinderivaten und Aminosäuregemischen als N-quellen. *Zeitschrift für vergleichende Physiologie* 63:165–181.

Jurzitza, G. 1969c. Der Vitaminbedarf normaler und aposymbiontischer *Lasioderma serricorne* F. (Coleoptera, Anobiidae) und die Bedeutung der symbiontischen Pilze als Vitaminquelle für ihre Wirte. *Oecologia* 3:70–83.

Jurzitza, G. 1969d. Aufzuchtversuche an *Lasioderma serricorne* F. in drogen- und Holzpulvern im Hinblick auf die Rolle der hefeartigen Symbionten. *Zeitschrift für Naturforschung* 24b:760–763.

Jurzitza, G. 1970. Über Isolierung, Kultur und Taxonomic einiger Anobiidensymbionten (Insecta, Coleoptera). *Archiv für Mikrobiologie* 72:203–222.

Jurzitza, G. 1972. Rasterelektronenmikroskopische Untersuchungen über die Strukturen der Oberfläche von Anobiideneiern (Coleoptera) und über die Verteilung der Endosymbionten auf den Eischalen. *Forma et Functio* 5:75–88.

Jurzitza, G. 1976. Die Aufzucht von *Lasioderma serricorne* F. in holzhaltigen Vitaminmangeldiäten. Ein Beitrag zur Bedeutung der Endosymbiosen holzzerstörender Insekten als Vitaminquellen für ihre Wirte. *Material und Organismen* 3:499–505.

Kagayama, K., N. Shiragami, T. Nagamine, T. Umehara, and T. Mitsui. 1993. Isolation and classification of intracellular symbiotes from the rice brown planthopper, *Nilaparvata lugens*, based on analysis of 18S-ribosomal DNA. *Journal of Pesticide Science* 18:231–237.

Keilin, D., and P. Tate. 1943. The larval stages of the celery fly (*Acidia heracleri* L.) and of the braconid *Adelura apii* (Curtis), with notes upon an associated parasitic yeast-like fungus. *Parasitology* 35:27–36.

Klemm, D., H.-P. Schmauder, and T. Heinze. 2002. Cellulose. In *Biopolymers*, Vol. 6 of *Polysaccharides II. Polysaccharides from Eukaryotes*, ed. E. J. Vandamme, S. De Baets and A. Steinbuchel, pp. 275–319. New York: John Wiley & Sons.

Koch, A. 1954. Symbionten als vitaminquellen der insekten. *Forschungen und Fortschritte* 28:33–37.

Kuhlwein, H., and G. Jurzitza. 1961. Studien an der Symbiose der Anobiiden. 1. Mitteilung: Die Kultur des Symbionten von *Sitodrepa panicea* L. *Archiv für Mikrobiologie* 40:247–260.

Kurtzman, C. P., and C. J. Robnett. 1998a. Identification and phylogeny of ascomycetous yeasts from analysis of nuclear large subunit (26S) ribosomal DNA partial sequences. *Antonie van Leeuwenhoek* 73:331–371.

Kurtzman, C. P., and C. J. Robnett. 1998b. Three new insect-associated species of the yeast genus *Candida*. *Canadian Journal of Microbiology* 44:965–973.

Kusumi, T., Y. Suwa, H. Kita, and S. Nasu. 1979. Symbiotes of planthoppers: I. The isolation of intracellular symbiotes from the smaller brown planthopper, *Laodelphax striatellus* Fallén (Hemiptera: Delphacidae). *Applied Entomology and Zoology* 14:459–463.

Lachaise, D., M.-C. Pignal, and J. Roualt. 1979. Yeast flora partitioning by drosophilid species inhabiting a tropical African savanna of the Ivory Coast [Diptera]. *Annales de la Société Entomologique de France* (*Nouvelle Série*) 15:659–680.

Larran, S., A. Perello, M. R. Simón, and V. Moreno. 2002. Isolation and analysis of endophytic microorganisms in wheat (*Triticum aestivum* L.) leaves. *World Journal of Microbiology & Biotechnology* 18:683–686.

Le Fay, A. J., J.-E. Courtois, A. Thuillier, C. Chararas, and S. Lambin. 1969. Étude des osidases de l'insecte xylophage *Ips sexdentatus* et de sa flore microbienne. *Comptes Rendus de l'Académie des Sciences de Paris* 268:2968–2970.

Le Fay, A. J., J.-E. Courtois, A. Thuillier, C. Chararas, and S. Lambin. 1970. Étude des osidases de l'insecte xylophage *Ips sexdentatus* et de sa flore microbienne. I. Étude de la flore microbienne et comparison de ses osidases avec celles de l'insecte total. *Annales de l'Institute Pasteur* 119:483–491.

Lebeck, L. M. 1989. Extracellular symbiosis of a yeast-like microorganism within *Comperia merceti* (Hymenoptera: Encyrtidae). *Symbiosis* 7:1–66.

Lee, Y. H., and R. F. Hou. 1987. Physiological roles of yeast-like symbiote in reproduction and embryonic development of the brown planthopper, *Nilaparvata lugens* Stål. *Journal of Insect Physiology* 33:851–860.

Leufvén A., G. Bergström, and E. Falsen. 1984. Interconversion of verbenols and verbenone by yeasts isolated from the spruce bark beetle *Ips typographus. Journal of Chemical Ecology* 10:1349–1361.

Leufvén A., and L. Nehls.1986. Quantification of different yeasts associated with the bark beetle, *Ips typographus*, during its attack on a spruce tree. *Microbial Ecology* 12:237–243.

Lichtwardt, R. W., M. M. White, M. J. Cafaro and J. K. Misra. 1999. Fungi associated with passalid beetles and their mites. *Mycologia* 91:694–702.

Lienig, K. 1993. Oogenesis and symbiont transfer in *Sogatodes orizicola* Muir (Homoptera: Delphacidae) [in German]. *Beitrage zur Entomologie* 43: 445–449.

Lu, K. C., D. G. Allen, and W. B. Bollen. 1957. Association of yeasts with the Douglas-fir beetle. *Forest Science* 3:336–343.

Lushbaugh, W. B., E. D. Rowton, and R. Barclay McGhee. 1976. Redescription of *Coccidiascus legeri* Chatton, 1913 (Nematosporaceae: Hemiascomycetidae), an intracellular parasitic yeastlike fungus from the intestinal epithelium of *Drosophila melanogaster. Journal of Invertebrate Pathology* 28:93–107.

Mahdihassan, S.1928. Symbionts specific of wax and pseudo lac insects. *Archiv für Protistenkunde* 63:18–22.

Mahdihassan, S. 1929. The microorganisms of red and yellow lac insects. *Archiv für Protistenkunde* 68:613–624.

Malan, C. E., and A. Gandini. 1966. Blastomiceti della fonte di alimentazione, del nido pedotrofico e dell'apparato digerente di larve di scarabei coprofagi (Coleoptera: Scarabaeidae). *Centro Entomologico Alpina Forestale Consiglio Nazionale delle Ricerche, Publicatione No. 99.*

Margulis, L., and M. J. Chapman. 1998. Endosymbioses: Cyclical and permanent in evolution. *Trends in Microbiology* 6:342–345.

Martignoni, M. E., P. J. Iwai, and L. J. Wickerham. 1969. A candidiasis in larvae of the Douglas-fir tussock moth, *Hemerocampa pseudotsugata. Journal of Invertebrate Pathology* 14:108–110.

Meyer, S. A., R. W. Payne, and D. Yarrow. 1998. *Candida* Berkhout. In *The yeasts. A taxonomic study*, 4th ed., ed. C. P. Kurtzman and J. W. Fell, pp. 454–573. New York: Elsevier.

Middeldorf, J., and A. Ruthmann. 1984. Yeast-like endosymbionts in an ichneumonid wasp. *Zeitschrift für Naturforschung* 39c:322–326.

Miller, M. W., and E. M. Mrak. 1953. Yeasts associated with dried-fruit beetles in figs. *Applied Microbiology* 1:174–178.

Milne, D. L. 1961. The mechanism of growth retardation by nicotine in the cigarette beetle, *Lasioderma serricorne. South African Journal of Agricultural Science* 4:277–278.

Milne, D. L. 1963. A study of the nutrition of the cigarette beetle, *Lasioderma serricorne* F. (Coleoptera: Anobiidae) and a suggested new method for its control. *Journal of the Entomological Society of Southern Africa* 26:43–63.

Mishra, S. C., and P. K. Sen-Sarma. 1985. Carbohydrases in xylophagous coleopterous larvae (Cerambycidae and Scarabaeidae) and their evolutionary significance. *Material und Organismen* 20:221–230.

Mishra, S. C., and P. Singh. 1978. Polysaccharide digestive enzymes in the larvae of *Stromatium barbatum* (Fabr.), a dry wood borer (Coleoptera: Cerambycidae). *Material und Organismen* 13:115–122.

Mitsuhashi, J. 1975. Cultivation of intracellular yeast-like organisms in the smaller brown planthopper, *Laodelphax striatellus* Fallén (Hemiptera: Delphacidae). *Applied Entomology and Zoology* 10:243–245.

Monk, K. A., and G. J. Samuels. 1990. Mycophagy in grasshoppers (Orthoptera: Acrididae) in Indo-Malayan rain forests. *Biotropica* 22:16–21.

Moore, G. E. 1972. Microflora from the alimentary tract of healthy southern pine beetles *Dendroctonus frontalis* (Scolytidae), and their possible relationship to pathogenicity. *Journal of Invertebrate Pathology* 19:72–75.

Morais, P. B., A. N. Hagler, C. A. Rosa, and L. C. Mendonça-Hagler. 1992. Yeasts associated with *Drosophila* in tropical forests of Rio de Janeiro, Brazil. *Canadian Journal of Microbiology* 38:1150–1155.

Morais, P. B., C. A. Rosa, A. N. Hagler, and L. C. Mendonça-Hagler. 1994. Yeast communities of the cactus *Pilosocereus arrabidae* as resources for larval and adult stages of *Drosophila serido*. *Antonie van Leeuwenhoek* 66:313–317.

Nardon, P., and A. M. Grenier. 1989. Endocytobiosis in Coleoptera: Biological, biochemical, and genetic aspects. In *Insect Endocytobiosis*, ed. W. Schwemmler and G. Gassner, pp. 175–215. Boca Raton, FL: CRC Press.

Nardon, P., and C. Nardon. 1998. Morphology and cytology of symbiosis in insects. *Annales de la Société Entomologique de France* (*Nouvelle Série*) 34:105–134.

Nasu, S. 1963. Studies on some leafhoppers and planthoppers which transmit virus disease of plants in Japan [in Japanese]. *Bulletin of the Kyushu Agricultural Experiment Station* 8:153–349.

Nasu, S., T. Kusumi, Y. Suwa, and H. Kita. 1981. Symbiotes of planthoppers: II. Isolation of intracellular symbiotic microorganisms from the brown planthopper, *Nilaparvata lugens* Stål, and immunological comparison of the symbiotes associated with rice planthoppers (Hemiptera: Delphacidae). *Applied Entomology and Zoology* 16:88–93.

Nes, W. D., M. Lopez, W. Zhou, D. Guo, P. F. Dowd, and R. A. Norton. 1997. Sterol utilization and metabolism by *Heliothis zea*. *Lipids* 32:1317–1323.

Noda, H. 1974. Preliminary histological observation and population dynamics of intracellular yeast-like symbiotes in the smaller brown planthopper, *Laodelphax striatellus* (Homoptera: Delphacidae). *Applied Entomology and Zoology* 9:275–277.

Noda, H. 1977. Histological and histochemical observation of intracellular yeastlike symbiotes in the fat body of the smaller brown planthopper, *Laodelphax striatellus* (Homoptera: Delphacidae). *Applied Entomology and Zoology* 12:134–141.

Noda, H., and K. Kodama. 1995. Phylogenetic position of yeast-like symbiotes of rice planthoppers based on partial 18S rDNA sequences. *Insect Biochemistry and Molecular Biology* 25:639–646.

Noda, H., and K. Kodama. 1996. Phylogenetic position of yeastlike endosymbionts of anobiid beetles. *Applied and Environmental Microbiology* 62:162–167.

Noda, H., N. Nakashima, and M. Koizumi. 1995. Phylogenetic position of yeast-like symbiotes of rice planthoppers based on partial 18S rDNA sequences. *Insect Biochemistry and Molecular Biology* 25:639–646.

Noda, H., and T. Saito. 1979a. The role of intracellular yeastlike symbiotes in the development of *Laodelphax striatellus* (Homoptera: Delphacidae). *Applied Entomology and Zoology* 14:453–458.

Noda, H., and T. Saito. 1979b. Effects of high temperature on the development of *Laodelphax striatellus* (Homoptera: Delphacidae) and on its intracellular yeastlike symbiotes. *Applied Entomology and Zoology* 14:64–75.

Noda, H., K. Wada, and T. Saito. 1979. Sterols in *Laodelphax striatellus* with special reference to the intracellular yeastlike symbiotes as a sterol source. *Journal of Insect Physiology* 25:443–447.

Pant, N. C., and M. Anand. 1985. Qualitative vitamin requirement of *Lasioderma serricorne* (Fab.) larvae. *Journal of Entomological Research* 9:165–169.

Pant, N. C., and Fraenkel, G. 1950. The function of the symbiotic yeasts of two insect species, *Lasioderma serricorne* F. and *Stegobium (Sitodrepa) paniceum* L. *Science* 112:498–500.

Pant, N. C., and G. Fraenkel. 1954. Studies on the symbiotic yeasts of two insect species, *Lasioderma serricorne* F. and *Stegobium paniceum* L. *Biological Bulletin* 107:420–432.

Pant, N. C., P. Gupta, and J. K. Nayar. 1959. Physiology of intracellular symbionts of *Stegobium paniceum* L., with special reference to amino acid requirements of the host. *Experientia* 16:311–312.

Pant, N. C., and S. Kapoor. 1963. Physiology of intracellular symbiotes of *Lasioderma serricorne* F. with special reference to the cholesterol requirements of the host. *Indian Journal of Entomology* 25:311–315.

Pérez, J., F. Infante, F. E. Vega, F. Holguín, J. Macías, J. Valle, G. Nieto, S. W. Peterson, C. P. Kurtzman, and K. O'Donnell. 2003. Mycobiota associated with the coffee berry borer *Hypothenemus hampei* (Coleoptera, Scolytidae) in Chiapas, Mexico. *Mycological Research* 107:879–887.

Phaff, H. J., M. W. Miller, J. A. Recca, M. Shifrine, and E. M. Mrak. 1956. Studies on the ecology of *Drosophila* in the Yosemite region of California. *Ecology* 37:533–538.

Phaff, H. J., and W. T. Starmer. 1987. Yeasts associated with plants, insects and soil. In *Biology of yeasts*, Vol. 1, 2nd ed., *The yeasts*, ed. A. H. Rose and J. S. Harrison, pp. 123–180. London: Academic Press.

Pignal, M.-C., C. Chararas, and M. Bourgeay-Causse. 1987. Isolement et étude de levures du tractes digestif d'*Ips sextendatus*, Coléoptère parasite des conifères. *Comptes Rendus de l'Académie des Sciences de Paris* 304:449–452.

Pignal, M.-C., C. Chararas, and M. Bourgeay-Causse. 1988. Yeasts from *Ips sexdentatus* (Scolytidae). *Mycopathologia* 103:43–48.

Potrykus, I. 2001. Golden rice and beyond. *Plant Physiology* 125:1157–1161.

Přibram, E. 1925. Über "schwarze Hefen" (*Zymonemata nigra* und eine *Torula variabilis*). *Ergebnisse der Physiologie* 24:95–106.

Prillinger, H., R. Messner, H. König, R. Bauer, K. Lopandic, O. Molnar, P. Dangel, F. Weigang, T. Kirisits, T. Nakase, and L. Sigler. 1996. Yeast associated with termites: A phenotypic and genotypic characterization and use of coevolution for dating evolutionary radiations in asco- and basidiomycetes. *Systematic and Applied Microbiology* 19:265–283.

Ralet, M.-C., E. Bonnin, and J.-F. Thibault. 2002. Pectins. In *Biopolymers*, vol. 6, *Polysaccharides II. Polysaccharides from Eukaryotes*, ed. E. J. Vandamme, S. De Baets and A. Steinbuchel, pp. 345–380. New York: John Wiley & Sons.

Riba, G. 1977. Étude ultrastructurale de la multiplication et de la dégénérescence des symbiontes des larves de *Criocephalus rusticus* [Coleoptera: Cerambycidae]; influence du jeune. *Annales de la Société Entomologique de France (Nouvelle Série)* 13:153–157.

Rosa, C. A., P. B. Morais, A. N. Hagler, L. C. Mendonça-Hagler, and R. F. Monteiro. 1994. Yeast communities of the cactus *Pilosocereus arrabidae* and associated insects in the Sandy Coastal Plains of Southeastern Brazil. *Antonie van Leeuwenhoek* 65:55–62.

Rosa, C. A., A. Norton Hagler, L. C. S. Mendonça-Hagler, P. Benevides de Morais, N. C. M. Gomes, and R. F. Monteiro. 1992. *Clavispora opuntiae* and other yeasts associated with the moth *Sigelgaita* sp. in the cactus *Pilosocereus arrabidae* of Rio de Janeiro, Brazil. *Antonie van Leeuwenhoek* 62:267–272.

Sang, J. H. 1978. The nutritional requirements of *Drosophila*. In *The genetics and biology of Drosophila*, vol. 2, ed. M. Ashburner and T. R. F. Wright, pp. 159–192. New York: Academic Press.

Sasaki, T., M. Kawamura, and H. Ishikawa. 1996. Nitrogen recycling in the brown planthopper, *Nilaparvata lugens*: Involvement of yeast-like endosymbionts in uric acid metabolism. *Journal of Insect Physiology* 42:125–129.

Schäfer, A., R. Konrad, T. Kuhnigk, P. Kampfer, H. Hertel, and H. Konig. 1996. Hemicellulose degrading bacteria and yeasts from the termite gut. *Journal of Applied Bacteriology* 80:471–478.

Schmidt, O., U. Theopold, and M. R. Strand. 2001. Innate immunity and its evasion and suppression by hymenopteran endoparasitoids. *BioEssays* 23:344–351.

Shankar, G., and P. Baskaran. 1987. Distribution of endosymbiotes among the insect fauna of Annamalainagar. *Current Science* 56:970–971.

Shankar, G., and P. Baskaran. 1992. Regulation of yeast-like endosymbiotes in the rice brown planthopper *Nilaparvata lugens* Stål (O:Homoptera, F:Delphacidae). *Symbiosis* 14:161–173.

Shen, S. K., and P. F. Dowd. 1989. Xenobiotic induction of esterases in cultures of the yeastlike symbiont from the cigarette beetle. *Entomologia Experimentalis et Applicata* 52:179–184.

Shen, S. K., and P. F. Dowd. 1991a. Detoxification spectrum of the cigarette beetle symbiont *Symbiotaphrina kochii* in culture. *Entomologia Experimentalis et Applicata* 60:51–59.

Shen, S. K., and P. F. Dowd. 1991b. 1–Naphthyl acetate esterase activity from cultures of the symbiont yeast of the cigarette beetle (Coleoptera: Anobiidae). *Journal of Economic Entomology* 84:402–407.

Shifrine, M., and H. J. Phaff. 1956. The association of yeasts with certain bark beetles. *Mycologia* 48:41–55.

Shihata, A. M. El-Tabey Awad, and E. M. Mrak. 1952. Intestinal yeast floras of successive populations of *Drosophila*. *Evolution* 6:325–332.

Shigenobu, S., H. Watanabe, M. Hattori, Y. Sakaki, and H. Ishikawa. 2000. Genome sequence of the endocellular bacterial symbiont of aphids *Buchnera* sp. APS. *Nature* 407:81–86.

Six, D. L. 2003. Bark beetle-fungus symbiosis. In *Insect symbiosis*, ed. K. Bourtzis and T. A. Miller, pp. 97–114. Boca Raton, FL: CRC Press.

Slaytor, M. 2000. Energy metabolism in the termite and its gut microbiota. In *Termites: Evolution, sociality, symbiosis, ecology*, ed. T. Abe, D. E. Bignell and M. Higashi, pp. 307–332. Boston: Kluwer Academic.

Southwood, T. R. E. 1973. The insect/plant relationship: An evolutionary perspective. In *Insect/plant interactions*, ed. H. F. van Emden, pp. 3–30. London: Blackwell.

Starmer, W. T., J. S. F. Barker, H. J. Phaff, and J. C. Fogleman. 1986. Adaptations of *Drosophila* and yeasts: Their interactions with the volatile 2-propanol in the cactusmicroorganism-*Drosophila* model system. *Australian Journal of Biological Sciences* 39:69–77.

Starmer, W. T., and J. C. Fogleman. 1986. Coadaptation of *Drosophila* and yeasts in their natural habitat. *Journal of Chemical Ecology* 12:1037–1053.

Starmer, W. T., W. B. Heed, M. Miranda, M. Miller and H. J. Phaff. 1976. The ecology of yeast flora associated with cactophilic *Drosophila* and their host plants in the Sonoran Desert. *Microbial Ecology* 3:11–30.

Starmer, W. T., H. J. Phaff, M. Miranda, M. W. Miller, and W. B. Heed. 1982. The yeast flora associated with the decaying stems of columnar cactus and *Drosophila* in North America. *Evolutionary Biology* 14:269–295.

Steinhaus, E. A. 1947. *Insect microbiology.* New York: Comstock Publishing.

Steinhaus, E. A. 1949. *Principles of insect pathology.* New York: McGraw-Hill.

Stern, D. L., S. Aoki, and U. Kurosu. 1997. Determining aphid taxonomic affinities and life cycles with molecular data: A case study of the tribe Cerataphidini (Hormaphididae: Aphidoidea: Hemiptera). *Systematic Entomology* 22:81–96.

Suh, S.-O., C. J. Marshall, J. V. McHugh, and M. Blackwell. 2003. Wood ingestion by passalid beetles in the presence of xylose-fermenting gut yeasts. *Molecular Ecology* 12:3137–3146.

Suh, S.-O., H. Noda, and M. Blackwell. 2001. Insect symbiosis: Derivation of yeast-like endosymbionts within an entomopathogenic filamentous lineage. *Molecular Biology and Evolution* 18:995–1000.

Suomi, D. A., and R. D. Akre. 1993. Biological studies of *Hemicoelus gibbicollis* (LeConte) (Coleoptera: Anobiidae), a serious structural pest along the Pacific Coast: Larval and pupal stages. *Pan-Pacific Entomologist* 69:221–235.

Tamas, I., L. Klasson, B. Canback, A. K. Naslund, A.-S. Eriksson, J. J. Wernegreen, J. P. Sandstrom, N. A. Moran, and S. G. E. Andersson. 2002. 50 million years of genomic stasis in endosymbiotic bacteria. *Science* 296:2376–2379.

Tremblay, E. 1989. Coccoidea endocytobiosis. In *Insect endocytobiosis: Morphology, physiology, genetics, evolution,* ed. W. Schwemmler and G. Gassner, pp. 145–163. Boca Raton, FL: CRC Press.

Turunen, S. 1985. Absorption. In *Regulation, digestion, nutrition and excretion,* vol. 4, *Comprehensive insect physiology, biochemistry and pharmacology,* ed. G. A. Kerkut and L. I. Gilbert, pp. 241–278. New York: Pergamon Press.

van der Walt, J.P. 1959. *Endomycopsis wickerhamii* nov. spec., a new heterothallic yeast. *Antonie van Leeuwenhoek* 25:344–348.

van der Walt, J. P. 1961. The mycetome symbiont of *Lasioderma serricorne. Antonie van Leeuwenhoek* 27:362–366.

van der Walt, J. P., and W.C. van der Klift. 1972. *Pichia melissophila* sp. n., a new osmotolerant yeast from apiarian sources. *Antonie van Leeuwenhoek* 38:361–364.

Vega, F. E., M. B. Blackburn, C. P. Kurtzman, and P. F. Dowd. 2003. Identification of a coffee berry borer-associated yeast: Does it break down caffeine? *Entomologia Experimentalis et Aplicata* 107:19–24.

Verrett, J. M., K. B. Green, L. M. Gamble, and F. C. Crochen. 1987. A hemocoelic *Candida* parasite of the American cockroach (Dictyoptera: Blattidae). *Journal of Economic Entomology* 80:1205–1212.

Wetzel, J. M., M. Ohnishi, T. Fujita, K. Nakanishi, Y. Naya, H. Noda, and M. Sugiura. 1992. Diversity in steroidogenesis of symbiotic microorganisms from planthoppers. *Journal of Chemical Ecology* 18:2083–2094.

Wilkinson, T. L., and H. Ishikawa. 2001. On the functional significance of symbiotic microorganisms in the Homoptera: a comparative study of *Acyrthosiphon pisum* and *Nilaparvata lugens. Physiological Entomology* 26:86–93.

Woolfolk, S. W., and G. Douglas Inglis. 2004. Microorganisms associated with field-collected *Chrysoperla rufilabris* (Neuroptera: Chrysopidae) adults with emphasis on yeast symbionts. *Biological Control* 29:155–168.

Xu, J., M. K. Bjursell, J. Himrod, S. Deng, L. K. Carmichael, H. C. Chiang, L. V. Hooper, and J. I. Gordon. 2003. A genomic view of the human-*Bacteroides thetaiotamicron* symbiosis. *Science* 299:2074–2076.

Zhao, J.-H., F.-Y. Bai, L.-D. Guo, and J.-H. Jia. 2002. *Rhodotorula pinicola* sp. nov., a basidiomycetous yeast species isolated from xylem of pine twigs. *FEMS Yeast Research* 2:159–163.

Zhongxian, L., Y. Xiaoping, C. Jianming, Z. Xusong, and X. Hongxing. 2001. Effects of endosymbiote on feeding, development, and reproduction of brown planthopper, *Nilaparvata lugens*. *Chinese Rice Research Newsletter* 9:11–12.

Zoberi, M. H., and J. K. Grace. 1990. Fungi associated with the subterranean termite *Reticulitermes flavipes* in Ontario. *Mycologia* 82:286–294.

10

The Beetle Gut as a Habitat
for New Species of Yeasts

Sung-Oui Suh
Meredith Blackwell

The study of endosymbioses continues to accentuate the extraordinary impact that microorganisms have had on the ecology and evolution of invertebrates, especially insects. One only has to learn of the studies of *Buchneria* and *Wolbachia* and their interactions with a wide variety of arthropods to wonder at the dramatic effect of endosymbionts on their hosts (van Meer et al. 1999; Dale et al. 2001; Funk et al. 2001). The study of eukaryotic endosymbionts, however, has lagged somewhat behind that of the prokaryotes. Nevertheless, fungi are increasingly being recognized as important endosymbionts of insects. It is difficult to predict which insects might be associated with fungi because even close relatives feeding on similar nutrient resources may vary in whether they are associated with bacterial or fungal symbionts, or for that matter with any symbiont (Buchner 1965; Martin 1987).

Some of the eukaryotic endosymbionts that have been discovered have obligate roles in the lives of their hosts. These include true yeasts (Ascomycota: Saccharomycotina: Saccharomycetes) as well as reduced yeastlike symbionts (YLSs) derived from several groups of filamentous ascomycetes (Pezizomycotina), the most complex and largest subphylum of ascomycetes by three orders of magnitude. True yeasts have been reported in associations with termites (Prillinger et al. 1996; chapter 8, this volume) and with certain cerambycid and anobiid beetles (Dowd 1991; chapter 9, this volume). A more recently discovered association between passalid beetles and true yeasts is suspected of facilitating wood degradation (Suh et al. 2003, 2004b). Two independent clades of derived YLSs occur in association with anobiid beetles (in contrast to true yeast endosymbionts) and with planthoppers and other aphid groups (Dowd 1989, 1991; Fukatsu and Ishikawa 1996; Noda and Kodama 1996; Suh et al. 2001). The exact placement of the derived anobiid YLS

Symbiotaphrina has not been determined, but phylogenetic analysis now places the two *Symbiotaphrina* species examined among the more basal lineages of poorly resolved speciose filamentous ascomycete groups (discomycetes and loculoascomycetes) of the Pezizomycotina (Jones and Blackwell 1996; Jones et al. 1999). The other group of derived yeasts, unnamed YLSs, are implicated in sterol utilization and nitrogen recycling in certain planthoppers (Homoptera: Delphacidae; Wetzel et al. 1992; Sasaki et al. 1996; Hongoh and Ishikawa 1997, 2000; Wilkinson and Ishikawa 2001; Noda and Koizumi 2003; see chapter 9, this volume, for more detail). Planthopper YLSs are derived from among a clade of *Cordyceps* species, one of several groups of a large entomopathogenic genus (Suh et al. 2001). The origin of an insect symbiont from within a clade of necrotrophic pathogens contributes to the theory of the origin of symbionts (whether from insect pathogens or from saprobes), in this case supporting the origin from a deadly pathogen (see chapter 9).

During our continuing studies of fungi associated with insects, we have been especially interested in dispersal interactions. We originally began a survey of mushroom-feeding beetles, not in search of undescribed yeasts inhabitating the gut of the beetles, but to determine if insects might be dispersers of the blastosporic basidiomycete yeasts that had been reported from basidiocarps (Prillinger 1987). Originally, these yeasts had been mistaken for basidiospores when basidiocarps were set up to drop spores on agar for culture. We wanted to discover which insects might be the dispersers of Prillinger's (1987) yeasts, but we also wondered if any of the beetles might be associated with endosymbiotic yeasts in this habitat. In addition to their taxonomic diversity, the beetles that feed and breed in basidiocarps have specificity for particular basidiocarps and can be targeted for recollection.

Although we did not isolate a large number of the ballistosporic yeasts that we originally sought, many of the targeted basidiocarp-feeding beetles were the source of an unexpected diversity of new species of true yeasts. We also discovered interesting yeasts in the gut of certain other beetles using different nutritional resources. Thus far, ours has been a study of species diversity, but in some cases there is circumstantial evidence for symbiosis. The suggestion that a symbiotic association exists is based on (1) reports of beetle-associated enzymes that might be of microbial origin (Martin 1987), (2) localization of certain yeasts within specialized gut cecal pouches of some beetles, (3) the co-occurrence of certain beetles and yeasts across a broad intercontinental geographical distribution, and (4) the quality of the mushroom substrate.

The Basidiocarp as a Beetle Habitat

Some organisms use a narrow range of nutritional resources that may be low in certain essential nutrients. A classical example is wood-ingesting insects that lack the enzymes necessary for degradation of wood taken into the gut; in addition, wood as a food source is low in available nitrogen (Martin 1987). The yeasts we have discovered are associated with insects that lack variety in their diets. At present, we have no direct evidence indicating that the yeasts are truly endosymbiotic and contribute to the nutrition of the beetles, although we will discuss circumstantial evidence suggesting that dietary supplements could be supplied by yeasts. The gut of such insects would seem, therefore, a profitable habitat in which to look for microorganisms that might aid the insects.

Many of the beetles from which we have isolated yeasts start life as eggs laid in a basidiocarp. They pass all the stages of their life histories from first instar to egg-laying adult in one or more of the basidiocarps upon which they rely for their total intake of external nutrients. Variation in basidiocarp nutritional value is correlated with several factors, including basidiomycete species, basidiocarp age and condition, microhabitat, environmental factors, and individual differences of the fungus. Although some information on nutritional content of commercial and wild edible mushrooms is available (Crisan and Sands 1978), less is known about basidiocarps such as polypores that usually are not normally eaten by humans. Just as a steady diet of basidiocarps probably would not supply all the nutrients needed for a human diet, it would not provide a healthy insect diet. For example, many of the yeasts from the gut of beetles synthesize a broader spectrum of vitamins than do most mushrooms (Crisan and Sands 1978; Suh and Blackwell, unpublished data). More information is needed on the exact nutrients provided by specific insect-containing basidiocarps.

The Beetle as a Yeast Habitat

The precise location of insect-gut yeasts varies from the crop at the anterior end of the gut to the midgut (including its cecal pouches) and the hindgut. Adult lacewings (*Chrysoperla* spp.) consistently possess yeasts of unknown function in their crops (Woolfolk and Inglis 2004; Suh et al. 2004a; N. H. Nguyen, pers. comm.). Yeasts associated with mushroom-feeding beetles inhabit the midgut, sometimes located in the cecum of the midgut. The endosymbionts known from several phytophagous beetles also occupy cecal pouches. The midgut of most insects has a pH between 6 and 8 and is the site of high enzyme activity due to excretion of enzymes from midgut cells. This region is also the site of the digestion of proteins, carbohydrates, and lipids in most insects (Chapman 1998). The correlation between gut position and type of food material appears to be tied to gut function, and digestion of intractable foodstuffs such as woolen fiber and wood often occurs in the hindgut. The hindgut habitat usually is a little more acidic than that of the midgut and is the site of uptake of most water and salts by beetles (Chapman 1998). Unrelated microorganisms associated with the degradation of cellulose and other wood components inhabit the hindgut in different groups of unrelated insects. These microorganisms include flagellated fermentative microorganisms associated with several wood-ingesting groups (e.g., termites, wood roaches), but also yeasts. Wood-ingesting passalid beetles have xylose-fermenting and assimilating yeasts that also occur in the hindgut, where they can be seen attached to the gut wall (Suh et al. 2003, 2004b).

Insects require a variety of nutrients, similar to those required by other animals: amino acids, lipids, carbohydrates, vitamins, and minerals. The importance of associated microorganisms comes from their wide variety of enzymes capable of modifying the raw materials ingested by the insects for conversion to higher quality nutrient, especially lipids, vitamins, and nitrogen-containing compounds that can be used by the insect (Martin 1987). Most of the yeasts discovered in this study synthesize a wide variety of B-complex vitamins, and these and other products as well as enzymes such as those that degrade xylose could supplement the insect diet and ability to digest food.

In some cases wood-ingesting insects benefit from substrates predigested by free-living microorganisms, including wood-inhabiting fungi that break down polysaccharides to benefit the insects (Martin 1987; chapter 11). The microorganisms living within an insect, however, have an advantage over those that are free-living because within the gut the yeasts are bathed by a regular supply of nutrients. Another benefit might be direct dispersal. Free-living fungi usually deplete their substrate, and dispersal to a new substrate occurs by wind, water, or animals. In the case of organisms that live within insects, however, dispersal shifts to become highly dependent on the insects, and the probability that an appropriate new substrate will be chosen by the insect is high. We have not yet been successful in determining the complete life histories of the yeasts, but in several cases there is evidence that vertical transmission from parent to offspring occurs in some beetles, although horizontal transmission almost certainly occurs in others as well (see Specificity of Beetle–Yeast Associations).

How Many Yeasts Species Are in the Gut of Fungus-Feeding Beetles?

The beetle gut is a rich source of undescribed yeasts. Much of the success in discovering the presence of new yeast species from nature was due to the availability of a database of about 600 bp of the large subunit (LSU) rRNA gene (rDNA) for more than 650 previously described species of true yeasts (Kurtzman and Robnett 1998). As each newly acquired beetle gut yeast was isolated, it was sequenced, and the sequence was used in BLAST searches and phylogenetic analyses to determine its status. The numbers we report in this study are *genotypes* or *groups* based on unique LSU rDNA sequences. In addition, the small subunit (SSU) rDNA sequences and information from the yeast standard description were available to help evaluate the DNA groups as potential taxa. The characters of the standard description include about 80 physiological and 20 morphological traits. This strategy has resulted in the isolation and characterization of many new yeasts from the insect gut habitat.

More than 650 yeasts were cultured from the gut of beetles, and of these about 75% were undescribed isolates that could be placed in about 150 DNA-based groups of true yeasts (Ascomycota: Saccharomycetes) and a small number of basidiomycete yeasts (Tremellales). No reduced yeastlike fungi derived from filamentous ascomycetes (e.g., similar to planthopper YLS or *Symbiotaphrina*) were isolated. Beetles in 25 families and 3 distinctive previously independent families, recently subsumed within other families on the basis of paraphyletic relationships, were hosts for the yeasts. These included Erotylidae, Tenebrionidae, Passalidae, Endomychidae, Nitidulidae, Scarabaeidae, Curculionidae (Scolytinae), Ciidae, Staphylinidae (Scaphidiinae), Mycetophagidae, Cucujoidae, Curculionidae

(Curculioninae), Anthribidae, Cerambycidae, Histeridae, Staphylinidae, Elateridae, Carabidae, Derodontidae, Trogossitidae, Anobiidae, Melandryidae, Clambidae, Dermestidae, Cerylonidae, Latridiidae, and Mordellidae. Also, Chrysopidae (Neuroptera) was included. A few of these beetles were not basidiocarp feeders, but rather predators in the basidiocarp habitats or plant feeders, and some were collected at light traps (table 10.1).

Our discovery of about 150 undescribed yeast taxa from the poorly explored beetle gut habitat adds 21% to the total number of known yeasts. The large number of new species gains even more significance with the realization that fewer than 700 species of yeasts have been described previously from all of the earth's habitats and geographical regions (Barnett et al. 2000). Our study was conducted over 4 years in two regions, the United States, mostly in the southeastern states, and Barro Colorado Island, Panama. Somewhat daunting is the prediction of many more yeasts to come based on an analysis of our data. A Bayesian analysis (Pollock and Larkin, pers. comm.) provided an estimate of yeast species yet to be discovered in this unique ecosystem. The best supported model indicated that approximately 1.5–2 times as many taxa remain undiscovered in the localities already sampled (D. Pollock, pers. comm.). Those who study yeasts estimate that only a small fraction of the yeasts that exist have been discovered (C. Kurtzman, pers. comm.), and our studies certainly support this view.

A well-defined species concept is necessary to a discussion of species numbers. We used a conservative measure of species that considers DNA sequence, physiological differences, and beetle host differences. Previous studies of yeasts arbitrarily set a difference of more than 5 bp or 1% in the D1/D2 domain of the LSU rDNA as a cut-off point (Kurtzman and Robnett 1998). This phenetic measure sometimes has been correlated with a biological species concept (C. Kurtzman, pers. comm.) and, when multiple isolates have been included in analyses, a phylogenetic species concept. In fact, it appears to be a somewhat conservative measure, especially when more variable internal transcribed spacer (ITS) sequences are considered. From the data presented in table 10.1, it is clear that DNA sequences from the majority of all the gut yeasts are more than 5 bp different from the closest known species both by BLAST search and phylogenetic analysis. Although it is not expressed in the table because we do not have exact numbers, the number of yeast samples obtained is roughly proportional to the total number of insects dissected from each family because our sampling success approached 100%.

Phylogenetic Relationships and Distribution of the Beetle-associated Yeasts

About 150 insect-associated yeasts were distributed widely in clusters throughout the phylogenetic tree depicting the position of the new yeast species from this study (bold branches, fig. 10.1). Although several new species of basidiomycete yeasts (fig. 10.2) have been discovered, the overwhelming number of new species were ascomycete yeasts (fig. 10.3; Saccharomycetes). It is interesting that no additional YLSs were discovered, although two different clades previously have been known

Table 10.1. Yeast groups by host and host nutritional resource for number of basepair differences in D1/D2 domain of LSU rDNA from the nearest known species.

Family	0 bp	1–5 bp	>5 bp	No. of LSU Groups	No. of Samples
Basidiocarp-feeding beetles					
Erotylidae	7	8	36	51	158
Tenebrionidae	1	4	32	37	119
Endomychidae	4	6	19	29	54
Nitidulidae	6	8	15	29	48
Curculionidae (Scolytinae)	3	5	3	11	25
Ciidae	0	2	7	9	17
Staphylinidae (Scaphidiinae)	0	3	4	7	11
Mycetophagidae	1	0	4	5	10
Derodontidae	0	0	2	2	4
Melandryidae	0	0	2	2	3
Clambidae	0	0	1	1	2
Cerylonidae	0	1	0	1	1
Lathridiidae	1	0	0	1	1
Anthribidae	0	1	3	4	6
Subtotal	23	38	128	189	459
Predaceous beetles					
Cucujidae	2	1	4	7	8
Histeridae	1	1	4	6	6
Staphylinidae	0	0	6	6	6
Carabidae	0	0	3	3	4
Subtotal	3	2	17	22	24
Plant-feeding beetles					
Scarabaeidae	1	5	14	20	36
Curculionidae (Curculioninae)	2	2	4	8	8
Chrysomelidae	0	0	5	5	7
Cerambycidae	0	1	4	5	6
Elateridae	0	0	4	4	5
Trogossitidae	0	0	3	3	4
Anobiidae	0	0	1	1	3
Dermestidae	2	0	0	2	2
Passalidae	5	8	12	25	58
Mordellidae	0	0	1	1	1
Subtotal	10	16	48	74	130
Total	36	56	193	285	613

The number of groups based on large subunit (LSU) rDNA and the total number of samples are given. Most of the beetle hosts feed and breed in basidiocarps. Other groups of beetles feeding on insects in basidiocarps and on plants also were the source of yeasts. The yeasts that are more than 5 bp (D1/D2 domain of LSU rDNA) different probably are all undescribed taxa by even the most conservative species concept that could be used.

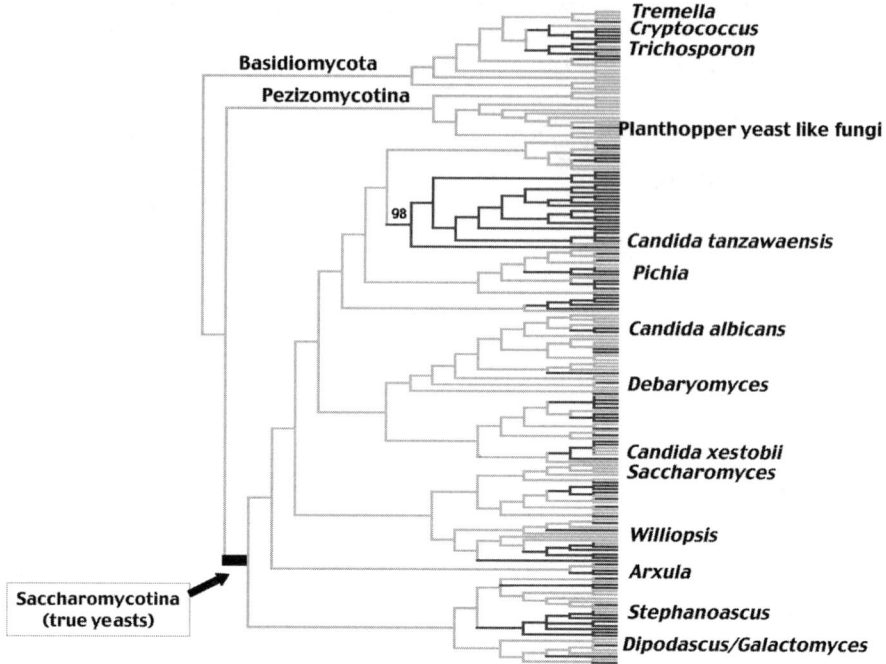

Figure 10.1. Phylogenetic tree depicting the distribution of beetle-gut yeasts and yeastlike fungi (bold lines) scattered among all fungi (stippled lines). Most are true yeasts (thick branch, Ascomycota: Saccharomycotina), but some are filamentous ascomycetes (Ascomycota: Pezizomycotina) and Basidiomycota (Tremellales). Several well-known marker taxa are indicated. Note the large clade of gut yeasts of which *Candida tanzawaensis* is basal. The majority of yeasts in this clade were isolated from the gut of beetles and several other groups of insects. The tree is a strict consensus tree derived from 1349 most parsimonious trees based on 857 phylogenetically informative characters from the combined dataset of small subunit and large subunit rDNA sequences and obtained from heuristic tree searches executed using the tree bisection-reconnection branch-swapping algorithm with random sequence analysis. Bootstrap values (only one shown) of the most parsimonious tree were obtained from 1000 replications. Maximum parsimony analyses were performed using PAUP 4.0b10 (Swofford, 2002).

as endosymbionts of anobiid beetles and planthoppers, but these occur in different habitats from the mushroom-feeding beetles.

In addition to the new species illustrated in the tree, several new clades were discovered that are composed almost entirely of insect-associated yeasts. One of these clades (the *Candida tanzawaensis* clade) previously was known from a single species, that of *C. tanzawaensis* (Barnett et al. 2000), but it now has been expanded by the discovery of 162 isolates and more than 16 undescribed taxa (see below).

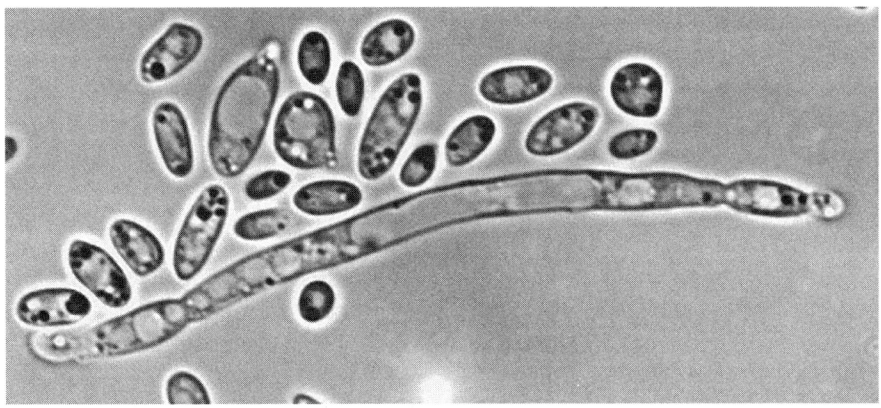

Figure 10.2. Many yeasts produce filaments made up of hyphae or pseudohyphae (formed by buds that do not separate and elongate after formation. This yeast is collected from a large, red scaphidiid beetle at Barro Colorado Island, Panama. This basidiomycete yeast is closely related to *Cryptococcus humicola*.

Specificity of Beetle–Yeast Associations

Usually, one species of yeast was isolated in culture from the gut of each beetle. Several possibilities could explain this phenomenon. Other yeasts may have been present, but rare yeasts may have been overgrown by a predominant yeast during the process of isolation and purification. Although the streak-plate method is supposed to prevent this outcome, it still is possible that overgrowth could have occurred. A second possibility is that some yeasts could not be cultivated under the conditions used and failed to grow. A third possibility is that only one yeast was

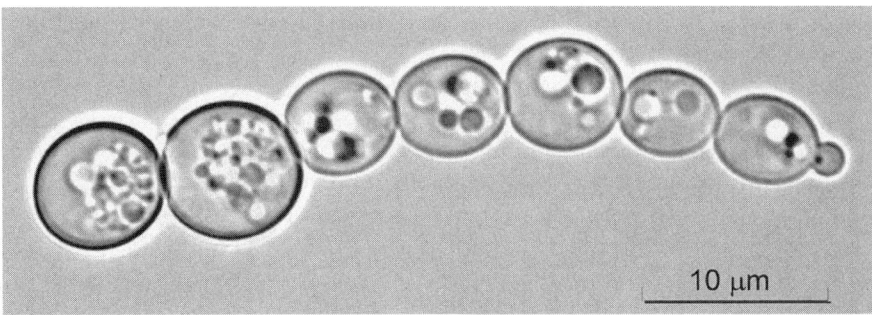

Figure 10.3. Example of an undescribed yeast isolated from the gut of a plant-eating scarabaeid beetle collected at Barro Colorado Island, Panama. Although this ascomycete yeast is genetically quite distant from *Candida albicans*, that well known human pathogen is its closest known relative. A short chain of cells, the product of successive budding, shows the hallmark of yeasts, buds.

present in the gut of each beetle. This last result was supported by our cultural studies and limited evidence from molecular cloning of the LSU rRNA gene from the gut of five species in four families of basidiocarp-feeding beetles (Zhang et al. 2003). In this study only one yeast that could not be cultured on the routinely used media was discovered. Some interesting questions remain unanswered: If a single or few yeasts are present, what mechanism accounts for exclusion of other yeasts? Does the yeast modify the habitat to exclude other yeasts, or does the insect modify the habitat to select a particular yeast?

When we consider the question of specificity in yeast–beetle associations, we find varying degrees of host specificity. The *Candida tanzawaensis* clade mentioned earlier provides a broad range of specificity examples. This group accounted for almost one-third of all beetle-associated yeasts isolated from mushroom-feeding beetles in our study, but the clade was virtually unknown, represented only by *C. tanzawaensis*. Over the last 3 years, 6 more species in the clade have been described (Kurtzman 2001), and 16 more are currently under study (Suh et al. in press). The taxa of this large clade are associated with 11 families of beetles and a geometrid.

Within the clade there are several degrees of host specificity, but strict specificity occurs with certain tenebrionid hosts. In the most extreme example, an undescribed yeast was isolated from *Bolitotherus cornutus* (Tenebrionidae) every time this beetle was collected from its broad range from Vermont, Georgia, and widely separated localities in Louisiana. The yeast had identical LSU rDNA and ITS sequences, appearing to be clonal. Another example of strict specificity occurs among beetles that ingest wood (e.g., Passalidae). One of the yeast isolates was cultured on more than 20 occasions in association with the wood-ingesting passalid beetle *Odontotaenius disjunctus*, from Pennsylvania, South Carolina, Georgia, and Louisiana. The markers (LSU rDNA and ITS) were identical throughout the collected range of the beetle from Pennsylvania to Louisiana. A third line of evidence for specificity comes from the independent isolation of the same yeast species from more than one beetle life-history stage (e.g., larva and adult, pupa and adult), which is indicative not only of host-species-level specificity, but also of early if not vertical transmission from the parents. It is also possible, however, that close association of all stages in a habitat literally smeared with a yeast would result in horizontal transfer (see below).

Specificity is also seen at high taxonomic levels, including that of beetle family. In several instances, clusters of closely related but not identical yeasts of the *C. tanzawaensis* clade are present in species of different genera of Tenebrionidae. It is highly unlikely that the yeasts are all descendents of an ancient common ancestor that once was associated with all tenebrionids. It is more likely that yeasts have been horizontally transmitted, perhaps within the basidiocarp habitat, and this view is further supported by the association of these closely related yeasts with insects besides tenebrionids—the 11 families of beetles and a geometrid mentioned above (Suh et al. in press).

None of the yeasts we cultured from the gut of basidiocarp-feeding beetles reproduced sexually under our cultural conditions, although they may have the capacity to undergo sexual reproduction. The widespread yeast associated with *Bolitotherus cornutus* from Vermont to Georgia and Louisiana appeared to be propagated clonally by vertical transmission. The other yeast used as an example of strict specificity was isolated from *Odontotaenius disjunctus*, a wood-ingesting passalid collected in the eastern United States; all of its yeast isolates had identical ITS sequences throughout the range in which they were collected. This yeast was unusual because it reproduced sexually, a condition that usually is correlated with horizontal symbiont transfer (Frank 1996, 2003; chapter 8, this volume). If transmission is not vertical (as a superficial consideration of the broad range might suggest), the occurrence of a single genotype over the broad range needs an explanation. It is possible that the beetle gut provides a habitat that supports growth of one species to the exclusion of others or that the yeast is free-living in the environment and invades the beetle. Xylose seldom occurs in a free form in nature (but see van Wyk and Nicolson 1995; Jackson and Nicolson 2002), and the beetle gut may provide a unique nutrient-rich habitat for a yeast with the rare or relatively rare traits of xylose fermentation and assimilation.

Summary and Future Directions

Sampling the gut of various beetles has resulted in the discovery of about 150 undescribed yeasts. The new yeasts species add almost 20% more species to the number of known ascomycete yeasts (ca. 700). A new model predicts that 1.5–2 times more yeasts will be discovered at localities already collected and the number will likely be much higher when collecting is done at previously unsampled localities.

The beetles associated with yeasts are not closely related, but many share a common habitat. Many of the beetle food sources (e.g., mushroom, wood) do not provide complete nutritional requirements. Most of the yeasts discovered in this study synthesize a wide variety of B-complex vitamins; these and other products, including enzymes, could supplement the insect diet and its ability to digest food. There is low yeast species diversity within individual beetles. Species-level specificity is found among the associated organisms. Transmission appears to be vertical in some examples, but this is not yet confirmed. The occurrence of a single genotype of a sexually-reproducing yeast over a broad range appears clonal, but early transmission from the environment is possible.

The results from our yeast–beetle system study have raised many questions that need to be addressed in subsequent studies. Some of these questions may enlighten the development of symbiosis theory. For example, how are yeasts transmitted in cases of specificity? Is vertical transmission usually the mode? If vertical transmission does not occur, how are apparent clonal forms maintained in sexually reproducing yeasts over broad intercontinental distributions?

In addition to discovering additional biodiversity and testing the model-based prediction of many more yeasts that are undiscovered, the basis of the association should be investigated. There is a lot of circumstantial evidence to suggest benefit to the beetles and the yeasts, but this needs to be tested. This system should be used as a model to discover other symbiotic relationships by examining other situations in which animals rely on a one-resource diet that might be lacking in some essential nutrients.

Some of the yeasts discovered in this study may have potential uses in biotechnology. The most promising of these are the xylose-fermenting yeasts because xylose, a backbone component of a major plant carbohydrate, hemicellulose, decreases the efficiency of fermentations aimed at the inexpensive production of fuel alcohols. In another case, some of the yeasts we isolated are excellent fermenters of glucose under anaerobic or severely oxygen-limited conditions (T. Jeffries, pers. comm.); these yeasts are members of a lineage not previously known for their ability to grow anaerobically.

This study was conceived as an investigation of fungal biodiversity. For the present time, the contribution of yeasts to overall fungal biodiversity could continue to occupy our research efforts for many years, and the total numbers of yeasts continues to be a fascinating question. If new yeasts continue in association, sometimes in highly specific interactions with particular beetle taxa, this aspect of the study will continue to be productive.

Acknowledgments　We are grateful to our colleague Joseph McHugh, who is involved in this study; in particular he helped us collect and identify the beetles discussed in this chapter. Undergraduate students Amy Whittington, Christine Ackerman, Katie Brillhart, Cennet Erbil, and Nhu Nguyen helped culture and characterize the yeasts using both molecular and cultural techniques. The resources of the Smithsonian Tropical Research Institute, Barro Colorado Island, Panama, and the efforts Dr. Donald Windsor and other staff members, especially Oris Acevedo and Maria Leone, are gratefully acknowledged. Dr. Cletus Kurtzman, USDA, Peoria, opened the way for nonspecialists to study and discover new yeasts by making available an extensive DNA sequence database. Our study was supported by a grant from the National Science Foundation (NSF DEB-0072741) and two REU supplements.

Literature Cited

Barnett, J. A., R. W. Payne, and D. Yarrow. 2000. *Yeasts—characteristics and identification*. Cambridge: Cambridge University Press.

Buchner, P. 1965. *Endosymbiosis of animals with plant microorganisms*. New York: John Wiley & Sons.

Chapman, R. E. 1998. *The insects. Structure and function*, 4th ed. Cambridge: Cambridge University Press.

Crisan, E. V., and A. Sands. 1978. Nutritional value. In *The biology and cultivation of edible mushrooms*, ed. S. T. Chang and W. A. Hayes, pp. 137–168. New York: Academic Press.

Dale, C., S. A. Young,, D. T. Haydon, and S. C. Welburn. 2001. The insect endosymbiont *Sodalis glossinidius* utilizes a type III secretion system for cell invasion. *Proceedings of the National Academy of Sciences of the USA* 98:1883–1888.

Dowd, P. F. 1989. *In situ* production of hydrolytic detoxifying enzymes by symbiotic yeasts in the cigarette beetle (Coleoptera: Anobiidae). *Journal of Economic Entomology* 82:396–400.

Dowd, P. F. 1991. Symbiont-mediated detoxification in insect herbivores. *In Microbial mediation of plant-herbivore interactions*, ed. P. Barbosa, V. A. Krischik and C. G. Jones, pp. 411–440. New York: John Wiley & Sons.

Frank, S. A. 1996. Host symbiont conflict over the mixing of symbiotic lineages. *Proceedings of the Royal Society of London B* 263:339–344.

Frank, S. A. 2003. Perspective: Repression of competition and the evolution of cooperation. *Evolution* 57:693–705.

Fukatsu, T., and H. Ishikawa. 1996. Phylogenetic position of yeast-like symbiont of *Hamiltonaphis styraci* (Homoptera, Aphididae) based on 18S rDNA sequence. *Insect Biochemistry and Molecular Biology* 25:639–646.

Funk, D. J., J. J. Wernegreen, and N. A. Moran. 2001. Intraspecific variation in symbiont genomes: Bottlenecks and the aphid-*Buchnera* association. *Genetics* 157:477–489.

Hongoh, Y., and H. Ishikawa. 1997. Uric acid as a nitrogen resource for the brown planthopper, *Nilaparvata lugens*: Studies with synthetic diets and aposymbiotic insects. *Zoological Science* 14:581–586.

Hongoh, Y., and H. Ishikawa. 2000. Evolutionary studies on uricases of fungal endosymbionts of aphids and planthoppers. *Journal of Molecular Evolution* 51:265–277.

Jackson, S., and S. W. Nicolson. 2002. Xylose as a nectar sugar: From biochemistry to ecology. *Comparative Biochemistry and Physiology B, Biochemistry and Molecular Biology* 131:613–620.

Jones, K. G., and M. Blackwell. 1996. Ribosomal DNA sequence analysis excludes *Symbiotaphrina* from the major lineages of ascomycete yeasts. *Mycologia* 88:212–218.

Jones, K. G., P. F. Dowd, and M. Blackwell. 1999. Polyphyletic origins of yeast-like endocytobionts from anobiid and cerambycid beetles. *Mycological Research* 103:542–546.

Kurtzman, C. P. 2001. Six new anamorphic ascomycetous yeasts near *Candida tanzawaensis*. *FEMS Yeast Research* 1:177–185.

Kurtzman, C. P., and C. J. Robnett. 1998. Identification and phylogeny of ascomycetous yeasts from analysis of nuclear large subunit (26S) ribosomal DNA partial sequences. *Antonie van Leeuwenhoek* 73:331–371.

Martin, M. M. 1987. Invertebrate-microbe interactions. Ithaca, NY: Cornell University Press.

Noda, H., and K. Kodama. 1996. Phylogenetic position of yeastlike endosymbionts of anobiid beetles. *Applied and Environmental Microbiology* 62:162–167.

Noda, H., and Y. Koizumi. 2003. Sterol biosynthesis by symbiotes: Cytochrome P450 sterol C-22 desaturase genes from yeastlike symbiotes of rice planthoppers and anobiid beetles. *Insect Biochemistry and Molecular Biology* 33:649–658.

Prillinger, H. 1987. Are there yeasts in Homobasidiomycetes? *Studies in Mycology* 30:33–59.

Prillinger, H., R. Messner, H. Konig, R. Bauer, K. Lopandic, O. Molnar, P. Dangel, F. Weigang, T. Kirisits, T. Nakase, and L. Sigler. 1996. Yeasts associated with termites: A phenotypic and genotypic characterization and use of coevolution for dating evolutionary radiations in asco- and basidiomycetes. *Systematic and Applied Microbiology* 19:265–283.

Sasaki, T., M. Kawamura, and H. Ishikawa. 1996. Nitrogen recycling in the brown planthopper, *Nilaparvata lugens*: Involvement of yeast-like endosymbionts in uric acid metabolism. *Journal of Insect Physiology* 42:125–129.

Suh, S.-O., H. Noda, and M. Blackwell. 2001. Insect symbiosis: Derivation of yeast-like endosymbionts within an entomophathogenic filamentous lineage. *Molecular Biology and Evolution* 18:995–1000.

Suh, S.-O., C. Marshall, J. V. McHugh, and M. Blackwell. 2003. Wood ingestion by passalid beetles in the presence of xylose-fermenting gut yeasts. *Molecular Ecology* 12:3137–3146.

Suh, S.-O., C. Gibson, and M. Blackwell. 2004a. *Metschnikowia chrysoperlae* sp. nov., *Candida picachoensis* sp. nov. and *Candida pimensis* sp. nov., isolated from the green lacewings *Chrysoperla comanche* and *Chrysoperla carnea* (Neuroptera: Chrysopidae). *International Journal of Systematic and Evolutionary Microbiology* 54:1883–1890.

Suh, S.-O., M. M. White, N. H. Nguyen, and M. Blackwell. 2004b. The identification of *Enteroramus dimorphus*: a xylose-fermenting yeast attached to the gut of beetles. *Mycologia* 96:756–760.

Suh, S.-O., J. V. McHugh, and M. Blackwell. Expansion of the *Candida tanzawaensis* clade: 16 new *Candida* species from basidiocarp-feeding beetles. *International Journal of Systematic and Evolutionary Microbiology*, in press.

Swofford, D. L. 2002. *PAUP. Phylogenetic analysis using parsimony (*and other methods)*, version 4.0b10. Sunderland, MA: Sinauer Associates.

van Meer, M. M. M., J. Witteveldt, and R. Stouthamer. 1999. Phylogeny of the arthropod endosymbiont *Wolbachia* based on the *wsp* gene. *Insect Molecular Biology* 8:399–408.

van Wyk, B. E., and S. W. Nicolson. 1995. Xylose is a major nectar sugar in *Protea* and *Faurea*. *South African Journal of Science* 91:151–153.

Wetzel, J. M., M. Ohnishi, T. Fujita, K. Nakanishi, Y. Naya, H. Noda, and M. Sugiura. 1992. Diversity in steroidogensis of symbiotic microorganisms from planthoppers. *Journal of Chemical Ecology* 18:2083–2094.

Wilkinson, T. L., and H. Ishikawa. 2001. On the functional significance of symbiotic microorganisms in the Homoptera: A comparative study of *Acyrthosiphon pisum* and *Nilaparvata lugens*. *Physiological Entomology* 26:86–93.

Woolfolk, S. W., and G. D. Inglis. 2004. Microorganisms associated with field-collected *Chrysoperla rufilabris* (Neuroptera: Chrysopidae) adults with emphasis on yeast symbionts. *Biological Control* 29:155–168.

Zhang, N., S.-O. Suh, and M. Blackwell. 2003. Gut organisms in beetles: Evidence from cloning. *Journal of Invertebrate Pathology*.

11

Ecology and Evolution
of Mycophagous Bark Beetles
and Their Fungal Partners

Thomas C. Harrington

Associations between bark beetles (Coleoptera: Curculionidae: Scolytinae, or family Scolytidae, depending on the classification used; Bright 1993; Farrell et al. 2001; Marvaldi et al. 2002) and fungi are varied and well known, but mycophagy (fungal feeding) by bark beetles has received relatively little attention. This may be due to the rarity or relative unimportance of fungal feeding by bark beetles, which feed in a nutrient-rich substrate, the inner bark of trees, or it may be due to a bias in research toward the possible importance of plant pathogenic fungi carried by a few coniferous bark beetles. The best studied of the more than 3500 species of bark beetles (Wood 1982; Farrell et al. 2001) construct their egg galleries in the inner bark (secondary phloem) of living trees, especially conifers in the family Pinaceae (Raffa et al. 1993). These bark beetles kill their host and are among the most economically important of forest insects. Although much has been said about the role of fungi in aiding bark beetles in killing their tree hosts, there are inconsistencies in such associations (Harrington 1993b; Paine et al. 1997), and mycophagy may be the more important symbiosis between some of the most important tree-killing bark beetles and fungi. Other species of bark beetles feed in thin-barked branches or treetops and also exploit fungi to supplement their diet.

Probably all bark beetles feed at least briefly on plant tissue colonized by fungi and could thus be considered mycophagous. However, I am here restricting the term mycophagy to grazing by larvae or young adults on fungal spores, fruiting structures, or hyphae (Lawrence 1989) on the surface of galleries or pupal chambers. Mycophagy does not appear to be obligatory for bark beetles, but I hypothesize that fungal feeding allows for more efficient use of the inner bark and gives these bark beetles a competitive edge over other species of bark beetles and phloeophagous (phloem feeding) wood borers. This scenario is consistent with the hypothesized evolution of xylomycetophagous (wood and fungal feeding) ambrosia beetles from phloeophagous bark beetles (Farrell et al. 2001). Mycophagy appears to have evolved many times in the bark beetles, as it has in the xylem-feeding ambrosia beetles (Farrell et al. 2001). There are also parallels and interesting contrasts between mycophagous bark beetles and ambrosia beetles in the way they carry their fungal symbionts, the range of fungi that have been exploited by the beetles, and the morphological adaptations that some of the fungi have made to maintain the symbioses.

Ambrosia Beetles

Approximately 3400 species of ambrosia beetles are found in 10 tribes of two subfamilies of the Curculionidae, the Platypodinae and the Scolytinae (Farrell et al. 2001). The adults of most ambrosia beetles lay their eggs in the wood, where the larvae develop. Lignified cellulose, the principal component of wood, is not readily digested by insects, and fungi serve as the ambrosia of these beetles.

Many ambrosia beetles have specific symbiotic fungi that colonize the wood and produce special spores or modified hyphal endings for insect grazing (Hartig 1844; Hubbard 1897; Neger 1909; Baker 1963; Batra 1963, 1967; Francke-Grosmann 1967; Kok 1979; Norris 1979; Beaver 1989). The ambrosia fungi typically produce fruity volatiles in culture (Neger 1909; Francke-Grosmann 1967), and perhaps these odors direct adults and larvae to actively growing and sporulating areas in the dark galleries. The fungi are probably a richer source of protein than wood, and they may also supply sterols and B-group vitamins important to beetle development (Kok 1979; Beaver 1989). In many of the ambrosia beetles, special spore-carrying sacs, called mycangia (Batra 1963), are found in one or both sexes of the adults, and the specific fungal symbionts are transported from one tree to the next in these sacs (Nunberg 1951; Batra 1963; Francke-Grosmann 1966, 1967; Beaver 1989). Glandular secretions into the mycangium may facilitate growth of specific ambrosia fungi (Norris 1979).

Coniferous bark beetles are basal in the Scolytinae (Farrell et al. 2001; Sequeira and Farrell 2001), and the xylomycetophagous habit is thought to have evolved at least seven times in the subfamily, each origin following a shift to angiosperms (Farrell et al. 2001). The subfamily Platypodinae forms a monophyletic group of ambrosia feeders within the Scolytinae or is sister to it (Farrell et al. 2001; Marvaldi et al. 2002), and this clade and the tribes Corthylini and Xyleborini account for 98% of the ambrosia beetles (Farrell et al. 2001). The Corthylini tribe contains seed eaters, pith borers, and cone borers, as well as bark and ambrosia beetles (Wood and Bright 1992). The bark beetle genus *Dryocoetes*, species of which are found on both conifers and angiosperms, is basal to a monophyletic group of more than 1300 ambrosia beetle species in the tribe Xyleborini, which was thought to have arisen about 20 million years ago (Jordal et al. 2000; Farrell et al. 2001).

The fungal symbionts of only a small percentage of the ambrosia beetles have been identified (Baker 1963; Batra 1963, 1967; Francke-Grosmann 1967), and with many of these it is not clear if the identified fungus is the primary symbiont or a contaminating fungus in the system. An unidentified basidiomycete associated with *Xyleborus dispar* (Happ et al. 1976b) was found to be near *Antrodia* (Hsiau and

Harrington 2003), a genus of brown rot fungi, but the basidiomycete associated with *X. dispar* may not be common or important to the beetle. Batra (1972) reported a *Tulasnella* sp. as an ambrosia fungus of *Trypodendron rufitarsus*. Aside from these two basidiomycetes, the identified ambrosia fungi are in the Ascomycota, though most are known only by their asexual states. Some of the symbionts are yeasts (Batra 1967; Francke-Grosmann 1967). Neger (1908, 1909) suggested and then questioned that the filamentous ambrosia fungi were derived forms of *Ophiostoma* species. Batra (1967) placed many of the ambrosia fungi in the anamorph (asexual) genera *Ambrosiella* or *Raffaelea*. Later phylogenetic studies (Cassar and Blackwell 1996; Jones and Blackwell 1998) have found that these species are closely related to the ascomycetous genera *Ceratocystis* and *Ophiostoma*, both of which have many sexual species associated with insects, especially bark beetles (Harrington 1993a,b; Harrington and Wingfield 1998).

Ceratocystis and *Ophiostoma* are members of the pyrenomycetes, which probably arose more than 200 million years ago (Berbee and Taylor 2001), but the two genera are not closely related, and their ancestors may have diverged more than 170 million years ago (Farrell et al. 2001). The biology of *Ceratocystis* differs substantially from that of *Ophiostoma*, but they have converged on long-necked perithecia with sticky ascospore masses for insect dispersal (Harrington 1981, 1987). Most *Ophiostoma* species are saprophytes on wood and inner bark, often in association with coniferous bark beetles, whereas *Ceratocystis* species are principally plant pathogens on angiosperms.

Phylogenetic analyses place most of the ambrosia fungi within the large genus *Ophiostoma* (Cassar and Blackwell 1996; Jones and Blackwell 1998; Farrell et al. 2001; Rollins et al. 2001), which may be more than 85 million years old (Farrell et al. 2001). *Ambrosiella* and *Raffaelea* species are each polyphyletic within *Ophiostoma* (Cassar and Blackwell 1996; Jones and Blackwell 1998; Farrell et al. 2001; Rollins et al. 2001). *Ceratocystis* may be younger than *Ophiostoma*, perhaps less than 40 million years old (Farrell et al. 2001), and the three known *Ambrosiella* species within *Ceratocystis* appear to be closely related, but not necessarily monophyletic (Cassar and Blackwell 1996; Paulin-Mahady et al. 2002). Species of the relatively young tribe Xyleborini have ambrosia fungi from both the *Ophiostoma* and *Ceratocystis* groups (Farrell et al. 2001). It appears that various species of ambrosia beetles have independently acquired their symbionts from numerous species in these relatively old, insect-associated genera.

In contrast to the ambrosia beetles, most bark beetles lay their eggs in the inner bark of trees, where their larvae feed on a relatively rich substrate. The inner bark may be nutritionally complete for the developing brood, but there is a great deal of competition among bark beetles and wood borers (Coleoptera: Cerambycidae and Buprestidae) for this substrate, and some fungi quickly colonize the inner bark and may render it unsuitable for beetle brood development. Ambrosia beetles have avoided this competition in the inner bark by feeding in the xylem on symbiotic fungi (Farrell et al. 2001). Although little reported compared to mycophagy by ambrosia beetles, some bark beetle species are also known to supplement their diet with fungal hyphae and spores, some carry fungi in highly developed mycangia, and some may depend on fungi for optimal development.

Associations between Fungi and Bark Beetles

A wealth of fungi, principally ascomycetes, can be found in the egg and larval galleries and pupal chambers of coniferous bark beetles (Francke-Grosmann 1967; Graham 1967; Dowding 1973, 1984; Whitney 1982; Harrington, 1993a,b). Many of the fungi show adaptations for insect dispersal; that is, they produce asexual or sexual spores in sticky drops at the tip of fungal fruiting structures (Dowding 1984; Malloch and Blackwell 1993). Typically, the larvae construct a chamber for pupation in the inner bark at the terminus of the larval gallery, and sporulation in the pupal chamber is of particular importance for dissemination of the fungus to a new host tree. After pupation, the young (teneral or callow) adult may first go through a period of maturation feeding, perhaps feeding on fungi sporulating in the chamber as well as on bark tissue (fig. 11.1). Then the adult bores through the outer bark and searches out new trees for breeding and egg laying, or the adults may first go through a period of hibernation or maturation feeding on trees. When the young adults emerge from their pupal chambers, they usually carry spores of many different fungi on their exoskeleton or in their gut, and they introduce the fungi into new host trees during the construction of egg galleries.

The most conspicuous and first noted associations between bark beetles and fungi were between conifer bark beetles and ascomycetous bluestain fungi such as *Ophiostoma minus* (fig. 11.2; Hartig 1878; Münch 1907). The bluestain fungi have melanized hyphae and get their name from the bluish-gray color of the sapwood they colonize. Species of *Ophiostoma* are the most common of the bluestain fungi on the Pinaceae, and both staining and nonstaining *Ophiostoma* species and their anamorphs (asexual genera such as *Leptographium* and *Pesotum*) are conspicuous associates of conifer bark beetles (Harrington 1988, 1993a,b; Harrington et al. 2001; Jacobs and Wingfield 2001). Seven species of the morphologically similar but unrelated genus *Ceratocystis* are also capable of causing bluestain in conifer sapwood, but only three of these species are known associates of conifer bark beetles (Harrington and Wingfield 1998). Yeasts are intimately associated with bark beetles, and various molds, mycoparasites, and some basidiomycetous wood decay fungi are also commonly found in bark beetle galleries (Whitney 1982; Harrington 1993a; Six 2003). Although it is clear that the fungi benefit from these relationships by transportation to fresh phloem, the benefits to the vector are not always apparent.

Given the niche available—that is, the highly nutritious, moist, and uncolonized phloem—it is not surprising that many fungi have evolved stalked, sticky spore masses for acquisition by bark beetles. Based on observations of beetle galleries and isolations from adult beetles, the most successful genera of filamentous fungi associated with conifer bark beetles are *Ophiostoma* and its anamorphic genus *Leptographium* (Harrington 1988, 1993a). *Ophiostoma* may have a late Cretaceous origin (Farrell et al. 2001). The bark beetles are also believed to have their origins in the late Cretaceous (estimated at 67–93 million years before present), feeding on the coniferous genus *Araucaria*, and there may have been a diversification of bark beetles on the Pinaceae sometime near the Cretaceous/Paleocene border (Sequeira and Farrell 2001). It could be hypothesized that the diversification of *Ophiostoma* followed the advance of the coniferous bark beetles, whose tunnels

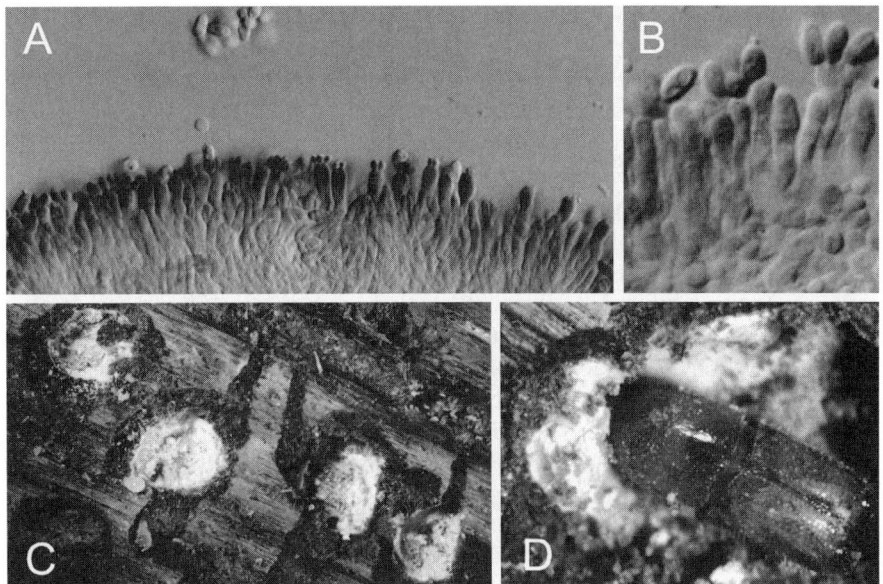

Figure 11.1. *Ips avulsus* and *Entomocorticium* sp. I. (A, B) Hymenium, basidia, and basidiospores of *Entomocorticium* sp. I. (C) Egg gallery (top), short larval galleries (descending), and pupal chambers of *I. avulsus* with white hymenia and basidiospores of *Entomocorticium* sp. I. (D) Teneral adult of *I. avulsus* feeding on the basidiospores and hymenium of *Entomocorticium* sp. I.

Figure 11.2. Loblolly pine trees attacked and killed by *Dendroctonus frontalis*. (A) Two trees in foreground with bark stripped to show the small, dark patches with black microsclerotia and perithecia of *Ophiostoma minus*, a bluestain fungus commonly associated with *D. frontalis*. The sapwood below these dark patches is stained blue-gray. (B) Winding egg galleries (EG) and short larval galleries (LG) in the inner bark of loblolly pine.

provided an avenue to uncolonized phloem. The alternative hypothesis is that plant-pathogenic *Ophiostoma* species allowed the bark beetles to colonize conifers by attacking the defensive resin canal system of the plant host (Farrell et al. 2001).

Some bark beetle associates are capable of causing lesions in the inner bark and/or can invade the sapwood of living trees, and coalition of lesions or sapwood occlusion may contribute to the death of bark beetle-attacked trees (Paine et al. 1997). However, the fungal associates of most tree-killing bark beetles show little capability to colonize a living host plant, and some aggressive tree-killing bark beetles are only inconsistently associated with plant-pathogenic fungi (Harrington 1993b; Paine et al. 1997). The pattern of host colonization by fungi in artificial inoculations is not consistent with the natural colonization of living trees following bark beetle attack (Parmeter et al. 1992). The *Ophiostoma* species apparently invade the sapwood only after the tissue has ceased to conduct water—that is, after the host has died (Hobson et al. 1994). Antagonistic associations also have been noted between bark beetles and fungi capable of causing lesions in trees (Barras 1970; Yearian et al. 1972). Thus, a few species of fungi may in some cases benefit aggressive bark beetle species by attacking the host and providing a suitable substrate for brood development. However, most bark beetles attack dead or severely weakened trees, and it is unlikely that plant-pathogenic *Ophiostoma* species have contributed greatly to the diversification of bark beetles.

Some fungi may be involved in the production of bark beetle pheromones, but this has only been investigated in the laboratory (Brand et al. 1976; Brand and Barras 1977; Hunt and Burden 1990). Other fungi are pathogens of bark beetles (Moore 1971; Whitney 1982; Whitney et al. 1984), or they may compete with the bark beetles for the phloem tissue and render the inner bark unsuitable for brood development (Barras 1970, 1973; Yearian 1972; Fox et al. 1992; Klepzig 2001a). The vast majority of bark beetle fungi, however, appear to be neither beneficial nor detrimental to the beetles (Harrington 1993a), and it would seem that the bark beetles are generally hapless taxis for the fungi.

Only a small fraction of the bark beetles have been seriously studied, and these investigations are heavily biased toward the bark beetles that attack living conifers—more specifically, toward the few bark beetles that attack living Pinaceae, species of *Pinus, Picea,* and *Abies.* Even with this small sampling, it is apparent that there are hundreds, if not thousands, of fungal species tightly associated with conifer bark beetles. With such a small number of studied beetles and such a large number of fungal associates, it would be difficult to make many generalizations about the associations between the beetles and their mycoflora. I feel safe in saying, however, that in most cases the beetles are indifferent to their fungal partners. Nonetheless, there are at least a few lineages of bark beetles known to use fungi as an important supplement to their diet, and most of these bark beetles have evolved special means of carrying fungi.

Bark Beetle Mycangia

Adults of all bark beetles have pits and crevices in their exoskeleton that may contain fungal spores, but only nine bark beetle species are known to have well-developed mycangia (table 11.1; Francke-Grosmann 1967; Beaver 1989; Bright 1993). As originally defined by Batra (1963), mycangia of ambrosia beetles are invaginations of the exoskeleton lined with secretory gland cells. The secretions into mycangia of ambrosia beetles are believed to favor the growth of specific symbiotic fungi (Norris 1979). However, the term "mycangium" has become much more broadly used with bark beetles and includes even mere pits in the exoskeleton that may accumulate fungal spores (Whitney 1982). Such simple pits and other crevices are found in virtually all bark beetles and may or may not be important in transporting fungal spores. Six (2003) recently suggested that these be considered nonglandular pit mycangia to distinguish them from the complex mycangia lined with secretory cells. However, even glandular pits may have a function other than fungal transport, and here I discuss only setal brush mycangia and sac mycangia (Six 2003) because of their obvious role in fungal transport.

Female adults of *Pityoborus* spp. have a unique pubescent, prothoracic mycangium (Furniss et al. 1987), which could be considered a setal brush mycangium in the terminology of Six (2003). The pubescence acts as a comb to collect basidiospores from the galleries, and the fungi probably do not grow within the mycangium.

Relatively few bark beetles have sac mycangia, composed of pockets or tubes that may be glandular or nonglandular (Six 2003), and there is a remarkable coincidence of sac mycangia and mycophagy in the bark beetles. There is at least a suggestion of mycophagy in seven of the eight bark beetle species with sac mycangia (table 11.1). It is not known if *Dryocoetes confusus* is mycophagous. Of the known mycophagus bark beetles, only *Ips avulsus* and *Tomicus minor* do not have either a sac or a setal brush mycangium (table 11.1). However, fungal spores can be seen in the median suture and lateral folds of the elytra of *T. minor* (Francke-Grosmann 1963).

Two types of saclike mycangia have been described for bark beetles: oral mycangia and prothoracic mycangia. Oral mycangia were first described for *Ips acuminatus*, in which there are small pouches behind the mandibles in both males and females (Francke-Grosmann 1967). *Dryocoetes confusus* also has oral sac mycangia at the base of the mandibles of male and female adults (Farris 1969). The prothoracic mycangium of *Dendroctonus* species is glandular, and the secretions allow for growth of fungi within the mycangium (fig. 11.3).

Most of the known mycangial bark beetles (table 11.1) are near-obligate parasites of pines and are within the genus *Dendroctonus*. This genus is in the tribe Tomicini, which is at the base of the bark beetle tree (Farrell et al. 2001), and the genus is relatively old, estimated at 30–50 million years (Sequeira and Farrell 2001). Two types of mycangia, oral and prothoracic, have been described in *Dendroctonus* (table 11.1), and the inferred phylogeny of *Dendroctonus* (Kelley and Farrell, 1998; Sequeira and Farrell 2001) shows that these respective mycangial types are synapomorphic (fig. 11.4), suggesting their importance in the evolution of *Dendroctonus*.

Male and female adults of two sister species of *Dendroctonus* (*D. ponderosae* and *D. jeffreyi*, the Jeffrey pine beetle) have oral mycangia in the form of simple pouches in the maxillary cardines (Whitney and Farris 1970; Six and Paine 1997). Well-developed prothoracic mycangia were reported in female adults of *D. brevicomis*, *D. frontalis*, and *D. adjunctus* (Francke-Grosmann 1965, 1966, 1967), and

Table 11.1. Species of bark beetles thought to be mycophagous, their hosts, mycangial types, and the principal ascomycetous and basidiomycetous fungi upon which they feed or carry.

Bark Beetle	Tribe	Principal Plant Hosts	Mycangial Type	Ascomycete Associates	Basidiomycete Associates
Dendroctonus frontalis	Tomicini	Pinus spp.	Prothoracic, glandular	Ceratocystiopsis ranaculosus	Entomocorticium sp. A
D. brevicomis	Tomicini	P. ponderosae, P. coulteri	Prothoracic, glandular	C. brevicomis	Entomocorticium sp. B
D. approximatus	Tomicini	Pinus spp.	Prothoracic, glandular	Unknown	Phlebiopsis gigantea
D. adjunctus	Tomicini	Pinus spp.	Prothoracic, glandular	Leptographium pyrinum	Unknown
D. ponderosae	Tomicini	Pinus spp.	Maxillary	Ophiostoma clavigerum, O. montium	Entomocorticium dendroctoni, E. sp. D, E, F, G, H, P. gigantea
D. jeffreyi	Tomicini	P. jeffreyi	Maxillary	O. clavigerum	Entomocorticium sp. E
Tomicus minor	Tomicini	Pinus spp.	Unknown	O. canum, Ambrosiella tingens	Unknown
Ips acuminatus	Ipini	Pinus spp.	Mandibular	O. clavatum, A. macrospora	Unknown
I. avulsus	Ipini	Pinus spp.	Unknown	O. ips	Entomocorticium sp. I
Pityoborus comatus	Corthylini	Pinus spp.	Prothoracic, pubescent	Unknown	Entomocorticium sp. C

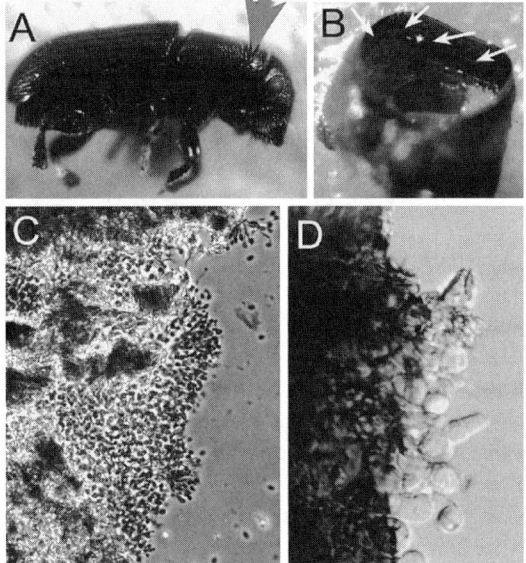

Figure 11.3. Mycangia of *Dendroctonus frontalis*. (A) Adult, female *D. frontalis* with a bulge (arrow) in the prothorax where the mycangium resides. (B) Prothorax removed from a decapitated female adult to show the mycangium, a pale tubelike structure indicated by white arrows. (C) Squashed mycangium with numerous, small conidia and conidiogenous cells of *Ceratocystiopsis ranaculosus*. (D) Prothorax of *D. frontalis* torn apart to show the budding growth of *Entomocorticium* sp. A within the mycangium.

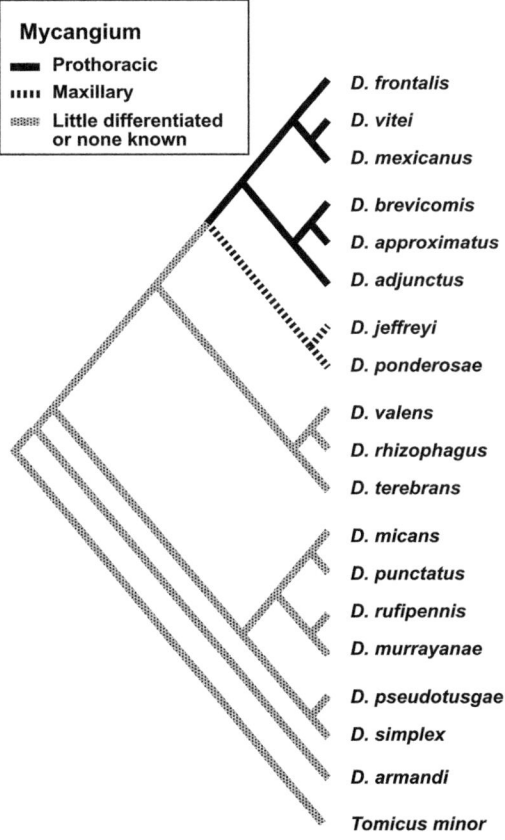

Mycangium
- ▬ **Prothoracic**
- ⅏ **Maxillary**
- ▦ **Little differentiated or none known**

D. frontalis
D. vitei
D. mexicanus
D. brevicomis
D. approximatus
D. adjunctus
D. jeffreyi
D. ponderosae
D. valens
D. rhizophagus
D. terebrans
D. micans
D. punctatus
D. rufipennis
D. murrayanae
D. pseudotusgae
D. simplex
D. armandi
Tomicus minor

Figure 11.4. Mycangial types in a cladogram of the inferred phylogeny of *Dendroctonus* species based on cytochrome oxidase I mitochondrial gene sequences (Kelley and Farrell 1998). The tree includes all recognized *Dendroctonus* species except the nonmycangial *D. parallelocollis,* which is near *D. terebrans.*

265

D. approximatus probably has a prothoracic mycangium (Hsiau and Harrington 2003). Similar prominent bulges in the prothorax of female *D. vitei* and *D. mexicanus* suggest that these species have functional mycangia similar to those of their close relatives (Wood 1982; Lanier et al. 1988; fig. 11.4). Secretory cells associated with the mycangium of *D. frontalis* (Happ et al. 1971), *D. brevicomis* (Francke-Grosmann 1967), and *D. adjunctus* (Barras and Perry 1971) suggest that material produced by these cells allows for the selective growth of special symbionts (Happ et al. 1976a; Paine and Birch 1983). As the fungi grow in the mycangium of the egg-laying females, the spores ooze out of the mycangium and inoculate the egg gallery.

Highly developed mycangia are not essential for the transport of fungi by bark beetles, and it is likely that all coniferous bark beetles at least occasionally carry *Ophiostoma* species. The nonmycophagous and nonmycangial *D. valens*, for instance, consistently carries *Leptographium terebrantis, L. procerum*, and/or *O. ips* (Whitney 1982; Harrington 1988). Even bark beetles that carry specific fungi in a well-developed mycangium also carry fungi on their exoskeleton and in their gut. It would appear that mycangia have evolved to carry specific beneficial fungi. With the exception of *O. clavigerum*, the fungi listed in table 11.1 have not been pathogenic to trees in inoculation experiments. Rather, mycangial fungi appear to be sources of nutrition for the beetles.

Fungal Feeding by Bark Beetles

Mycophagous associations between phloem-feeding bark beetles and fungal symbionts appear to be relatively uncommon, presumably because the inner bark tissues of trees are rich in nutrients. In some cases of mycophagy by bark beetle species, the larval galleries are short and quickly widen to form a large pupal chamber, where the teneral adults can remain and feed on conspicuous fungal growth (fig. 11.1) before emergence. In other cases, the larval galleries extend out into the xylem or the outer bark (fig. 11.2), which would be dry and very low in nutrients, and thus probably unsuitable for larval growth if not for the presence of fungi.

To date, mycophagy is known among only four genera of bark beetles, and the species are all conifer feeders, primarily on species of pine, and most of the mycophagous beetles are capable of attacking living trees (table 11.1). Species of *Ips* and *Dendroctonus* are the best known mycophagous bark beetles, and a species of *Tomicus,* the lesser pine shoot borer, and another of *Pityoborus* also are thought to feed on fungi. The western balsam bark beetle, *Dryocoetes confusus*, has a well-developed mycangium (Farris 1969), but it is not known if it feeds on fungi.

The benefits of fungal feeding have been studied only in a few species, mostly in *Dendroctonus frontalis*, the southern pine beetle, which may carry one or both of two specific fungal symbionts in a well-developed mycangium (fig. 11.3; table 11.1). Experimental manipulations with the mycoflora of bark beetles in a realistic environment have proven difficult but, as will be discussed later, such studies with *D. frontalis*, the mountain pine beetle (*D. ponderosae*), and the small southern pine engraver (*I. avulsus*) have shown that feeding on fungi or fungal-colonized phloem can be beneficial to the bark beetle. The specific nutritional benefits conferred to the beetles are not clear, but they may be similar to the benefits suggested for ambrosia beetles (Kok 1979; Beaver 1989). Vitamins produced by the fungi may aid beetle development. Six (2003) speculated that the fungi produce essential sterols. Nutrients from the inner bark are likely concentrated by the fungi into larval galleries and pupal chambers, and the lowered carbon–nitrogen ratio is likely beneficial to the beetles (Barras and Hodges 1969; Ayres et al. 2000).

Fungal feeding does not appear to be obligatory for the completion of the life cycle of any of these bark beetles. Rather, the concentration and production of nutrients, sterols, and vitamins in fungal hyphae and spores may supplement the diet and allow for shorter larval galleries in the inner bark, which may reduce intraspecific competition and the intense interspecific competition that exists with other bark beetles and wood borers. Such interspecific competition is particularly acute with subcortical insects (Denno et al. 1995), especially in *Pinus* spp. Five of the six mycophagous *Dendroctonus* species attack *P. ponderosa* (table 11.1), which hosts more than 75 species of Scolytidae, including many species of *Ips*, and at least 40 species of wood borers (Farrell et al. 2001). In such a competitive environment, effective transport and introduction of particularly nutritious fungi could give a distinct advantage.

Mycophagy by Ips *spp.*

The genus *Ips* is economically important and well known, and there are a number of reports of fungal feeding by *Ips* species, or at least of the nutritional superiority of phloem colonized by fungi. For instance, Fox et al. (1992) reported that larvae of *I. paraconfusus* could develop successfully to adults without fungi, but the presence of associated fungi in the phloem led to shorter galleries for first and second instar larvae, and larvae were larger when compared to fungus-free larvae. In contrast, Colineau and Lieutier (1994) found little difference between brood of *I. sexdentatus* raised from axenic adults and those from naturally contaminated adults.

Leach et al. (1934) observed young adults of both *Ips pini* and *I. grandicollis* feeding on conidia and hyphae of *Ambrosiella ips*, which was sporulating heavily in some pupal chambers. When *A. ips* was present, the young adults fed upon the growth and consumed it before exiting the tree. No special transport vesicles have been characterized in adults of these two species of *Ips*, and because *A. ips* was not frequently seen in the galleries, Leach et al. (1934) did not consider fungal feeding important to these *Ips* species.

Later work with three species of *Ips* that commonly compete in freshly killed southern pine trees tends to support the claim that fungal feeding is not important for *I. grandicollis*. Wild adults of *I. grandicollis, I. calligraphus*, and *I. avulsus* were introduced into pine bolts, and the developing broods were compared to those from fungus-free adults and adults with only bluestain fungi (*O. ips*) added (Yearian et al. 1972). In general, there was little difference among the treatments for *I. grandicollis* and *I. calligraphus*. However, wild *I. avulsus* adults constructed longer egg galleries, laid more eggs, and the broods matured faster and weighed more than broods from fungus-free adults or from adults with only *O. ips* added.

Ips avulsus, unlike the other *Ips* and *Dendroctonus* species attacking southern yellow pines, is relatively small, can have a life cycle as short as 18–25 days, and can breed in small branches (Berisford 1980; Drooz 1985), perhaps because of its short larval galleries and mycophagy. The larval galleries of *I. avulsus* soon open out into a feeding/pupal chamber (Drooz 1985) that contains luxurious fungal growth (fig. 11.1; Yearian et al. 1972). Yearian (1966) reported that the fungal growth in the pupal chambers was that of a *Tuberculariella* species, and Gouger (1971) referred to the fungus as an *Ambrosiella* sp. Klepzig et al. (2001b) concluded that these observations were instead of the basidiomycete observed by Brian Sullivan (unpublished obs.), which is now known as an *Entomocorticium* (Hsiau and Harrington 2003). *Ophiostoma ips* also sporulates in these pupal chambers (Yearian et al. 1972). Teneral adults of various *Ips* species feed on perithecia of *O. ips*, and ascospores of *O. ips* are commonly present in the gut and feces of adult *I. avulsus* and other *Ips* species (Leach et al. 1934; Yearian et al. 1972; Gouger et al. 1975). No mycangium was found in *I. avulsus* (Gouger et al. 1975).

Francke-Grosmann (1952) found an intimate association between *I. acuminatus* and its fungal associates. Mycangia stuffed with spores were identified behind the mandibles of the adults. Mathiesen-Käärik (1951) and Francke-Grosmann (1952) found *O. clavatum* to be tightly associated with *I. acuminatus*. Francke-Grosmann (1952) also noted yeasts and a second symbiont unique to *I. acuminatus*, *Trichosporium tingens* var. *macrospora*, later placed in *Ambrosiella* by Batra (1967). *Ambrosiella macrospora* may be the most important fungal source of nutrition for *I. acuminatus* (Francke-Grosmann 1952), but more work is needed on this beetle and its fungal symbionts.

Mycophagy by Tomicus minor

Mathiesen-Käärik (1950) found that *O. canum* and *Trichosporium tingens* (syn. *Ambrosiella tingens*) were tightly associated with *Tomicus minor* and not found with other bark beetles in Sweden. Francke-Grosmann (1952) observed luxuriant fungal growth of *O. canum* and *A. tingens* in the galleries of *T. minor* and speculated that feeding on fungi allowed the beetle to develop normally in small branches and under very thin bark. The larvae leave their relatively short larval galleries after the second molt and enter the nutritionally poor xylem, where they become strictly fungal feeders. Thus, this beetle species is both a bark beetle and a xylomycetophagous ambrosia beetle. *Ophiostoma canum*, *A. tingens*, and some yeasts are found in the median suture and the lateral folds of the elytra of *T. minor* (Francke-Grosmann 1963).

Mycophagy by Pityoborus comatus

Pityoborus is within the omnivorous tribe Corthylini, which includes species of pith borers, cone borers, seed borers, and ambrosia beetles (Wood and Bright 1992). Although not economically important and thus little studied, *Pityoborus* species are interesting in that they breed in small branches (1–8 cm diameter) of *Pinus* spp. (Furniss et al. 1987). *Pityoborus comatus* adults are less than 2 mm long and breed in the underside of living pine branches, deeply engraving the xylem (Drooz 1985).

Mycophagy may allow *Pityoborus* species to feed in the thin bark of these branches (Furniss et al. 1987). Adult females of *Pityoborus* species have a unique pubescent mycangium in the prothorax (Furniss et al. 1987). Spores are apparently combed into the mycangium of the adults and transported to new branches. Furniss et al. (1987) isolated a fungus from the basidiospore-filled mycangium of *P. comatus*, and Hsiau and Harrington (2003) placed this fungus in the genus *Entomocorticium*.

Mycophagy by Dendroctonus *spp.*

The *Dendroctonus* genus of bark beetles is arguably the most economically important of forest insects in North America and also has the greatest number of mycophagous species. *Dendroctonus*, meaning "tree killer," is a relatively old genus of largely North American bark beetles restricted to the Pinaceae. The tribe Tomicini is basal to the Scolytinae, and xylomycetophagy has not been reported in the tribe (Farrell et al. 2001; Sequeira and Farrell 2001), except for that by the later instar larvae of *Tomicus minor*. As discussed above, well-developed mycangia are not known for most species of *Dendroctonus*, but six species have been shown to carry fungi consistently in their mycangia (table 11.1), and two others (*D. mexicanus* and *D. vitei*) likely have mycangia (fig. 11.4). *Dendroctonus frontalis* has a rich mycoflora and well-developed prothoracic mycangium (fig. 11.3). The related *D. brevicomis* is also a major economic pest and has a biology and fungal relationships similar to those of *D. frontalis*. In addition to similar mycangial fungi, *D. brevicomis* and *D. frontalis* are associated with *O. minus* and *O. nigrocarpum* (table 11.1; Whitney and Cobb 1972; Paine and Birch 1983; Harrington and Zambino 1990; Hsiau and Harrington 1997, 2003; de Beer et al. 2003). *Dendroctonus adjunctus* and *D. approximatus* also have prothoracic mycangia, and fungal feeding is probably important for these beetles, though it appears that the fungal symbionts of the latter two *Dendroctonus* species differ from those of the better-studied *D. frontalis* and *D. brevicomis* (table 11.1). Two other mycophagous *Dendroctonus* species, *D. ponderosae* and *D. jeffreyi*, form a sister clade to the *Dendroctonus* species with prothoracic mycangia (fig. 11.4).

The Mycoflora of *Dendroctonus ponderosae* and *D. jeffreyi*

The maxillary mycangia, mycoflora, and biology of the sister species *D. ponderosae* and *D. jeffreyi* are similar, except that *D. jeffreyi* is restricted to *Pinus jeffreyi* and *D. ponderosae* attacks most *Pinus* species in its range, except *P. jeffreyi* (Wood and Bright 1992). There is less genetic variation in *D. jeffreyi* than in *D. ponderosae* (Kelley et al. 2000), suggesting that the geographically and host-limited *D. jeffreyi* population has gone through a genetic bottleneck. It could be speculated that the specialist *D. jeffreyi* diverged from the more generalist *D. ponderosae* (Kelley et al. 2000).

The larvae of *D. ponderosae* primarily feed on phloem devoid of fungal growth, as the fungi lag behind the larvae in emanating from the egg gallery (Whitney 1971). During pupation, the fungi continue to advance and colonize the pupal chamber, where they form copious amounts of spores in an ambrosial growth. Teneral adults of *D. ponderosae* consume the fungal growth and enlarge the pupal chamber by eating phloem and phellem before emergence (Whitney 1971; Whitney et al. 1987). Fungal spores enter the oral mycangium during grazing on fungi, and in this manner the fungi are transported to the new host trees (Whitney 1971).

Many fungi have been found sporulating in the pupal chambers of *D. ponderosae* (table 11.1; Whitney 1971; Whitney et al. 1987; Hsiau and Harrington 2003). Special isolation techniques and media would be needed to isolate basidiomycetes from mycangia of *D. ponderosae*, but spores of the basidiomycetes listed in table 11.1 are copious in pupal chambers and likely enter the maxillary mycangia during grazing. Yeasts, molds, *Ophiostoma montium*, and *O. clavigerum* were isolated from the mycangia of *D. ponderosae* (Whitney and Farris 1970), suggesting that the mycangia do not select for the growth of specific fungi. Only *O. clavigerum* was isolated from the mycangia of *D. jeffreyi* (Six and Paine 1997). More work needs to be done to identify the mycoflora of *D. jeffreyi*, but if this bark beetle is recently derived from a *D. ponderosae* population, then it might be expected that the diversity of its mycoflora also has gone through a bottleneck.

Ophiostoma clavigerum appears to be more beneficial in the *D. ponderosae* system compared to *O. montium*. Adults of *D. ponderosae* with *O. clavigerum* produced more progeny, and emergence of the brood occurred sooner than with *O. montium* (Six and Paine 1998). No progeny were produced by fungus-free *D. ponderosae*. However, larvae of *D. ponderosae* developed into apparently normal adults in axenic pine phloem (Whitney 1971), so fungal feeding is not obligatory. *Entomocorticium dendroctoni* is also a likely nutritional symbiont of *D. ponderosae*, apparently superior to the bluestain fungi *O. clavigerum* and *O. montium* (Whitney et al. 1987), but the numerous basidiomycetous fungi found sporulating in pupal chambers of *D. ponderosae* and *D. jeffreyi* (table 11.1) have not been well studied.

The Mycoflora of *Dendroctonus frontalis*

Most of the investigations and observations of the mycoflora of the southern pine beetle have focused on three primary associates of the beetle, the bluestain fungus *O. minus* and two mycangial fungi (Barras and Perry 1972), now known as *Ceratocystiopsis ranaculosus* and *Entomocorticium* sp. A (Klepzig et al. 2001a,b). However, *O. nigrocarpum* is also a common associate of *D. frontalis* (Harrington and Zambino 1990). Further, there has been considerable confusion over the identity of the ascomycetous mycangial fungus *C. ranaculosus*, which has been known as SJB 133, *Sporothrix* sp., *Raffaelea* sp., and *Ceratocystis minor* var. *barrasii* (Harrington and Zambino 1990).

Because of the wealth of information and the confusion concerning fungi associated with *D. frontalis*, its mycoflora need to be characterized more clearly. In an unpublished study, Bob Bridges, the late Thelma Perry of the U.S. Forest Service, and I collected bark infested by *D. frontalis* in eight outbreak regions from east Texas to North Carolina. At each national forest site sampled, bark from two heights of 3–12 trees was collected, and adults were reared from the samples in the laboratory. From each bark sample, 10 males and 10 females were ground in sterile water, and dilution-plating techniques on cycloheximide-amended medium were used to quantify the presence of fungi (Harrington 1992; Hsiau and Harrington 1997). Aside from some yeasts, especially *Candida* spp., which were present in relatively small numbers, three filamentous fungi were recovered. *Ophiostoma minus* and *O. nigrocarpum* were recovered in equal frequency from males and females and in similar frequency to each other, although their relative incidence varied greatly among the eight national forests (fig. 11.5). *Ceratocystiopsis ranaculosus* was rarely isolated from the male beetles, but it was very numerous in the females, often in the tens of thousands of colony forming units per beetle, in some cases in concen-

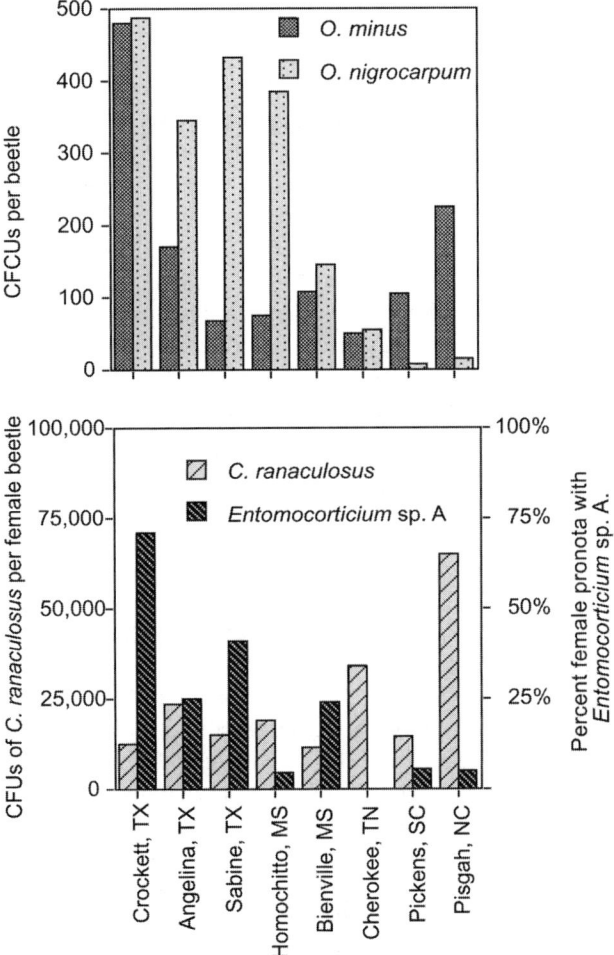

Figure 11.5. Incidence of *Ophiostoma minus*, *O. nigrocarpum*, *Ceratocystiopsis ranaculosus* (colony forming units, CFUs), and *Entomocorticium* sp. A. (percent female pronota colonized) on populations of *Dendroctonus frontalis* from eight national forests in the southeastern United States.

trations hundreds of times higher than the concentrations of *O. nigrocarpum* and *O. minus*.

The numerous small conidia of *C. ranaculosus* found in female mycangia (fig. 11.3C) are the result of conidiogenesis within the mycangium. The basidiomycete *Entomocorticium* sp. A. also grows in the mycangium (Happ et al. 1976a), but mostly as hyphae with swollen cells (fig. 11.3D), and dilution plating failed to recover the fungus, even with media lacking cycloheximide. To quantify the presence of the basidiomycete, the pronota of 21–70 females per national forest site were excised, surface sterilized, and plated on medium amended with benomyl (Harrington et al. 1981). The surface sterilization procedure killed much of the fungal material in the mycangium. In spite of the severe surface sterilization, however, *Entomocorticium* sp. A was isolated from nearly 75% of the pronota at Crockett National Forest, but it was not isolated from the 21 pronota from Cherokee National Forest (fig. 11.5).

It is clear that the relative abundance of the four main associates of *D. frontalis* can vary greatly, with little correlation between the relative frequencies of the various fungi. Among the populations from 13 sites, only two significant ($P < .05$) correlations were found among the frequencies of the four fungal species. The strongest was a negative correlation ($r = -.379$) between the incidence of *C. ranaculosus* and *Entomocorticium* sp. A. The mycangial fungi probably compete with each other within the pupal chambers, where the propagules enter the mycangia of exiting adults, and a negative correlation of the two fungi might be expected. Bridges (1985) found that most individual mycangia had either *C. ranaculosus* or *Entomocorticium* sp. A, but rarely were both species present.

Although it has been speculated that competition between *O. minus* and the mycangial fungi is important to the biology of the beetle (Bridges and Perry 1985; Harrington 1993a; Klepzig 2001a,b), there was a positive correlation ($r = .360$) between the incidence of *Entomocorticium* sp. A and *O. minus*. However, this significant correlation was primarily due to the population at Crockett National Forest, which had the highest incidence of *O. minus*, *O. nigrocarpum*, and *Entomocorticium* sp. A (fig. 11.5). *Ophiostoma minus* primarily contaminates the beetles via the sporothecae of fungal-feeding mites that are phoretic on the adult beetles (Bridges and Moser 1983; Moser and Bridges 1986), and *O. nigrocarpum* may be acquired by adult beetles in a similar manner. Thus, the frequencies of *O. minus* and *O. nigrocarpum* many not be directly correlated with the frequencies of the mycangial fungi, as there may not be much direct competition between mycangial and nonmycangial fungi in the pupal chambers.

Ophiostoma minus is the most conspicuous associate of *D. frontalis* because it causes an intense bluing of the sapwood of the beetle-attacked trees. The role of *O. minus* in the biology of *D. frontalis* has received much speculation (Harrington 1993b; Paine et al. 1997; Klepzig et al. 2001a,b), and it was once thought that the fungus aided the beetle in killing the host tree. I inoculated loblolly pine seedlings and large trees with *O. minus*. A few seedlings died within 3 months of inoculation, and small lesions developed in the larger trees. The lesions were much smaller than those caused by the more aggressive fungus *L. terebrantis*, which was used as a positive check and killed one of the trees and more seedlings than did *O. minus*.

No seedling mortality or lesions greater than controls were found in inoculations with *O. nigrocarpum, C. ranaculosus,* or *Entomocorticium* sp. A (Harrington, unpublished data). These results are similar to those using closely related fungi from *D. brevicomis* (Owen et al. 1987; Parmeter et al. 1989) and suggest that *O. minus* is not an aggressive plant pathogen.

Often *O. minus* colonizes only a small portion of the inner bark and sapwood (fig. 11.2A), usually around the entrance holes of the egg galleries. When the developing larvae encounter *O. minus*, it is clear that it is detrimental to the development of the beetle brood (Bridges 1983; Bridges and Perry 1985; Goldhammer et al. 1990; Klepzig 2001a). It has been speculated that the habit of migration of late instar larvae to the dry outer bark evolved as a way to avoid phloem colonized by *O. minus* (Harrington 1993b). Because of the low moisture content and nutritional value of the outer bark, the larvae might only survive there through mycophagy, and the mycangium of *D. frontalis* would assure that only beneficial fungi dominate in the galleries and pupal chambers.

Broods developing with either *Entomocorticium* sp. A or *Ceratocystiopsis ranaculosus* are larger, have higher lipid contents, and are more fecund than beetles developing without these fungi, and *Entomocorticium* sp. A gives a greater benefit than *C. ranaculosus* (Barras 1973; Bridges 1983; Goldhammer et al. 1990; Coppedge et al. 1995). Because *Entomocorticium* sp. A is more beneficial to the beetle brood than is *C. ranaculosus*, and *C. ranaculosus* is well adapted to dispersal by mites phoretic on bark beetles (Moser et al. 1995), the evolution of mycangia in *D. frontalis* may have initially been to carry *Entomocorticium* sp. A. However, *C. ranaculosus* competes effectively with the basidiomycete in entering and sporulating in the mycangium of *D. frontalis*.

Fungi Fed upon by Bark Beetles

Ascomycetes and basidiomycetes are both fed upon by bark beetles and carried in mycangia (table 11.1). All of the known examples of mycophagy by bark beetles are by conifer bark beetles, which have well-known associations with members of *Ophiostoma* and their *Leptographium* and *Pesotum* anamorphs. Thus, it is not surprising that the ascomycetes fed upon by coniferous bark beetles or carried in their mycangia are all members of the genus *Ophiostoma* or their close relatives (table 11.1). As discussed earlier, *Ophiostoma* apparently predates *Dendroctonus* and other mycophagous bark beetle groups. If a bark beetle species were to begin to feed on fungi and in this manner shorten the feeding galleries and lessen competition with other beetles, then the conspicuous, large masses of conidia of numerous *Ophiostoma* species would be readily available. Minor morphological adaptations, such as spreading conidiogenous aparati and larger conidia, could greatly facilitate feeding on *Ophiostoma* species by beetles. Such minor adaptations can be seen in many specialized *Ophiostoma* species associated with mycophagous bark beetles. However, *Ophiostoma* species may not be the most nutritious fungi.

The known basidiomycete associates of mycophagus bark beetles are *Phlebiopsis gigantea* and species in the genus *Entomocorticium* (Hsiau and Harrington 2003).

The relatives of these bark beetle associates are wood-decaying basidiomycetes that degrade complex, lignified carbohydrates in the secondary xylem and produce macroscopic, sexual fruiting bodies on the outside of trees (Hibbett and Thorn 2001). The basidiospores are borne on basidia formed in a dense palisade and are forcibly discharged from distinctive sterigmata on the basidia, and the basidiospores fall free from the hymenium to be carried passively in wind currents. Spores of *Peniophora aurantiaca* have been trapped more than 1000 km away from the nearest fruit body (Hallenberg and Kuffer 2001). Some of the wood-decaying fungi also produce conidial states that might form an ambrosial layer in beetle galleries. One such basidiomycete is *Phlebiopsis gigantea*, which forms arthroconidia as well as airborne basidiospores (Hsiau and Harrington 2003). In contrast, species of *Entomocorticium* have apparently lost the capacity to produce airborne basidiospores and have more dramatic, irreversible adaptations that provide for symbioses with bark beetles (Hsiau and Harrington 2003).

Ophiostoma *and* Leptographium *as Food for Beetles*

All of the ascomycete species listed in table 11.1 are species of *Ophiostoma*, their anamorphs, or other closely related genera. *Ophiostoma* is a relatively old group of more than 85 species that have slimy ascospore masses at the tip of perithecial necks (Seifert et al. 1993). *Leptographium* and *Pesotum* have conidia produced in a slimy mass at the tip of stalks, and the mass of sticky spores readily adheres to the exoskeleton of insects. *Pesotum* species have synnemata, which are stalks composed of many parallel strands of conidiophores, and most *Pesotum* species have known *Ophiostoma* sexual states (Harrington et al. 2001). *Leptographium* species have mononematous (individual) conidiophores but are similarly associated with insects (Harrington 1988; Jacobs and Wingfield 2001). In addition to named *Ophiostoma* species with *Leptographium* anamorphs, there are 28 species of *Leptographium* (Jacobs and Wingfield 2001), most of which are likely anamorphs of *Ophiostoma*.

As mentioned above, it had been speculated that some *Ophiostoma* species aid beetles in killing host trees, but these fungi are not particularly strong plant pathogens, with the exception of the species that cause Dutch elm disease and black stain root disease (Harrington 1993b; Harrington et al. 2001). The most aggressive of the bark beetles are not necessarily associated with aggressive *Ophiostoma* species (Paine et al. 1997). Harrington (1993b) suggested that some *Ophiostoma* species associated with tree-killing bark beetles are pathogenic to their conifer host because of the selective advantage over other *Ophiostoma* species that cannot colonize living phloem and sapwood.

Many of the *Ophiostoma* species listed in table 11.1 show interesting characteristics that may indicate evolutionary adaptations to insect feeding. *Ophiostoma clavigerum*, among the most important food sources for young adults of *D. ponderosae* and its sibling species, *D. jeffreyi*, has a *Leptographium* state that is frequently found in the pupal chambers of these beetles. The conidia are extremely long and often septate (fig. 11.6), unusual characters for *Leptographium* species. Also, the conidiophores tend to branch at the base and form a shaving-brush cluster of conidiogenous cells at the apex, producing columns of waxy spore masses.

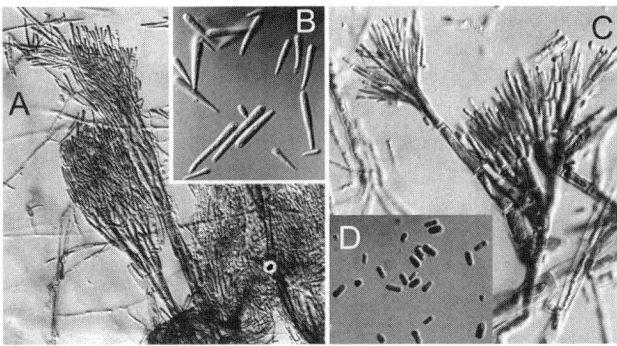

Figure 11.6. Conidiophores and conidia of the mycangial fungus *Ophiostoma clavigerum* (A, B) and its nonmycangial, generalist sister species *Leptographium terebrantis* (C, D). A and C are shown at the same scale, as are B and D.

The waxy columns of elongated conidia appear well suited for consumption and for entering the opening of the maxillary mycangia. Surprisingly, genetic analyses (Six et al. 2003) showed that *O. clavigerum* is very closely related to *L. terebrantis*, which is loosely associated with a number of bark beetles and weevils (Harrington 1988; Jacobs and Wingfield 2001) and does not appear to serve as food for bark beetles. In contrast to the ambrosialike *O. clavigerum*, *L. terebrantis* produces only short, single-celled conidia and does not produce the shaving-brush type of conidial apparatus (fig. 11.6C, 6D). In addition to the typical ambrosial-type conidiophores and clavate conidia, degenerate cultures of *O. clavigerum* also produce simpler conidiophores and small, single-celled conidia (Tsuneda and Hiratsuka 1984), thus reverting to a *L. terebrantis* morphology. Six et al. (2003) speculated that *O. clavigerum* evolved from a *L. terebrantis*-like ancestor and filled a new niche created when the *D. ponderosae*/*D. jeffreyi* lineage of mycophagous bark beetles evolved. *Ophiostoma clavigerum* is capable of causing lesions in inoculated ponderosa pine seedlings and large trees (Owen et al. 1987; Parmeter et al. 1989; Yamaoka et al. 1990, 1995), as does *L. terebrantis*, though the unique conidia and conidiophores of *O. clavigerum* suggest its role as food rather than as a plant pathogen.

Leptographium pyrinum, a mycangial fungus from *D. adjunctus* (Six and Paine 1996), is also related to *L. terebrantis* and *O. clavigerum* (Six et al. 2003). Mycophagy by *D. adjunctus* has not been well studied, but the well-developed mycangium of *D. adjunctus* suggests a biologically important role for fungi. The conidia and conidiophores *L. pyrinum* do not differ greatly from those of the nonambrosial *L. terebrantis* (Six et al. 2003).

In addition to *O. clavigerum*, *O. montium* is a common associate of *D. ponderosae*. Based on morphology and DNA sequence comparisons, *O. montium* is closely related to *O. ips* (Kim et al. 2003), a widespread species found on many species of *Pinus* and associated with many species of bark beetles (Francke-Grosmann 1963; Whitney 1982). Again, it could be speculated that the specialized mycangial fungus *O. montium* evolved from a generalist ancestor like *O. ips*.

Another intriguing evolutionary relationship between a generalist *Ophiostoma* and its specialized sister species is that of *O. piceae* and *O. canum*. *Ophiostoma canum* is believed to be a food source for *Tomicus minor* (Francke-Grosmann 1952) and is only found with this beetle (Mathiesen-Käärik 1950). In rDNA sequence and morphology of the teleomorph, *O. canum* is indistinguishable from *O. piceae* (Harrington et al. 2001), a widespread species loosely associated with many conifer insects and a common colonizer of sapwood in the apparent absence of bark beetles. The distinguishing feature of *O. canum* is the masses of large, round conidia, which resemble the ambrosia growth of many *Ambrosiella* species, except the conidia are found at the tip of synnemata (fig. 11.7). The large conidia could be a morphological adaptation for serving as food for *T. minor*. *Ophiostoma canum* and *O. piceae* also differ in the nutrients they use (Mathiesen-Käärik 1960).

Francke-Grosmann (1952) reported that larvae of *Ips acuminatus* fed upon *O. clavatum*, although there also is a species of *Ambrosiella* found in association with this mycophagous bark beetle. It is not clear which of these fungi is most common in the oral mycangium of *I. acuminatus*. The conidia of *O. clavatum* are relatively small compared to other ascomycetes that serve as food for bark beetles, but the ramifying synnemata of this species may compensate for small spore size by producing large, copious conidial masses for grazing (Mathiesen-Käärik 1951). The nearest relatives of *O. clavatum* are not known, but morphologically it is quite similar to *O. brunneo-ciliatum* (Mathiesen-Käärik 1953), a common associate of *Ips sexdentatus*. *Ophiostoma brunneo-ciliatum* was reported to be associated with *I. acuminatus* in France (Lieutier et al. 1991). Guerard et al. (2000) inoculated Scots pine with *O. brunneo-ciliatum* and concluded that it was not very pathogenic to the tree host.

Ceratocystiopsis *as Food for Beetles*

Although considered synonymous with *Ophiostoma* by some (Hausner et al. 1993, 2003), *Ceratocystiopsis* species may prove to be a monophyletic group with distinctive long, thin ascospores and small perithecia and necks, perhaps adaptations for dispersal by mites phoretic on bark beetles. However, the phylogenetic picture of *Ceratocystiopsis*, especially of the variable *C. minuta*, is cloudy (Hausner et al. 1993, 2003). Ascospores of the best known of these species, *C. minuta*, *C. ranaculosus*, and *C. brevicomis*, have been found in the sporothecae of *Tarsonemus* spp. and on

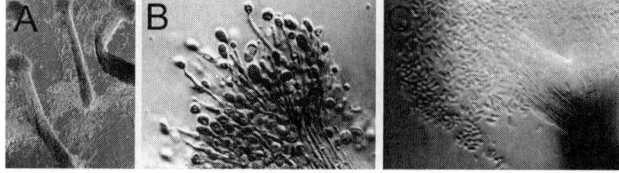

Figure 11.7. Synnemata and conidia of *Ophiostoma canum* (A, B), a species associated solely with the mycophagous *Tomicus minor*, and conidia of its generalist sister species *Ophiostoma piceae* (C). B and C are shown at the same scale.

other genera of mites (Moser et al. 1989, 1995; Moser and Macias-Samano 2000). *Ceratocystiopsis collifera*, associated with *D. valens*, may be a synonym of *C. ranaculosus*, or it is a very closely related species (Hsiau and Harrington 1997).

Ceratocystiopsis ranaculosus and *C. brevicomi* produce large conidia in the pupal chambers of their associated beetles, *D. frontalis* and *D. brevicomis*, respectively. The large conidia are believed to be the ambrosial growth on which the beetles feed and are the likely propagules for entering the mycangium (Harrington and Zambino 1990; Hsiau and Harrington 1997). It is clear that *C. ranaculosus* can proliferate within the mycangium of *D. frontalis* as small conidia (fig. 11.3C) and occur in thousands of colony-forming units per mycangium (fig. 11.5). Small conidia of *C. ranaculosus* and *C. brevicomi* are also present in phloem tissue and may serve as spermatia in these heterothallic ascomycetes (Harrington and Zambino 1990; Moser et al. 1995; Hsiau and Harrington 1997). However, ascospores of these species seem particularly effective for dispersal by mites. For example, ascospores of *C. ranaculosus* were abundant on *Tarsonemus* mites phoretic on *D. frontalis* in Chiapas, Mexico (Moser and Macias-Samano 2000).

It is possible that *C. ranaculosus* and *C. brevicomis* share a most recent common ancestor, perhaps a mite-dispersed generalist species such as *C. minuta*, which is frequently found in old bark-beetle galleries (Mathiesen-Käärik 1960). *Ceratocystis ranaculosus* and *C. brevicomis* appear to be mycangial interlopers, adapted to produce large conidia for beetle grazing and entering mycangia, proliferate in prothoracic mycangia, and compete with *Entomocorticium* sp. A. and sp. B, respectively. It is not clear how nutritionally beneficial *C. ranaculosus* and *C. brevicomi* are to their beetles. As discussed earlier in the section on *D. frontalis*, *C. ranaculosus* is less beneficial to its beetle than is *Entomocorticium* sp. A, with which it competes (fig. 11.5), though *C. ranaculosus* may be important in competition with the detrimental bluestain fungus *O. minus*.

Ambrosiella *as Food for Beetles*

Batra (1967) erected *Ambrosiella* to accommodate the asexual states of a number of fungi associated with ambrosia beetles. *Ambrosiella* species produce sporodochial-like masses of undifferentiated conidiophores and large masses of large conidia in monilioid chains, presumably adaptations that allow for easy grazing by beetles. No teleomorphs of these fungi have been discovered, but rDNA analysis (Cassar and Blackwell 1996) separated *Ambrosiella* species associated with ambrosia beetles into two groups, one clade closely related to *Ceratocystis* teleomorphs and the other related to *Ophiostoma* teleomorphs. Not surprisingly, the *Ambrosiella* species associated with bark beetles are all within the *Ophiostoma* group (Rollins et al. 2001).

There are three described species of *Ambrosiella* associated with coniferous bark beetles: *A. ips*, *A. macrospora*, and *A. tingens* (Batra 1967). Isolates of two undescribed *Ambrosiella* species from two other conifer bark beetles were recently found to be related to *A. macrospora* (Rollins et al. 2001), but these undescribed *Ambrosiella* species may not be important sources of nutrition for their bark beetles, *Polygraphus polygraphus* and *Hylurgops palliatus*. Phylogenetic relationships among the *Ophiostoma* species are not clear, but the *Ambrosiella* species associated with bark beetles are related to each other, though they are not necessarily mono-phyletic, and the bark beetle symbionts are distinct from the ambrosia beetle symbionts (Farrell et al. 2001; Rollins et al. 2001).

Leach et al. (1934) observed grazing of teneral adults of *Ips pini* and *I. grandicollis* on *A. ips*, but he did not consider fungal feeding important to these beetles. Francke-Grosmann (1967) concluded that *A. tingens* and *A. macrospora* are important food sources for *Ips acuminatus* and *Tomicus minor*, respectively, although mycophagy by these bark beetles has received little attention since that work. As discussed earlier, *O. clavatum* and *O. canum* also are specifically associated with these mycophagous bark beetles (table 11.1).

Phlebiopsis gigantea *as Food for Beetles*

Phlebiopsis gigantea is a well-known saprophytic wood-decaying fungus of conifers that produces wind-disseminated basidiospores from basidiocarps on the outside of logs or stumps. It is found in the polyphyletic family Corticiaceae (Hibbett and Thorn 2001). Like a minority of basidiomycetous wood-decaying fungi, it also produces a conidial state (Hibbett and Thorn 2001), consisting of hyphal tips that disarticulate into arthrospores. This conidial state is used commercially for the treatment of fresh stump tops as a biological control agent against the root rot fungus *Heterobasidion annosum* (Vainio and Hantula 2000).

It was shown that the arthrospore-producing basidiomycete noted in galleries of *D. ponderosae* (Tsuneda et al. 1993) is *P. gigantea*, and it was speculated that its conidial state forms a suitable ambrosial growth on which the young adults can feed (Hsiau and Harrington 2003). Isolations and polymerase chain reaction amplifications from DNA extracted from the pronota of *D. approximatus* showed that *P. gigantea* is a mycangial fungus of *D. approximatus* (Hsiau and Harrington 2003). Isolates of *P. gigantea* from basidiomes (basidiocarps) and isolates from bark beetles have identical sequences of the internal transcribed spacer regions of the nuclear rDNA, but the isolates from bark beetles have wider hyphae and conidia and grow slower in culture than do those from basidiomes (Hsiau and Harrington 2003). Thus, it is a possibility that the beetle isolates represent a newly derived taxon. The wider conidiogenous cells and conidia would seem to be particularly advantageous to mycophagous bark beetles. However, the number of isolates examined was low. It may be that the evolution and persistence of the conidial state by *P. gigantea* (and perhaps other wood-decaying basidiomycetes with conidial states) is an adaptation for insect dissemination that supplements the wind dissemination of basidiospores. Some species of *Amylostereum*, have a similar dual life history with arthroconidia or fragments of hyphae produced in the mycangia of horntails (*Sirex* spp.) but also free-living with windborne basidiospores (Tabata et al. 2000).

Entomocorticium *as Food for Beetles*

Most basidiomycetous species associated with bark beetles are of the genus *Entomocorticium*, a recently evolved, monophyletic genus within the paraphyletic genus *Peniophora* (fig. 11.8). *Peniophora* is a large and complex genus of wood-decaying

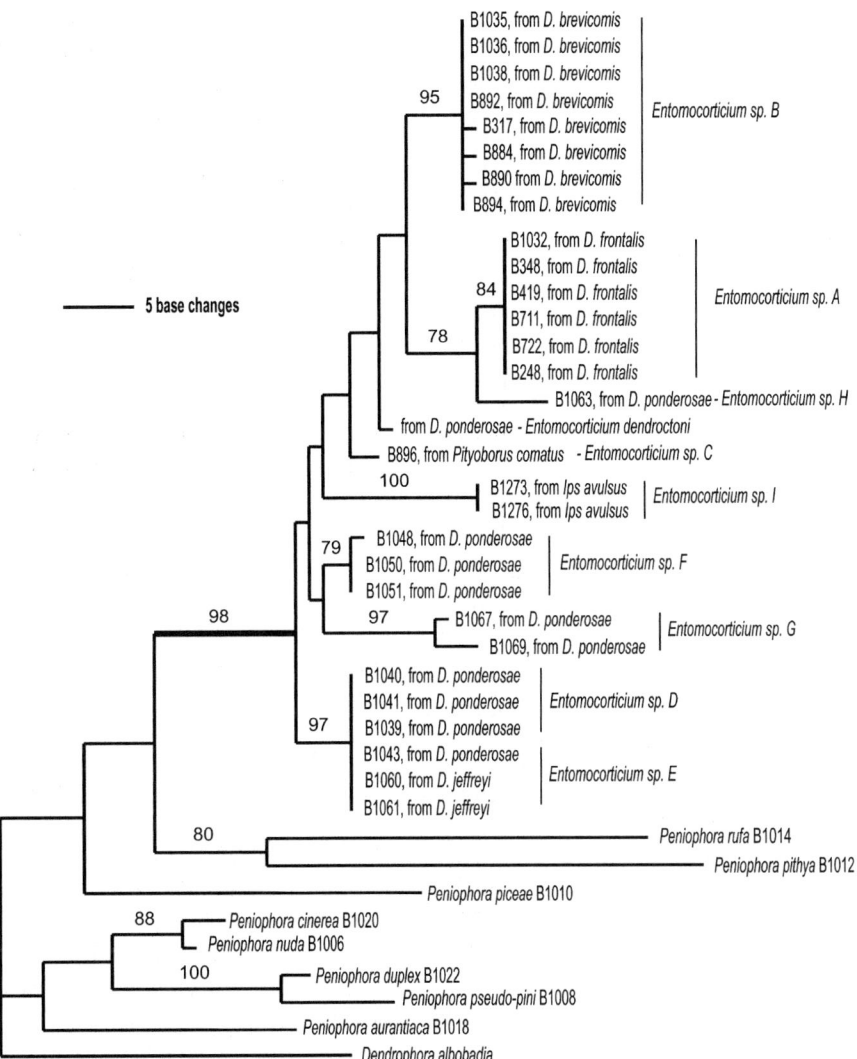

Figure 11.8. One of 8 most parsimonious trees of 254 steps from the internal transcribed spacers and 5.8S rDNA sequences of select *Peniophora* species and the known *Entomocorticium* species (consistency index = 0.6850, retention index = 0.8072). The tree is rooted to *Dendrophora albobadia*. Bootstrap values greater than 50% are shown above branches.

fungi with rather simple basidiocarps, classified in the Corticiaceae (Eriksson et al. 1978; Hallenberg et al. 1996), but phylogenetic analyses place *Peniophora* in the "Russuloid clade" of the "Euagarics" (Hibbett and Thorn 2001). Hallenberg and Parmasto (1998) suggested that *Peniophora* is near *Amylostereum*, the genus of horntail symbionts mentioned above.

Amylostereum and *Peniophora* species produce their basidiocarps on exposed bark or wood, the basidia forcibly discharge their basidiospores from the basidia by way of specialized sterigmata, and the spores are then dispersed by wind (Hallenberg and Kuffer 2001). In *Entomocorticium*, however, the sterigmata are merely pegs and do not discharge the basidiospores, and the sticky basidiospores accumulate on the surface of the hymenium for grazing by the beetles and for adhering to the exoskeleton of the adult beetles (Whitney et al. 1987; Hsiau and Harrington 2003). The loss of forcible discharge of basidiospores would appear to be an irreversible but successful adaptation. The phylogenetic analyses (fig. 11.8; Hsiau and Harrington 2003) strongly support the monophyly of *Entomocorticium* and suggest that this is a young clade with a rapid radiation of species, particularly in association with *D. ponderosae*.

Only one species of *Entomocorticium* has been formally described, and only a few others have been studied morphologically. Cultures of *Entomocorticium* species G, H, and I (fig. 11.8) form basidia with short sterigmata identical to those described for *E. dendroctoni* (Whitney et al. 1987), and some of these species produce cystidia similar to those found in some of the *Peniophora* species. Most of the *Entomocorticium* species were discovered through examinations of pupal chambers of *D. ponderosae* (table 11.1). The spores of these *Entomocorticium* species likely collect in the maxillary mycangia of *D. ponderosae* while the young adults are grazing, but no basidiomycetes have been isolated from the mycangia.

An undescribed basidiomycete was seen in pupal chambers of *Ips avulsus* (Klepzig et al. 1991b; Brian Sullivan, unpublished data). Small stems of loblolly pine were sent to me by Brian Sullivan, and many of the pupal chambers of *I. avulsus* had luxurious, white fungal growth, which were the hymenium and basidiospores of an *Entomocorticium*. When the bark was peeled away from the wood, teneral adults were seen grazing on this fungal growth (fig. 11.1).

Basidiospores have not been observed in two of the mycangial *Entomocorticium* species (A and B), and it is not clear if the cells forming in the pupal chambers of *D. frontalis* were asexual spores or basidiospores. However, Klepzig et al. (2001a) illustrated a basidiomycete from a pupal chamber of *D. frontalis* that had the characteristic basidiospores and cystidia of an *Entomocorticium* species, and that fungus may have been *Entomocorticium* sp. A. The mycangial species of *D. frontalis* and *D. brevicomis* (*Entomocorticium* sp. A and B, respectively) grow more slowly than the other *Entomocorticium* species, perhaps indicative of their more specialized habitat. The ambrosial-like growth in the pupal chambers of *D. frontalis* and *D. brevicomis*, whether due to the accumulation of basidiospores or conidia, likely serves as the propagules for entrance into the mycangium of the young adults. In the mycangium *Entomocorticium* sp. A grows in a yeastlike phase (fig. 11.3D; Happ et al. 1976a).

Other Basidiomycetes Associated with Bark Beetles

Gloeocystidium ipidophilum was described by Siemaszko (1939) as an associate of *Ips typographus*, though it is not apparently a food source for this beetle. *Gloeocystidium ipidophilum* was thought to be related to *E. dendroctoni* (Whitney et al. 1987), but this speculation was not supported by mitochondrial small subunit rDNA sequence data (Hsiau and Harrington 2003). *Heterobasidion annosum* was reported to be loosely associated with Curculionidae (Bakshi 1952; Himes and Skelly 1972; Hunt and Cobb 1982; Kadlec et al. 1992), although it would appear to be more detrimental than mutualistic with the beetles. The proposition (Castello et al. 1976) that exiting bark beetles can acquire fragments of mycelium of wood-decaying basidiomycetes was not supported by later studies (Harrington et al. 1981). *Cryptoporus volvatus* is commonly found producing fruiting bodies near the entrance and exit holes of bark beetles, but bark beetles do not acquire basidiospores or mycelial fragments of the fungus from beetle galleries (Harrington 1980; Harrington et al. 1981). A number of wind-disseminated basidiomycetes have been isolated from bark beetles that have been trapped in flight, but these appear to be casual contaminants of the bark beetles and are probably unimportant to the biology of the beetles (Harrington et al. 1981). A *Sebacina*-like species was reported from galleries of *Ips avulsus* (Moser and Roton 1971). Basidiomycetous yeasts and mycoparasites also have been found in galleries of bark beetles (Kirschner et al. 1999; Kirschner and Chen 2003).

Evolution of Mycophagy in Bark Beetles

Mycophagy appears to be important in many of the well-studied species of bark beetles, but the biology of few bark beetles is well known. Strict dependence on fungal feeding to complete the bark beetle life cycle has not been demonstrated, but the nutritional benefit of mycophagy has been suggested. The fungi could provide a source of sterols, B vitamins, and perhaps a concentration of nitrogen and readily digestible carbohydrates. This extra nutritional boost could make adults more robust and fit for emergence, flight, mating, gallery construction, and egg laying. It also could lessen the amount of phloem tissue needed to complete development and thus shorten the life cycle, which could prove advantageous in a highly competitive environment.

As discussed by Denno et al. (1995), interspecific competition tends to be most intense among mandibulate herbivores feeding in concealed niches, and such competition may be a major driving force in the evolution of mycophagy by bark beetles. The known mycophagous bark beetles all have pine hosts, which have an exceptionally large number of phloeophagous insects, especially bark beetles on some of the North American *Pinus* species like *P. ponderosa* (Farrell et al. 2001), a host of most of the mycophagous bark beetles. Two of the five highly competitive species of bark beetles on southern yellow pines, *D. frontalis* and *Ips avulsus*, also are mycophagous. A common characteristic of mycophagous bark beetles is their short larval galleries, which likely result in more efficient use of the tree for brood development and less competition with other species of bark beetles and wood borers.

Ambrosia beetles apparently escaped the intense interspecific competition in phloem by feeding on fungal symbionts in the xylem (Farrell et al. 2001). Mycophagous bark beetles feed in thin-barked branches or tops of pines, often deeply grazing the xylem, their later instars are found often in the xylem or the dry outer bark, or their teneral adults feed on fungi in pupal chambers before exiting. Fungal feeding probably allows for shorter larval galleries and for feeding in nutritionally poor plant tissues.

It is likely that many bark beetle species feed frequently on fungal spores and fruiting structures, and at times the fungal growth forms a thick, sporulating mass of tissue, well suited for grazing by the beetles. The most luxuriant fungal growth is often seen in the pupal chambers, and those adults that feed on fungi would most likely carry spores of fungi from this ambrosial growth, especially if the adult beetles have specialized mycangia near their mouthparts or head. Like many ambrosia beetles, some mycophagous bark beetles have mycangia into which excretions apparently provide nutrition for specific fungi to grow and sporulate, and bark beetles with such mycangia carry only one or two species of uniquely adapted fungi. Well-developed mycangia are not essential for transport of fungi, but almost all known mycophagous bark beetles have sac or setal brush mycangia. Various types of mycangia are found in such phylogenetically diverse genera as *Dendroctonus*, *Dryocoetes*, *Ips*, and *Pityoborus* (Farrell et al. 2001), suggesting that mycophagy has evolved many times among the phloeophagous bark beetles. Two related clades of *Dendroctonus* appear to have adapted to mycophagy, and the clades have their respective mycangial types. Dependence on fungi for nutrition is apparently a derived character in the bark beetles, and in *Dendroctonus* mycophagy and mycangia are found in what might be considered the two most advanced clades within the genus.

It is not surprising that the ascomycetous fungi fed upon by bark beetles have proven to be related to *Ophiostoma* species, which may have diversified with the diversification of the earliest conifer bark beetles. Thus, *Ophiostoma* predates the mycophagous bark beetles. *Ophiostoma* and *Ceratocystiopsis* species and their related anamorphic genera are generally the fungi most abundant and frequently found in galleries of coniferous bark beetles. Many of the species fed upon by bark beetles are associated with specific beetle species, and the fungal symbionts appear to have morphological adaptations that make them more suitable as food, more readily adherent to bark beetles, and, for some, able to grow in beetle mycangia. In most cases, these specialized fungi are close relatives of more generalist fungi, fungi associated with many different species of coniferous bark beetles. *Ophiostoma montium* is very closely related to the generalist *O. ips*, for instance. *Ophiostoma clavigerum* is genetically almost indistinguishable from its generalist sister species *L. terebrantis*; *O. canum* has an identical rDNA sequence with its generalist sister species *O. piceae*; and the mycangial *Ceratocystiopsis ranaculosus* and *C. brevicomis* are closely related to the mite-associated *C. minuta*. In most of these cases, large asexual spores are produced in an aggregated, ambrosial-like layer, while their generalist counterparts produce spores that are much smaller and accumulate on more discrete conidiophores.

Aside from associations with mycophagous bark beetles, basidiomycetes are not common associates of Scolytinae. However, basidiomycetes may be the preferred food source of some of the best studied mycophagous bark beetles. It appears that these basidiomycetes evolved at least twice from wind-disseminated wood-decaying fungi. An asexual basidiomycete associated with some *Dendroctonus* species is either the common wood-decaying fungus *Phlebiopsis gigantea*, which may have a dual life history, or the bark beetle symbiont is recently derived from *P. gigantea*. The other known associated basidiomycetes appear to reside in a young, recently derived, monophyletic lineage nested within the paraphyletic genus *Peniophora*. Loss of forcible discharge appears to be a recently derived, irreversible character in *Entomocorticium*, and DNA sequence analyses suggest that this adaptation has led to a rapid radiation of species. Most of the known *Entomocorticium* species are associated with *D. ponderosae*.

It has been speculated that the phloeophagous, saprophytic habit is a primitive character in the Scolytinae and that xylomycetophagy is more advanced (Wood 1982; Kirkendall 1983; Beaver 1989; Berryman 1989; Farrell et al. 2001). The habit of feeding in the nutritionally poor xylem by ambrosia beetles likely evolved several times with the feeding on fungal fruiting structures and spores, mostly with shifts to angiosperm hosts (Farrell et al. 2001), and many of these ambrosia beetles have well-developed mycangia for transporting their symbionts. Similarly, but independently, mycophagous bark beetles appear to have evolved from nonmycophagous bark beetles, probably many times, and mycangia are a common characteristic of mycophagous bark beetles. Like the evolution of mycophagy within the ambrosia beetles, a number of independent evolutionary events are needed to explain the diversity of fungi adapted to being fed upon and transported by mycophagous bark beetles. Although the adaptations were often dramatic and perhaps irreversible in both the beetles and the fungi, there does not appear to be a clear case for the co-evolution of the mycophagous bark beetles and the fungi upon which they feed. Rather, mycophagous bark beetles have exploited a number of different fungi for their fungal partners, and based on the morphological adaptations evident in these fungi, they appear to be willing and successful symbionts.

Literature Cited

Ayres, M. P., R. T. Wilkens, J. J. Ruel, M. J. Lombardero, and E. Vallery. 2000. Nitrogen budgets of phloem-feeding bark beetles with and without symbiotic fungi. *Ecology* 81:2198–2210.

Baker, J. M. 1963. Ambrosia beetles and their fungi, with particular reference to *Platypus cylindrus* Fab. *Symposium of the Society for General Microbiology* 13:232–265.

Bakshi, B. K. 1952. *Oedocephalum lineatum* is a conidial state of *Fomes annosus*. *Transactions of the British Mycological Society* 35:195.

Barras, S. J. 1970. Antagonism between *Dendroctonus frontalis* and the fungus *Ceratocystis minor*. *Annals of the Entomological Society of America* 63:1187–1190.

Barras, S. J. 1973. Reduction in progeny and development in the southern pine beetle following removal of symbiotic fungi. *Canadian Entomologist* 105:1295–1299.

Barras, S. J., and J. D. Hodges. 1969. Carbohydrates of inner bark of *Pinus taeda* as affected by *Dendroctonus frontalis* and associated micro-organisms. *Canadian Entomologist* 101:489–493.

Barras, S. J., and T. J. Perry. 1971. Gland cells and fungi associated with prothoracic mycangium of *Dendroctonus adjunctus* (Coleoptera: Scolytidae). *Annals of the Entomological Society of America* 64:123–126.

Barras, S. J., and T. J. Perry. 1972. Fungal symbionts in the prothoracic mycangium of *Dendroctonus frontalis* (Coleoptera: Scolytidae). *Zeitschrift fur Angewandte Entomologie* 71:95–104.

Batra, L. R. 1963. Ecology of ambrosia fungi and their dissemination by beetles. *Transactions of the Kansas Academy of Science* 66:213–236.

Batra, L. R. 1967. Ambrosia fungi: A taxonomic revision, and nutritional studies of some species. *Mycologia* 59:976–1017.

Batra, L. R. 1972. Ectosymbiosis between ambrosia fungi and beetles. 1. *Indian Journal of Mycology and Plant Pathology* 2:165–169.

Beaver, R. A. 1989. Insect-fungus relationships in the bark and ambrosia beetles. In *Insect-fungus interactions,* ed. N. Wilding, N. M. Collins, P. M. Hammond and J. F. Webber, pp. 121–143. London: Academic Press.

Berbee, M. L., and J. W. Taylor. 2001. Fungal molecular evolution: Gene trees and geologic time. In *The Mycota.* vol. VIIB, ed. D. J. McLaughlin, E. McLaughlin, and P. A. Lemke, pp. 229–246. Heidelberg: Springer Verlag.

Berisford, C. W. 1980. Natural enemies and associated organisms. In *The southern pine beetle,* ed. R. C. Thatcher, J. L. Searcy, J. E. Coster, and G. D. Hertel, pp. 31–54. USDA Forest Service Technical Bulletin 1631, Washington, DC.

Berryman, A. A. 1989. Adaptive pathways in Scolytid-fungus associations. In *Insect-fungus interactions,* ed. N. Wilding, N. M. Collins, P. M. Hammond and J. F. Webber, pp. 145–159. London: Academic Press.

Brand, J. M., and S. J. Barras. 1977. The major volatile constituents of a basidiomycete associated with the southern pine beetle. *Lloydia* 40:318–399.

Brand, J. M., J. W. Bracke, L. N. F. Britton, A. J. Markovetz, and S. J. Barras. 1976. Bark beetle pheromones: Production of verbenone by a mycangial fungus of *D. frontalis.* *Journal of Chemical Ecology* 2:195–199.

Bridges, J. R. 1983. Mycangial fungi of *Dendroctonus frontalis* (Coleoptera: Scolytidae) and their relationship to beetle population trends. *Environmental Entomology* 12:858–861.

Bridges, J. R. 1985. Relationships of symbiotic fungi to southern pine beetle population trends. In *Integrated pest management research symposium: The proceedings,* ed. S. J. Branham and R. C. Thatcher, pp. 127–135. USDA Forest Service, General Technical Report SO-56, Asheville, NC.

Bridges, J. R., and J. C. Moser. 1983. Role of two phoretic mites in transmission of bluestain fungus, *Ceratocystis minor. Ecological Entomology* 8:9–12.

Bridges, J. R., and T. J. Perry. 1985. Effects of mycangial fungi on gallery construction and distribution of bluestain in southern pine beetle-infested pine bolts. *Journal of Entomological Science* 20:271–275.

Bright, D. E. 1993. Systematics of bark beetles. In *Beetle-pathogen interactions in conifer forests,* ed. R. D. Schowalter and G. M. Filip, pp. 23–33. New York: Academic Press.

Cassar, S., and M. Blackwell. 1996. Convergent origins of ambrosia fungi. *Mycologia* 88:596–601.

Castello, J. D., C. G. Shaw, and M. M. Furniss. 1976. Isolation of *Cryptoporus volvatus* and *Fomes pinicola* from *Dendroctonus pseudotsugae. Phytopathology* 66:1421–1434.

Colineau, B., and F. Lieutier. 1994. Production of *Ophiostoma*-free adults of *Ips sexdentatus* Boern (Coleoptera, Scolytidae) and comparison with naturally contaminated adults. *Canadian Entomologist* 126:103–110.

Coppedge, B. R., M. S. Frederick, and W. F. Gary. 1995. Variation in female southern pine beetle size and lipid content in relationship to fungal associates. *Canadian Entomologist* 127:145–153.

De Beer, Z. M., T. C. Harrington, H. F. Vismer, B. D. Wingfield, and M. J. Wingfield. 2003. Phylogeny of the *Ophiostoma stenoceras-Sporothrix schenckii* complex. *Mycologia* 95:434–441.

Denno, R. F., M. S. McClure, and J. R. Ott. 1995. Interspecific interactions in phytophagous insects: Competition reexamined and resurrected. *Annual Review of Entomology* 40:297–331.

Dowding, P. 1973. Effects of felling time and insecticide treatment on the interrelationships of fungi and arthropods in pine logs. *Oikos* 24:422–429.

Dowding, P. 1984. The evolution of insect-fungus relationships in the primary invasion of forest timber. In *Invertebrate-microbial interactions*, ed. J. M. Anderson, A. D. M. Raynor, and D. W. H. Walton, pp. 133–153. New York: Cambridge University Press.

Drooz, A. T. 1985. *Insects of eastern forests*. USDA Forest Service Miscellaneous Publication 1426, Washington, DC.

Eriksson, J., K. Hjortstam, and L. Ryvarden. 1978. *The Corticiaceae of North Europe*, vol. 5. Mycoaciella *to* Phanerochaete. Norway: Fungiflora.

Farrell, B. D., A. S. Sequeira, B. C. O'Meara, B. B. Normark, J. H. Chung, and B. H. Jordal. 2001. The evolution of agriculture in beetles (Curculionidae: Scolytinae and Platypodinae). *Evolution* 55:2011–2027.

Farris, S. H. 1969. Occurrence of mycangia in the bark beetle *Dryocoetes confusus* (Coleoptera: Scolytidae). *Canadian Entomologist* 101:527–532.

Fox, J. W., D. L. Wood, R. P. Akers, and J. R. Parmeter. 1992. Survival and development of *Ips paraconfusus* Lanier (Coleoptera, Scolytidae) reared axenically and with tree pathogenic fungi vectored by cohabiting *Dendroctonus* species. *Canadian Entomologist* 124:1157–1167.

Francke-Grosmann, H. 1952. Über die Ambrosiazucht der beiden Kiefernborkenkäfer *Mycelophilus minor* Htg. und *Ips acuminatus* Gyll. Meddelanden Från Statens Skogsforskningsinstitut 41:1–52.

Francke-Grosmann, H. 1963. Some new aspects in forest entomology. *Annual Review of Entomology* 8:415–436.

Francke-Grosmann, H. 1965. Ein Symbioseorgan bei dem Borkenkäfer *Dendroctonus frontalis* Zimm. (Coleoptera, Scolytidae). *Die Naturwissenschaften* 52:143.

Francke-Grosmann, H. 1966. Über Symbiosen von xylomycetophagen und phloeophagen Scolytoidea mit holzbewohnenden Pilzen. *Material und Organismen* 1:503–522.

Francke-Grosmann, H. 1967. Ectosymbiosis in wood-inhabiting insects. In *Symbiosis*, vol. II, ed. S. M. Henry, pp. 171–180. New York: Academic Press.

Furniss, M. M., J. Y. Woo, M. A. Deyrup, and T. H. Atkinson. 1987. Prothoracic mycangium on pine-infesting *Pityoborus* spp. (Coleoptera: Scolytidae). *Annals of the Entomological Society of America* 80:692–696.

Goldhammer, D. S., F. M. Stephen, and T. D. Paine. 1990. The effect of the fungi *Ceratocystis minor* (Hedgecock) Hunt, *Ceratocystis minor* (Hedgecock) Hunt var. *barrasii* Taylor, and SJB-122 on reproduction of the southern pine beetle, *Dendroctonus frontalis* Zimmermann (Coleoptera, Scolytidae). *Canadian Entomologist* 122:407–418.

Gouger, R. J. 1971. Interrelations of *Ips avulsus* (Eichh.) and associated fungi. PhD dissertation, University of Florida.

Gouger, R. J., W. C. Yearian, and R. C. Wilkinson. 1975. Feeding and reproductive behavior of *Ips avulsus*. *Florida Entomologist* 58:221–229.

Graham, K. 1967. Fungal-insect mutualism in trees and timber. *Annual Review of Entomology* 12:105–126.

Guerard, N., E. Dreyer, and F. Lieutier. 2000. Interactions between Scots pine, *Ips acuminatus* (Gyll.) and *Ophiostoma brunneo-ciliatum* (Math.): Estimation of the critical thresholds of attack and inoculation densities and effects on hydraulic properties in the stem. *Annals of Forest Science* 57:681–690.

Hallenberg, N., and N. Kuffer. 2001. Long-distance spore dispersal in wood-inhabiting Basidiomycetes. *Nordic Journal of Botany* 21:431–436.

Hallenberg, N., E. Larsson, and M. Mahlapuu. 1996. Phylogenetic studies in *Peniophora*. *Mycological Research* 100:179–187.

Hallenberg, N., and E. Parmasto. 1998. Phylogenetic studies in species of Corticiaceae growing on branches. *Mycologia* 90:640–654.

Happ, G. M., C. M. Happ, and S. J. Barras. 1971. Fine structure of the prothoracic mycangium, a chamber for the culture of symbiotic fungi, in the southern pine beetle, *Dendroctonus frontalis*. *Tissue and Cell* 3:295–308.

Happ, G. M., C. M. Happ, and S. J. Barras, 1976a. Bark beetle-fungal symbiosis. II. Fine structure of a basidiomycetous ectosymbiont of the southern pine beetle. *Canadian Journal of Botany* 54:1049–1062.

Happ, G. M., C. M. Happ, and J. R. J. French. 1976b. Ultrastructure of the mesonotal mycangium of an ambrosia beetle, *Xyleborus dispar* (F.) (Coleoptera: Scolytidae). *International Journal of Insect Morphology and Embryology* 5:381–391.

Harrington, T. C. 1980. Release of airborne basidiospores from the pouch fungus, *Cryptoporus volvatus*. *Mycologia* 72:926–936.

Harrington, T. C. 1981. Cycloheximide sensitivity as a taxonomic character in *Ceratocystis*. *Mycologia* 73:1123–1129.

Harrington, T. C. 1987. New combinations in *Ophiostoma* of *Ceratocystis* species with *Leptographium* anamorphs. *Mycotaxon* 28:39–43.

Harrington, T. C. 1988. *Leptographium* species, their distribution, hosts, and insect vectors. In *Leptographium root diseases on conifers*, ed. T. C. Harrington and F. W. Cobb, Jr., pp. 1–39. St. Paul, MN: The American Phytopathological Society Press.

Harrington, T. C. 1992. *Leptographium*. In *Methods for research on soilborne phytopathogenic fungi,* ed. L. L. Singleton, J. D. Mihail, and C. M. Rush, pp. 129–133. St. Paul, MN: The American Phytopathological Society Press.

Harrington, T. C. 1993a. Biology and taxonomy of fungi associated with bark beetles. In *Beetle-pathogen interactions in conifer forests,* ed. R. D. Schowalter and G. M. Filip, pp. 37–58. New York: Academic Press.

Harrington, T. C. 1993b. Diseases of conifers caused by species of *Ophiostoma* and *Leptographium*. In Ceratocystis *and* Ophiostoma: *Taxonomy, ecology, and pathogenicity,* ed. M. J. Wingfield, K. S. Seifert, and J. F. Webber, pp. 161–172. St. Paul, MN: The American Phytopathological Society Press.

Harrington, T. C., M. M. Furniss, and C. G. Shaw. 1981. Dissemination of hymenomycetes by *Dendroctonus pseudotsugae* (Coleoptera: Scolytidae). *Phytopathology* 71:551–554.

Harrington, T. C., D. McNew, J. Steimel, D. Hofstra, and R. Farrell. 2001. Phylogeny and taxonomy of the *Ophiostoma piceae* complex and the Dutch elm disease fungi. *Mycologia* 93:111–136.

Harrington, T. C., and M. J. Wingfield. 1998. The *Ceratocystis* species on conifers. *Canadian Journal of Botany* 76:1446–1457.

Harrington, T. C., and P. J. Zambino. 1990. *Ceratocystiopsis ranaculosus*, not *Ceratocystis minor* var. *barrasii*, is the mycangial fungus of the southern pine beetle. *Mycotaxon* 38:103–115.

Hausner, G., G. G. Eyjólfsdóttir, and J. Reid. 2003. Three new species of *Ophiostoma* and notes on *Cornuvesica falcata*. *Canadian Journal of Botany* 81:40–48.

Hausner, G., J. Reid, and G. R. Klassen. 1993. *Ceratocystiopsis*—a reappraisal based on molecular criteria. *Mycological Research* 97:625–633.

Hartig, R. 1878. *Die Zersetzungserscheinungen des Holzes der Nadelbäume und der Eiche forstlicher, botanischer und chemischer Richtung*. Berlin: Julius Springer.

Hartig, T. 1844. Ambrosia des *Botrichus dispar*. Allgemeine Forst- und Jagzeitung, n.s. 13:73–74.

Hibbett, D. S., and R. G. Thorn. 2001. Basidiomycota: Homobasidiomycetes. In *The Mycota*, vol. VIIB, ed. D. J. McLaughlin, E. McLaughlin, and P. A. Lemke, pp. 121–168. Heidelberg: Springer-Verlag.

Himes, W. E., and J. M. Skelly. 1972. An association of the black turpentine beetle, *Dendroctonus terebrans*, and *Fomes annosus* in loblolly pine. *Phytopathology* 62:670. (Abstr.)

Hobson, K. R., J. R. Parmeter, and D. L. Wood. 1994. The role of fungi vectored by *Dendroctonus brevicomis* Leconte (Coleoptera, Scolytidae) in occlusion of ponderosa pine xylem. *Canadian Entomologist* 126:277–282.

Hsiau, P. T. W., and T. C. Harrington. 1997. *Ceratocystiopsis brevicomi* sp. nov., a mycangial fungus from *Dendroctonus brevicomis* (Coleoptera: Scolytidae). *Mycologia* 89:659–661.

Hsiau, P. T. W., and T. C. Harrington. 2003. Phylogenetics and adaptations of basidiomycetous fungi fed upon by bark beetles (Coleoptera: Scolytidae). *Symbiosis* 34:111–131.

Hubbard, H. G. 1897. The ambrosia beetles of the United States. *United States Department of Agriculture Bulletin* 7:1–36.

Hunt, R. S., and J. H. Burden. 1990. Conversion of verbenols to vebenone by yeasts isolated from *Dendroctonus ponderosae* (Coleoptera: Scolytidae). *Journal of Chemical Ecology* 16:1385–1397.

Hunt, R. S. and F. W. Cobb, Jr. 1982. Potential arthropod vectors and competing fungi of *Fomes annosus* in pine stumps. *Canadian Journal of Plant Pathology* 4:247–253.

Jacobs, K., and M. J. Wingfield. 2001. *Leptographium* species. *Tree pathogens, insect associates, and agents of bluestain*. St. Paul, MN: The American Phytopathological Society Press.

Jones, K. G., and M. Blackwell. 1998. Phylogenetic analysis of ambrosial species in the genus *Raffaelea* based on 18S rDNA sequences. *Mycological Research* 102:661–665.

Jordal, B. H., B. B. Normark, and B. D. Farrell. 2000. Evolutionary radiation of a haplodiploid beetle lineage (Curculionidae, Scolytinae). *Biological Journal of the Linnean Society* 71:483–499.

Kadlec, Z., P. Stary, and V. Zumr. 1992. Field evidence for the large pine weevil, *Hylobius abietis* as a vector of *Heterobasidion annosum*. *European Journal of Forest Pathology* 22:316–318.

Kelley, S. T., and B. D. Farrell. 1998. Is specialization a dead end? The phylogeny of host use in *Dendroctonus* bark beetles (Scolytidae). *Evolution* 52:1731–1743.

Kelley, S. T., B. D. Farrell, and J. B. Mitton. 2000. Effects of specialization on genetic differentiation in sister species of bark beetles. *Heredity* 84:218–227.

Kim, J. J., S. H. Kim, S. Lee, and C. Breuil. 2003. Distinguishing *Ophiostoma ips* and *Ophiostoma montium*, two bark beetle-associated sapstain fungi. *FEMS Microbiology Letters* 222:187–192.

Kirkendall, L. R. 1983. The evolution of mating systems in bark and ambrosia beetles (Coleoptera: Scolytidae and Platypodidae). *Zoological Journal of the Linnean Society* 77:293–352.

Kirschner, R., R. Bauer, and F. Oberwinkler. 1999. *Atractocolax*, a new heterobasidiomycetous genus based on a species vectored by conifericolous bark beetles. *Mycologia* 91:538–543.

Kirschner, R., and C. J. Chen. 2003. A new record of *Rogersiomyces okefenofeensis* (Basidiomycota) from beetle galleries in pines in Taiwan. *Sydowia* 55:86–92.

Klepzig, K. D., J. C. Moser, M. J. Lombardero, M. P. Ayres, R. W. Hofstetter, and C. J. Walkinshaw. 2001a. Mutualism and antagonism: Ecological interactions among bark beetles, mites, and fungi. In *Biotic interactions in plant-pathogen associations,* ed. M. J. Jeger and N. J. Spence, pp. 237–269. New York: CABI Publishing.

Klepzig, K. D., J. C. Moser, M. J. Lombardero, R. W. Hofstetter, and M. P. Ayres. 2001b. Symbiosis and competition: complex interactions among beetles, fungi and mites. *Symbiosis* 30:83–96.

Kok, L. T. 1979. Lipids of ambrosia fungi and the life of mutualistic beetles. In *Insect-fungus symbiosis,* ed. L. R. Batra, pp. 33–52. Sussex, UK: Halsted Press.

Lanier, G. N., J. P. Hendrichs, and J. E. Flores. 1988. Biosystematics of the *Dendroctonus frontalis* (Coleoptera: Scolytidae) complex. *Annals of the Entomological Society of America* 81:403–418.

Lawrence, J. F. 1989. Mycophagy in the Coleoptera: Feeding strategies and morphological adaptations. In *Insect-fungus interactions,* ed. N. Wilding, N. M. Collins, P. M. Hammond and J. F. Webber, pp. 2–23. London: Academic Press.

Leach, J. G., L. Orr, and C. Christensen. 1934. The interrelationships of bark beetles and blue staining fungi in felled Norway pine timber. *Journal of Agricultural Research* 49:315–341.

Lieutier, F., J. Garcia, A. Yart, G. Vouland, M. Pettinetti, and M. Morelet. 1991. Ophiostomatales (Ascomycetes) associated with *Ips acuminatus* Gyll. (Coleoptera, Scolytidae) in Scots pine (*Pinus sylvestris* L) in southeastern France, and comparison with *Ips sexdentatus* Boern. *Agronomie* 11:807–817.

Malloch, D., and M. Blackwell. 1993. Dispersal biology of the ophiostomatoid fungi. In Ceratocystis *and* Ophiostoma. *Taxonomy, ecology, and pathogenicity,* ed. M. J. Wingfield, K. A. Seifert, and J. F. Webber, pp. 195–206. St. Paul, MN: The American Phytopathological Society Press.

Marvaldi, A. E., A. S. Sequeira, C. W. O'Brien, and B. D. Farrell. 2002. Molecular and morphological phylogenetics of weevils (Coleoptera: Curculionoidea): Do niche shifts accompany diversification? *Systematic Biology* 51:761–785.

Mathiesen-Käärik, A. 1950. Über einige mit Borkenkäfern assoziierte Bläupilze in Schweden. *Oikos* 2:275–308.

Mathiesen-Käärik, A. 1951. Einige neue *Ophiostoma*-arten in Schweden. *Swensk Botanisk Tidskrift* 45:203–232.

Mathiesen-Käärik, A. 1953. Eine Übersicht über die gewöhnlichsten mit Borkenkäfern assoziierten Bläuplize in Schweden und einige für Schweden neue Bläupilze. *Meddelanden Från Statens Skogsforskningsinstitut* 43:1–74.

Mathiesen-Käärik, A. 1960. Studies on the ecology, taxonomy and physiology of Swedish insect-associated blue stain fungi, especially the genus *Ceratocystis*. *Oikos* 11:1–25.

Moore, G. E. 1971. Mortality factors caused by pathogenic bacteria and fungi of the southern pine beetle in North Carolina. *Journal of Invertebrate Pathology* 17:28–37.

Moser, J. C., and J. R. Bridges. 1986. *Tarsonemus* (Acarina: Tarsonemidae) mites phoretic on the southern pine beetle (Coleoptera: Scolytidae): attachment sites and numbers of bluestain (Ascomycetes: Ophiostomataceae) ascospores carried. *Proceedings of the Entomological Society of Washington* 8:297–299.

Moser, J. C., and J. E. Macias-Samano. 2000. Tarsonemid mite associates of *Dendroctonus frontalis* (Coleoptera: Scolytidae): Implications for the historical biogeography of *D. frontalis*. *Canadian Entomologist* 132:765–771.

Moser, J. C., T. J. Perry, J. R. Bridges, and H. F. Yin. 1995. Ascospore dispersal of *Ceratocystiopsis ranaculosus*, a mycangial fungus of the southern pine beetle. *Mycologia* 87:84–86.

Moser, J. C., T. J. Perry, and H. Solheim. 1989. Ascospores hyperphoretic on mites associated with *Ips typographus*. *Mycological Research* 93:513–517.

Moser, J. C., and L. R. Roton. 1971. Mites associated with southern pine beetles in Allan Parish, Louisiana. *Canadian Entomologist* 103:1775–1798.

Münch, E. 1907. Die Blaufäule des Nadelholzes II. *Naturwissenschaftliche Zeitschrift für Forst-und Landwirtschaft* 62:297–323.

Neger, F. W. 1908. Ambrosiapilze. *Deutsche Botanische Gesellschaft* 26:735–745.

Neger, F. W. 1909. Ambrosiapilze II. *Deutsche Botanische Gesellschaft* 27:372–389.

Norris, D. M. 1979. The mutualistic fungi of xyleborine beetles. In *Insect-fungus symbiosis*, ed. L. R. Batra. pp. 53–63. Sussex, UK: Halsted Press.

Nunberg, M. 1951. Contribution to the knowledge of prothoracic glands of Scolytidae and Platypodidae (Coleoptera). *Annales Musei Zoologici Polonici* 14:261–265.

Owen, D. R., K. Q. Lindahl, Jr., D. L. Wood, and J. R. Parmeter. 1987. Pathogenicity of fungi isolated from *Dendroctonus valens*, *D. brevicomis*, and *D. ponderosae* to seedlings. *Phytopathology* 77:631–636.

Paine, T. D., and M. C. Birch. 1983. Acquisition and maintenance of mycangial fungi by *Dendroctonus brevicomis* Leconte (Coleoptera, Scolytidae). *Environmental Entomology* 12:1384–1386.

Paine, T. D., K. F. Raffa, and T. C. Harrington. 1997. Interactions among scolytid bark beetles, their associated fungi, and live host conifers. *Annual Review of Entomology* 42:179–206.

Parmeter, J. R., G. W. Slaughter, M. Chen, and D. L. Wood. 1992. Rate and depth of sapwood occlusion following inoculation of pines with bluestain fungi. *Forest Science* 38:34–44.

Parmeter, J. R., G. W. Slaughter, M. M. Chen, D. L. Wood, and H. A. Stubbs. 1989. Single and mixed inoculations of ponderosa pine with fungal associates of *Dendroctonus* spp. *Phytopathology* 79:768–772.

Paulin-Mahady, A. E., T. C. Harrington, and D. McNew. 2002. Phylogenetic and taxonomic evaluation of *Chalara*, *Chalaropsis*, and *Thielaviopsis* anamorphs associated with *Ceratocystis*. *Mycologia* 94:62–72.

Raffa, K. F., T. W. Phillips, and S. M. Salom. 1993. Strategies and mechanisms of host colonization by bark beetles. In *Beetle-pathogen interactions in conifer forests*, ed. R. D. Schowalter and G. M. Filip, pp. 103–128. New York: Academic Press.

Rollins, F., K. G. Jones, P. Krokene, H. Solheim, and M. Blackwell. 2001. Phylogeny of asexual fungi associated with bark and ambrosia beetles. *Mycologia* 93:991–996.

Sequeira, A. S., and B. D. Farrell. 2001. Evolutionary origins of Gondwanan interaction: How old are *Araucaria* beetle hervibores? *Biological Journal of the Linnean Society* 74:459–474.

Seifert, K. A., M. J. Wingfield, and W. B. Kendrick. 1993. A nomenclator for described species of *Ceratocystis, Ophiostoma, Ceratocystiopsis, Ceratostomella* and *Sphaeron-aemella*. In *Ceratocystis and Ophiostoma: Taxonomy, ecology, and pathogenicity,* ed. M. J. Wingfield, K. S. Seifert, and J. F. Webber, pp. 269–287. St. Paul, MN: The American Phytopathological Society Press.

Siemaszko, W. 1939. Zespoly grzybow towarzyszacych kornikom polskim. *Planta Polonica* 7:1–52.

Six, D. L. 2003. Bark beetle-fungus symbioses. In *Insect symbiosis,* ed. K. Bourtzis and T. A. Miller, pp. 99–116. Boca Raton, FL: CRC Press.

Six, D. L., T. C. Harrington, J. Steimel, D. McNew, and T. D. Paine. 2003. Genetic relationships among *Leptographium terebrantis* and the mycangial fungi of three western *Dendroctonus* bark beetles. *Mycologia* 95:781–792.

Six, D. L., and T. D. Paine. 1996. *Leptographium pyrinum* is the mycangial fungus of *Dendroctonus adjunctus. Mycologia* 88:739–744.

Six, D. L., and T. D. Paine. 1997. *Ophiostoma clavigerum* is the mycangial fungus of the Jeffrey pine beetle, *Dendroctonus jeffreyi* (Coleoptera: Scolytidae). *Mycologia* 89:858–866.

Six, D. L., and T. D. Paine. 1998. Effects of mycangial fungi and host tree species on progeny survival and emergence of *Dendroctonus ponderosae* (Coleoptera: Scolytidae). *Environmental Entomology* 27:1393–1401.

Tabata, M., T. C. Harrington, W. Chen, and Y. Abe. 2000. Molecular phylogeny of species in the genera *Amylostereum* and *Echinodontium. Mycoscience* 41:585–593.

Tsuneda, A., and Y. Hiratsuka. 1984. Sympodial and annelidic conidiation in *Ceratocystis clavigera. Canadian Journal of Botany* 62:2618–2624.

Tsuneda, A., S. Murakami, L. Sigler, and Y. Hiratsuka. 1993. Schizolysis of dolipore-parenthesome septa in an arthroconidial fungus associated with *Dendroctonus ponderosae* and in similar anamorphic fungi. *Canadian Journal of Botany* 71:1032–1038.

Vaino, E. J., and J. Hantula. 2000. Genetic differentiation between European and North American populations of *Phlebiopsis gigantea. Mycologia* 92:436–446.

Whitney, H. S. 1971. Association of *Dendroctonus ponderosae* (Coleoptera: Scolytidae) with blue stain fungi and yeasts during brood development in lodgepole pine. *Canadian Entomologist* 103:1495–1503.

Whitney, H. S. 1982. Relationships between bark beetles and symbiotic organisms. In *Bark beetles in North American conifers: A system for the study of evolutionary biology,* ed. J. B. Milton and K. B. Sturgeon, pp. 183–211. Austin: University of Texas Press.

Whitney, H. S., R. J. Bandoni, and F. Oberwinkler. 1987. *Entomocorticium dendroctoni* gen. et. sp. nov. (Basidiomycotina), a possible nutritional symbiote of the mountain pine beetle in lodgepole pine in British Columbia. *Canadian Journal of Botany* 65:95–102.

Whitney, H. S., and F. W. Cobb, Jr. 1972. Non-staining fungi associated with the bark beetle *Dendroctonus brevicomis* (Coleoptera: Scolytidae) on *Pinus ponderosa. Canadian Journal of Botany* 50:1943–1945.

Whitney, H. S., and S. H. Farris. 1970. Maxillary mycangium in the mountain pine beetle. *Science* 176:54–55.

Whitney, H. S., D. C. Ritchie, H. H. Borden, and A. J. Stock. 1984. The fungus *Beauveria bassiana* (Deuteromycotina: Hyphomycetaceae) in the western balsam bark beetle, *Dryocoetes confusus* (Coleoptera: Scolytidae). *Canadian Entomologist* 116:1419–1424.

Wood, S. L. 1982. The bark and ambrosia beetles of North and Central America (Coleoptera: Scolytidae), a taxonomic monograph. *Great Basin Naturalist Memoirs* 6:42–50.

Wood, S. L., and D. E. Bright. 1992. A catalog of Scolytidae and Platypodidae (Coleoptera), Part 2: Taxonomic index. *Great Basin Naturalist Memoirs* 13:1–1553.

Yamaoka, Y., Y. Hiratsuka, and P. J. Maruyama. 1995. The ability of *Ophiostoma clavigerum* to kill mature lodgepole pine trees. *European Journal of Forest Pathology* 25:401–404.

Yamaoka, Y., R. H. Swanson, and Y. Hiratsuka. 1990. Inoculation of lodgepole pine with four blue-stain fungi associated with mountain pine beetle, monitored by a heat pulse velocity (HPV) instrument. *Canadian Journal of Forest Research* 20:31–36.

Yearian, W. C. 1966. Relations of the blue stain fungus, *Ceratocystis ips* (Rumbold) C. Moreau, to *Ips* bark beetles (Coleoptera: Scolytidae) occurring in Florida. PhD dissertation, University of Florida.

Yearian, W. C., R. J. Gouger, and R. C. Wilkinson. 1972. Effects of the bluestain fungus, *Ceratocystis ips*, on development of *Ips* bark beetles in pine bolts. *Annals of the Entomological Society of America* 65:481–487.

Conclusion

Symbioses, Biocomplexity, and Metagenomes

Fernando E. Vega
Meredith Blackwell

A common thread in this book is symbiosis, organisms living together in an association, the outcome of which could be neutral, positive, or negative and that might even change from time to time. Ever since de Bary (1879) popularized this term, we have seen the reductionist approach of science in vogue, an approach that has been quite productive. This approach has resulted in the development of revolutionary technologies, such as the tools of molecular biology, which have revealed a tremendous level of biocomplexity in nature. For example, Hermsmeier et al. (2001) found that more than 500 genes in *Nicotiana attenuata* respond to attack by the lepidopteran *Manduca sexta*. This mind-boggling analysis points at the need for a concerted effort aimed at understanding the biocomplexity of insect–fungal associations. We have to start fitting the parts together in a new puzzle that uses parts based on molecular studies. It is not just a matter of examining tri-trophic interactions or of examining how insects deal with plant trichomes, allelochemicals, or a thick cuticle.

Finding more than 500 plant genes responding to herbivory by just one insect is significant. We need to be aware, however, that there are organisms residing in the insect and the plant, in addition to those found on the cuticle and on the phylloplane. When we consider the myriad endophytes in the plant, including yeasts, bacteria, and fungi, and how these might be influencing the plant, and consequently the insect, the picture becomes even more complex. There is an enormous complexity of interactions involving several different trophic levels.

To assess this complexity, it will be necessary to examine metagenomes in insects, an approach that considers a particular insect species as a community in which genomes belonging to various other organisms such as yeasts, fungi, and bacteria might be present. This concept has been used for microbial community analyses in soil (Handelsman et al. 1998; Torsvik and Øvreås 2002), hot springs (Barns et al. 1994), and marine environments (DeLong 1992; López-García et al. 2001), and it is based on various molecular techniques, including polymerase chain reaction (PCR) with primers for small subunit ribosomal RNA (rDNA) (Amann et al. 1995; Jarrell et al. 1999), gene cassette PCR (Stokes et al. 2001), and microarrays (Murray et al. 2001; Zhou 2003), among others (see Torsvik et al. 1998). These techniques allow a better understanding of the vastness of microbial community diversity, which in the past was not known simply because the microbes could not be cultured.

What will we find when we apply these or other related techniques to insects? Will insects reveal themselves to be reservoirs of unknown biodiversity, or will we find that we have been close to reality in terms of knowing what is present, and that the metagenome approach used in other systems is not applicable to insects? Using the metagenome approach, Reed and Hafner (2002) extracted total community DNA from chewing lice to study bacterial communities associated with the insect; their results revealed 35 bacterial lineages and was the first study documenting bacterial associations with chewing lice in the Trichodectidae. Ohkuma and Kudo (1996) used mixed-population DNA from the termite *Reculitermes speratus* to analyze the diversity of intestinal bacteria, revealing the presence of a wide array of previously unknown microorganisms. These results indicate that there is likely a huge knowledge gap in the breadth of insect–fungal associations that can be filled using various molecular techniques. The insect–fungal association field is likely to become much more complex.

It also would be interesting to find out how the "genomic islands" concept (*sensu* Doolittle 2002) applies to insect-associated fungi. Are there "pathogenicity islands" (large gene clusters correlated with virulence) or "symbiotic islands" (genes necessary for symbiosis)? Can these be identified and manipulated for pest control strategies?

Referring to molecular biology, the late Peter Medawar (1968) wrote: "It is simply not worth arguing with anyone so obtuse as not to realize that this complex of discoveries is the greatest achievement of science in the twentieth century" (p. 4). Are there further developments in molecular tools that can be used to advance the field? Certainly. Are there tools at present that are not being used to study insect–fungal association studies? Obviously. We need to pursue a broader approach to understanding the nature of insect–fungal associations knowing that Occam's razor will forever be with us, but with one important caveat: From the reductionist approach that has yielded powerful tools which in turn point at the importance of more holistic studies, we must become true naturalists as our scientific ancestors once were. It is not just a matter of being well versed in molecular biology, but also in ecology, pathology, mycology, and entomology. Once we have put these various fields together as individuals or, perhaps, as teams of reductionist biologists to comprise the new naturalist of our studies, we will be able to advance at gigantic steps in our understanding of insect–fungal associations.

Literature Cited

Amann, R. I., W. Ludwig, and K.-H. Schleifer. 1995. Phylogenetic identification and in situ detection of individual microbial cells without cultivation. *Microbiological Reviews* 59:143–169.

Barns, S. M., R. E. Fundyga, M. W. Jeffries, and N. R. Pace. 1994. Remarkable archaeal diversity detected in a Yellowstone National Park hot spring environment. *Proceedings of the National Academy of Sciences of the USA* 91:1609–1613.

de Bary, A. 1879. Die Erscheinung der Symbiose. Straßburg: Verlag Trubner.

DeLong, E. F. 1992. Archaea in coastal marine environments. *Proceedings of the National Academy of Sciences of the USA* 89:5685–5689.

Doolittle, R. F. 2002. Microbial genomes multiply. *Nature* 416:697–700.

Handelsman, J., M. R. Rondon, S. F. Brady, J. Clardy, and R. M. Goodman. 1998. Molecular biological access to the chemistry of unknown soil microbes: A new frontier for natural products. *Chemistry & Biology* 5:R245–R249.

Hermsmeier, D., U. Schittko, and I. T. Baldwin. 2001. Molecular interactions between the specialist herbivore *Manduca sexta* (Lepidoptera, Sphingidae) and its natural host *Nicotiana attenuata*. I. Large-scale changes in the accumulation of growth- and defense-related plant mRNAs. *Plant Physiology* 125:683–700.

Jarrell, K. F., D. P. Bayley, J. D. Correia, and N. A. Thomas. 1999. Recent excitement about the Archaea. *BioScience* 49:530–541.

López-García, P., F. Rodríguez-Valera, Carlos Pedrós-Alió, and D. Moreira. 2001. Unexpected diversity of small eukaryotes in deep-sea Antarctic plankton. *Nature* 409:603–607.

Medawar, P. B. 1968. Lucky Jim. Review of *The Double Helix* by James Watson. *The New York Review of Books*, March 28, 1968, pp. 3–5.

Murray, A. E., D. Lies, G. Li., K. Nealson, J. Zhou, and J. M. Tiedje. 2001. DNA/DNA hybridization to microarrays reveals gene-specific differences between closely related microbial genomes. *Proceedings of the National Academy of Sciences of the USA* 98:9853–9858.

Ohkuma, M., and T. Kudo. 1996. Phylogenetic diversity of the intestinal bacterial community in the termite *Reticulitermes speratus*. *Applied and Environmental Microbiology* 62:461–468.

Reed, D. L., and M. S. Hafner. 2002. Phylogenetic analysis of bacterial communities associated with ectoparasitic chewing lice of pocket gophers: A culture-independent approach. *Microbial Ecology* 44:78–93.

Stokes, H. W., A. J. Holmes, B. S. Nield, M. P. Holley, K. M. Helena Nevalainen, B. C. Mabbutt, and M. R. Gillings. 2001. Gene cassette PCR: Sequence-independent recovery of entire genes from environmental DNA. *Applied and Environmental Microbiology* 67:5240–5246.

Torsvik, V., and L. Øvreås. 2002. Microbial diversity and function in soil: From genes to ecosystems. *Current Opinion in Microbiology* 5:240–245.

Torsvik, V., F. L. Daae, R.-A. Sandaa, and L. Øvreås. 1998. Novel techniques for analysing microbial diversity in natural and perturbed habitats. *Journal of Biotechnology* 64:53–62.

Zhou, J. 2003. Microarrays for bacterial detection and microbial community analysis. *Current Opinion in Microbiology* 6:288–294.

Index